T0317775

STATISTICS FOR CENSORED ENVIRONMENTAL DATA
USING MINITAB® AND R

STATISTICS FOR CENSORED ENVIRONMENTAL DATA
USING MINITAB® AND R

SECOND EDITION

Dennis R. Helsel
Practical Stats
Denver, Colorado

A JOHN WILEY & SONS, INC., PUBLICATION

Library of Congress Cataloging-in-Publication Data:

Helsel, Dennis R.
  Statistics for censored environmental data using Minitab® and R / Dennis R. Helsel. – 2nd ed.
      p. cm. – (Wiley series in statistics in practice)
  Rev. ed. of: Nondetects and data analysis / Dennis R. Helsel. 2005.
  Includes bibliographical references and index.
  ISBN 978-0-470-47988-9 (cloth)
  1. Environmental sciences–Statistical methods. 2.   Pollution–Measurement–Statistical methods.
  3. Minitab. 4. R (Computer program language) I.   Helsel, Dennis R. Nondetects and data analysis.
  II. Title.
  GE45.S73H45 2012
  363.730285'53–dc23

                                                                                    2011028945

Printed in the United States of America

10 9 8 7 6 5 4 3 2 1

CONTENTS

PREFACE

This book introduces methods for censored data, some simple and some more complex, to potential users who until now were not aware of their existence, or perhaps not aware of their utility. These methods are directly applicable to air quality, water quality, soils, and contaminants in biota, among other types of data. Most of the methods come from the field of survival analysis, where the primary variable being investigated is length of time. Here they are instead applied to environmental measures such as concentration. The first edition (under the name *Nondetects And Data Analysis*) has influenced the methods used by scientists in several disciplines, as reflected in guidance documents and usage in journals. It is my hope that the second edition of this book will continue this progress, broadening the readership to statisticians who are just becoming familiar with environmental applications for these methods.

Within each chapter, examples have been provided in sufficient detail so that readers may apply these methods to their own work. Readily available software was used so that methods would be easily accessible. Examples throughout the book were computed using *Minitab*® *(version 16)*, one of several software packages providing routines for survival analysis, and using the freely available R statistical software system.

The web site linked with this book: http://practicalstats.com/nada contains material for the reader that augments this textbook. Located on the web site are

1. answers to exercises computed using *Minitab* and R,
2. *Minitab* macros and R scripts,
3. a link to the *NADA for R* package,
4. data sets used in this book, and
5. as necessary, an errata sheet listing corrections to the text.

Comments and feedback on both the web site and the book may be emailed to me at nada@practicalstats.com

I sincerely hope that you find this book helpful in your work.

DENNIS HELSEL
April 2011

ACKNOWLEDGMENTS

My sincere appreciation goes to Dr. Ed Gilroy and to a host of students in our *Nondetects And Data Analysis* short courses who have reviewed portions of notes and overheads, making many suggestions and improvements.

To A.T. Miesch, who led the way decades ago.

To my wife Cindy, for her patience and support during what seems to her a never-ending process.

Yesterday upon the stair
I saw a man who wasn't there
He wasn't there again today
Oh how I wish he'd go away.

Hughes Mearns (1875–1965)

Introduction to the First Edition: An Accident Waiting To Happen

On January 28, 1986 the space shuttle *Challenger* exploded 73 seconds after liftoff from Kennedy Space Center, killing all seven astronauts on board and severely wounding the US space program. In addition to career astronauts, on board was America's Teacher In Space, Christa McAuliffe, who was to tape and broadcast lessons designed to interest the next generation of children in America's space program. Her participation ensured that much of the country, including its school children, was watching.

What caused the accident? Would it happen again on a subsequent launch? Four months later the Presidential Commission investigating the accident issued its final report (Rogers Commission, 1986). It pinpointed the cause as a failure of O-rings to flex and seal in the 30°F temperatures at launch time. Rocket fuel exploded after escaping through an opening left by a failed O-ring. An on-camera experiment during the hearings by physicist Richard Feynman illustrated how a section of O-ring, when placed in a glass of ice water, failed to recover from being squeezed by pliers. The experiment's refreshing clarity contrasted sharply with days of inconclusive testimony by officials who debated what might have taken place.

The most disturbing part of the Commission's report was that the O-ring failure had been foreseen by engineers of the booster rockets' manufacturer, who were unable to convince managers to delay the launch. Rocket tests had previously shown evidence of thermal stress in O-rings when temperatures were 65°F and colder. No data were available for the extremely low temperatures predicted for launch time. Faxes sent to NASA on January 27th, the night before launch, presented a graph of damage incidents to one or more rocket O-rings as a function of temperature (Figure i1). This evidence given in the figure seemed inconclusive to managers—there were few data and no apparent pattern.

The Rogers Commission noted in its report that the above graph had one major flaw—flights where damage had not been detected were deleted. The Commission produced a modified graph, their assessment of what should have been (but was not) sent to NASA managers. Their graph added back in the censored values (Figure, i2). By including all recorded data, the Commission proved that the pattern was a bit more striking.

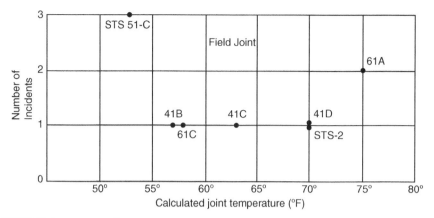

FIGURE i1   Plot of flights with incidents of O-ring thermal distress—"censored observations" deleted. (Figure 6 from Rogers Commission, 1986, p. 146.)

What type of graph could the engineers have used to best illustrate the risk they believed was present? The vast store of information in censored observations is contained in the *proportions* at which they occur. A simple bar chart could have focused on the proportion of O-rings exhibiting damage. For a possible total of three damage incidents in each rocket, a graph of the proportion of failure incidents by ranges of 5° in temperature is shown in Figure i3. The increase in the proportion of damaged O-rings with lower temperatures is clear.

In Figure i1, the information content of data below a (damage) detection threshold was discounted, and the data ignored. Not recognizing and recovering this information was a serious error by engineers. Today the same types of errors are being made by numerous environmental scientists. Deleting censored observations, concentrations

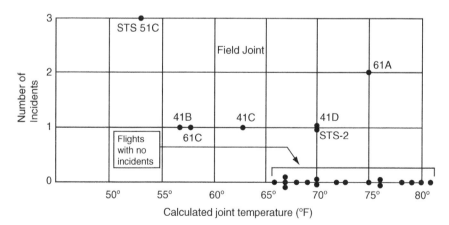

FIGURE i2   Plot of flights with and without incidents of O-ring thermal distress— "censored observations" included. (Figure 7 from Rogers Commission, 1986, p. 146.)

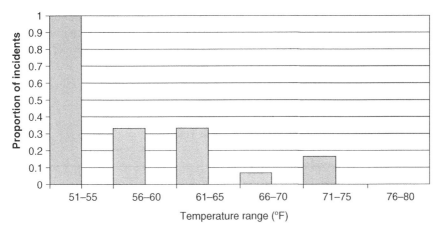

FIGURE i3   O-ring thermal distress data, re-expressed as proportions.

below a measurement threshold, obscures the information in graphs and numerical summaries. Statements such as the one below from the ASTM committee on intralaboratory quality control are all too common:

> Results reported as "less than" or "below the criterion of detection" are virtually useless for either estimating outfall and tributary loadings or concentrations for example.
>
> (ASTM D4210, 1983)

A second, equally serious error occurred prior to the *Challenger* launch when managers assumed that they possessed more information on launch safety than was contained in their data. They decided to launch without knowing the consequences of very low temperatures. According to Richard Feynman, their attitude had become "a kind of Russian roulette . . .. We can lower our standards a little bit because we got away with it the last time" (Rogers Commission, 1986, p. 148). A similar error is now frequently made by environmental programs that fabricate numbers, such as one-half the detection limit, to replace censored observations. Substituting a constant value is even mandated by some Federal agencies—it seemed to work the last time they used it. Its primary error lies in assuming that the scientist/regulator knows more information than what is actually contained in their data. This can easily result in the wrong conclusion, such as declaring that an area is "clean" when it really is not. For the *Challenger* accident, the consequences were a tragic one-time loss of life. For environmental sciences, the consequences are likely to be more chronic and continuous. The health effects of many environmental contaminants occur in the same ranges as current detection limits. Assuming that measurements are at one value when they could be at another is not a safe practice, and as we shall see, totally unnecessary. Fabricating numbers for concentrations could also lead to unnecessary expenditures for cleanup, declaring an area is worse than it actually is. With the large (but limited) amounts of funding now spent on environmental measurements and evaluations, it is incumbent on scientists to use the best available methodologies. In regards to deleting censored observations, or fabricating numbers for them, there are better ways.

When interpreting data that include values below a detection threshold, keep in mind three principles:

1. Never delete censored observations.
2. Capture the information in the proportions.
3. Never assume that you know more than you do.

This book is about what else is possible.

Introduction to the Second Edition: Invasive Data

In his satire *Hitchhiker's Guide To The Galaxy*, Douglas Adams wrote of his characters' search through space to find the answer to "the question of Life, The Universe and Everything." In what is undoubtedly a commentary on the inability of science to answer such questions, the computer built to process it determines that the answer is 42. Environmental scientists often provide an equally arbitrary answer to a different question—what to do with censored "nondetect" data?

The most common procedure within environmental chemistry to deal with censored observations continues to be substitution of some fraction of the detection limit. This method is better labeled as "fabrication", as it substitutes a specific value for concentration data even though a specific value is unknown (Helsel, 2006). Within the field of water chemistry, one-half is the most commonly- used fraction, so that 0.5 is used as if it had been measured whenever a <1 (detection limit of 1) occurs. For air chemistry, one over the square root of two, or about 0.7 times the detection limit, is commonly used. Douglas Adams might have chosen 0.42.

In addition to the environmental sciences where I work, the issue of correctly handling nondetect data has been of great interest in astronomy (Feigelson and Nelson, 1985), in risk assessment (Tressou, 2006), and in occupational health (Succop et al., 2004; Hewett and Ganser, 2007; Finkelstein, 2008; Krishnamoorthy et al., 2009; Flynn, 2010). We all deal with information overload, barely having time to read the relevant literature of our own discipline. It is next to impossible to keep up with work in other disciplines, even when they encounter the same issues as we do. Handling nondetect data is one example.

There is an incredibly strong pull for doing something that is simple and cheap, not to mention familiar. In 1990, I stated that techniques of survival analysis, statistical methods for handling right-censored data in medical and industrial applications, could be turned around and applied to censoring on the low end (Helsel, 1990). The 1990 article clearly states that substitution of values such as one-half the detection limit is generally a bad idea. Because I mention substitution in it, the article has since been referenced a myriad of times to justify using substitution! It makes me wonder whether they read the article at all. As I said, there is an incredibly strong pull for doing something simple and cheap.

The problem with substitution is what I have come to call "invasive data." Substitution is not neutral, but invasive—a pattern is being added to the data that may be quite different than the pattern of the data itself. It can take over and choke out

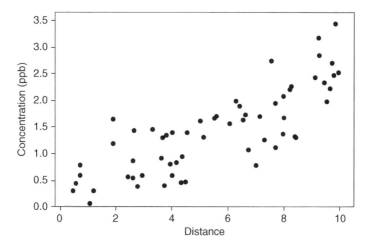

FIGURE i4   Original data prior to censoring. True correlation equals 0.81.

the native pattern. Consider the data of Figure i4, a straight-line relationship between two variables, Concentration (y) versus distance (x) downstream. The slope of the relationship is significant, with a strong positive correlation between the variables. Concentrations are increasing (perhaps with increasing urbanization) downstream. What happens when the data are reported using two detection limits of 1 and 3, and one-half the limit is substituted for the censored observations? The result (Figure i5) includes horizontal lines of substituted values, changing the slope and dramatically decreasing the correlation coefficient between the variables. Looking only at these numbers, the data analyst obtains the (wrong) impression that there is no correlation, no increase in concentration.

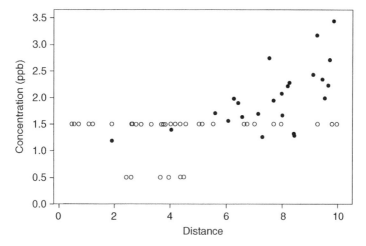

FIGURE i5   Data from Figure i4 after censoring at detection limits of 1 and 3 ppb and substituting $1/2$ DL (shown as open circles). These invasive data form flat lines at one-half the detection limits, lowering the correlation to 0.55.

There are many published articles where substitution was used prior to computing a correlation coefficient. It is cheap and simple. Tajimi et al. (2005), as just one example, calculated correlation coefficients between dioxin concentrations and possible causative factors after substituting one-half the detection limit for all censored observations. A low correlation coefficient was considered evidence that the factor was not the likely cause of the contamination. They found no significant correlations. Was this because there were none, or was it the result of their data substitutions? When adding an invasive flat line to the original data, the original relationship may easily be missed. Thankfully, there are better ways.

Finkelstein (2008) re-examined a study that compared asbestos in the lungs of automobile brake mechanics to a control group. The original study decided that no difference in tremolite asbestos was evident between the two groups, based on visually comparing group medians. The study was faced with many censored observations in the two groups, and was not sure how to best incorporate them into a statistical test. Finkelstein used censored maximum likelihood (see Chapter 9) to test for differences, finding that concentrations of tremolite asbestos were indeed elevated in the mechanics' lungs. The message of his paper is clear—ignoring methods that incorporate censored data leads to wrong decisions both economically and for human or ecosystem health. In the introduction to the first edition, I used the flawed decision to launch the Challenger shuttle as the example. Finkelstein's example of missing the elevated levels of asbestos in the lungs of brake mechanics is equally compelling. Simple, cheap, easy but ineffective methods today can often lead to expensive, heart-breaking, difficult consequences later.

Here are at three recommendations to consider while reading this book:

1. In general, do not use substitution. Journals should consider it a flawed method compared to the others that are available, and reject papers that use it. The lone exception might be when only estimating the mean for data with one censoring threshold, but not for other situations or procedures. Substitution is NOT imputation, which implies using a model such as the relationship with a correlated variable to impute (estimate) values. Substitution is fabrication. It may be simple and cheap, but its results can be noxious.

2. We should all become more familiar with the literature on censored data from survival/reliability analysis. There should be more widespread training in survival/reliability methods within university programs in both the environmental and public health disciplines.

3. Commercial software should more easily incorporate left- and interval-censored data into its survival/reliability routines. For example, plots and hypothesis tests of whether censored data fit a normal and other distributions, as requested by Hewett and Ganser (2007), already exist in many commercial software packages. But they are sometimes coded to handle only right-censored data. They usually do not return p-values for the test. They often incorrectly delete the highest point prior to plotting (see Chapter 5). These and similar considerations will not change until software users in both environmental sciences and public health loudly request that they be changed.

1 Things People Do with Censored Data that Are Just Wrong

Censored observations are low-level concentrations of organic or inorganic chemicals with values known only to be somewhere between zero and the laboratory's detection/reporting limits. The chemical signal on the measuring instrument is small in relation to the process noise. Measurements are considered too imprecise to report as a single number, so the value is commonly reported as being less than an analytical threshold, for example, "<1." Long considered second-class data, censored observations complicate the familiar computations of descriptive statistics, of testing differences among groups, and of correlation coefficients and regression equations.

Statisticians use the term "censored data" for observations that are not quantified, but are known only to exceed or to be less than a threshold value. Values known only to be below a threshold (less-thans) are left-censored data. Values known only to exceed a threshold (greater-thans) are right-censored data. Values known only to be within an interval (between 2 and 5) are interval-censored data. Techniques for computing statistics for censored data have long been employed in medical and industrial studies, where the length of time is measured until an event occurs, such as the recurrence of a disease or failure of a manufactured part. For some observations the event may not have occurred by the time the experiment ends. For these, the time is known only to be greater than the experiment's length, a right-censored "greater-than" value. Methods for incorporating censored data when computing descriptive statistics, testing hypotheses, and performing correlation and regression are all commonly used in medical and industrial statistics, without substituting arbitrary values. These methods go by the names of "survival analysis" (Klein and Moeschberger, 2003) and "reliability analysis" (Meeker and Escobar, 1998). There is no reason why these same methods should also not be used in the environmental sciences, but until recently their use has been relatively rare. Environmental scientists have not often been trained in survival analysis methods.

The worst practice when dealing with censored observations is to exclude or delete them. This produces a strong bias in all subsequent measures of location or hypothesis tests. After excluding the 80% of observations that are left-censored nondetects, for example, the mean of the top 20% of concentrations is reported. This provides almost no insight into the original data. Excluding censored observations removes the

Statistics for Censored Environmental Data Using Minitab® and R, Second Edition.
Dennis R. Helsel.
© 2012 John Wiley & Sons, Inc. Published 2012 by John Wiley & Sons, Inc.

primary information contained in them—the proportion of data in each group that lies below the reporting limit(s). And while better than deleting censored observations, fabricating artificial values as if these had been measured provides its own inaccuracies. Fabrication (substitution) adds an invasive signal to the data that was not previously there, potentially obscuring the information present in the measured observations.

Studies 25 years ago found substitution to be a poor method for computing descriptive statistics (Gilliom and Helsel, 1986). Numerous subsequent articles (see Chapter 6) have reinforced that opinion. Justifications for using one-half the reporting limit usually point back to Hornung and Reed (1990), who only considered estimation of the mean, and assumed that data below the single reporting limit follow a uniform distribution. Estimating the mean is not the primary issue. Any substitution of a constant fraction times the reporting limits will distort estimates of the standard deviation, and therefore all (parametric) hypothesis tests using that statistic. This is illustrated in a later section using simulations. Also, justifications for substitution rarely consider the common occurrence of changing reporting limits. Reporting limits change over time due to methods changes, change between samples due to changing interferences, amounts of sample submitted, and other causes. Substituting values that are tied to changing reporting limits introduces an external (exotic) signal into the data that was not present in the media sampled. Substituted values using a fraction anywhere between 0 and 0.99 times the detection limit are equivalently arbitrary, easy, and wrong.

There have been voices objecting to substitution. In 1967, a US Geological Survey report by Miesch (1967) stated that substituting a constant for censored observations created unnecessary errors, instead recommending Cohen's Maximum Likelihood procedure. Cohen's procedure was published in the statistical literature in the late 1950s and early 1960s (Cohen, 1957, 1961), so its movement into an applied field by 1967 is a credit indeed to Miesch. Two other early environmental pioneers of methods for censored data are Millard and Deverel (1988) and Farewell (1989). Millard and Deverel (1988) pioneered the use of two-group survival analysis methods in environmental work, testing for differences in metals concentrations in the groundwaters of two aquifers. Many censored values were present, at multiple reporting limits. They found differences in zinc concentrations between the two aquifers using a survival analysis method called a score test (see Chapter 9). Had they substituted one-half the reporting limit for zinc concentrations and run a t-test, they would not have found those differences. Farewell (1989) suggested using nonparametric survival analysis techniques for estimating descriptive statistics, hypothesis testing, and regression for censored water quality data. Many of his suggestions have been expanded in the pages of this book. Since that time, a guide to the use of censored data techniques for environmental studies was published by Akritas (1994) as a chapter in volume 12 of the *Handbook of Statistics*. In an applied setting, She (1997) computed descriptive statistics of organics concentrations in sediments using a survival analysis method called Kaplan–Meier. Means, medians, and other statistics were computed without substitutions, even though 20% of data were observations censored at eight different reporting limits.

Guidance documents have evolved over the years when recommending methods to deal with censored observations. In 1991 the *Technical Support Document for Water-Quality Based Toxics Control* (USEPA, 1991) recommended use of the delta-lognormal (also called Aitchison's or DLOG) method when computing means for censored data. Gilliom and Helsel (1986) had previously shown that the delta-lognormal method was essentially the same as substituting zeros for censored observations, and so its estimated mean was consistently biased low. Hinton (1993) found that the delta-lognormal method was biased low and had a larger bias than either Cohen's MLE or the parametric ROS procedure (see Chapter 6 for more information on the latter). The 1998 *Guidance for data quality assessment: Practical methods for data analysis* recommended substitution when there were fewer than 15% censored observations, otherwise using Cohen's method (USEPA, 1998a). Cohen's method, an approximate MLE method using a lookup table valid for only one reporting limit, may have been innovative when proposed by Miesch in 1967, but by 1998 there were better methods available. Minnesota's *Data Analysis Protocol for the Ground Water Monitoring and Assessment Program* presented an early adoption of some of the better, simpler methods for censored data (Minnesota Pollution Control Agency, 1999). In 2002, substitution of the reporting limit was still recommended in the *Development Document for theProposed Effluent Limitations Guidelines and Standards for the Meat and Poultry Products Industry Point Source Category* (USEPA, 2002c). States have forged their own way at times—in 2005 the California Ocean Plan recommended use of robust ROS when computing a mean and upper confidence limit on the mean (UCL95) for determining reasonable potential (California EPA, 2005, Appendix VI). More recently, the *2009 Stormwater BMP Monitoring Manual* (Geosyntec Consultants and Wright Water Engineers, 2009) states "It is strongly recommended that simple substitution is avoided," and instead recommends methods found in this book for estimating summary statistics. And the 2009 *Unified Guidance* on statistical methods for groundwater quality at RCRA facilities (USEPA, 2009) recommended the use of survival analysis methods, although they unfortunately allowed substitution for estimation and hypothesis testing when the proportion of censored observations was below 15%.

1.1   WHY NOT SUBSTITUTE—MISSING THE SIGNALS THAT ARE PRESENT IN THE DATA

Statisticians generate simulated data for much the same reasons as chemists prepare standard solutions—so that the starting conditions are exactly known. Statistical methods are then applied to the data, and the similarity of their results to the known, correct values provides a measure of the quality of each method. Fifty pairs of X,Y data were generated by Helsel (2006) with X values uniformly distributed from 0 to 100. The Y values were computed from a regression equation with slope = 1.5 and intercept = 120. Noise was then randomly added to each Y value so that points did not fall exactly on the straight line. The result is data having a strong linear relation between Y and X with a moderate amount of noise in comparison to that linear signal.

The noise applied to the data represented a "mixed normal" distribution, two normal distributions where the second had a larger standard deviation than the first. All of the added noise had a mean of zero, so the expected result over many simulations is still a linear relationship between X and Y with a slope $= 1.5$ and intercept $= 120$. Eighty percent of data came from the distribution with the smaller standard deviation, while 20% reflected the second distribution's increased noise level, to generate outliers. The 50 generated values are plotted in Figure 1.1a.

The 50 observations were also assigned to one of the two groups in a way that group differences should be discernible. The first group is mostly of early (low X) data and second of later (high X) data. The mean, standard deviation, correlation coefficient, regression slope of Y versus X, a t-test between the means of the two groups, and its p-value for the 50 generated observations in Figure 1.1a were then all computed and stored. These "benchmark" statistics are the target values to which later estimates are compared. The later estimates are made after censoring the points plotted as squares in Figure 1.1a.

Two reporting limits (at 150 and 300) were then applied to the data, the black dots of Figure 1.1a remaining as uncensored values with unique numbers, and the squares becoming censored observations below one of the two reporting limits. In total, 33 of 50 observations, or 66% of observations, were censored below one of the two reporting limits. This is within the range of amounts of censoring found in many environmental studies. Use of a smaller percent censoring would produce many of the same effects as found here, though not as obvious or as strong. All of the data between 150 and the higher reporting limit of 300 were censored as <300. In order to mimic laboratory results with two reporting limits, data below 150 were randomly selected and some assigned <150 while others became <300.

1.1.1 Results

Figure 1.1b–g illustrate the results of estimating a statistic or running a hypothesis test after substituting numbers for censored observations by multiplying the reporting limit value by a fraction between 0 and 1. Estimated values for each statistic are plotted on the Y-axes, with the fraction of the reporting limit used in substitution on the X-axes. A fraction of 0.5 on the X axis corresponds to substituting a value of 75 for all <150s, and 150 for all <300s, for example. On each plot is also shown the value for that statistic before censoring, as a "benchmark" horizontal line. The same information is presented in tabular form in Table 1.1.

Estimates of the mean of Y are presented in Figure 1.1b. The mean Y before censoring equals 198.1. Afterwards, substitution across the range between 0 and the detection limits (DL) produces a mean Y that can fall anywhere between 72 and 258. For this data set, substituting data using a fraction somewhere around 0.7 DL appears to mimic the uncensored mean. But for another data set with different characteristics, another fraction might be "best." And 0.7 is not the "best" for these data to duplicate the uncensored standard deviation, as shown in Figure 1.1c. Something larger or smaller, closer to 0.5 or 0.9 would work better for that statistic, for this set of data. Performance will also differ depending on the proportion of data censored, as

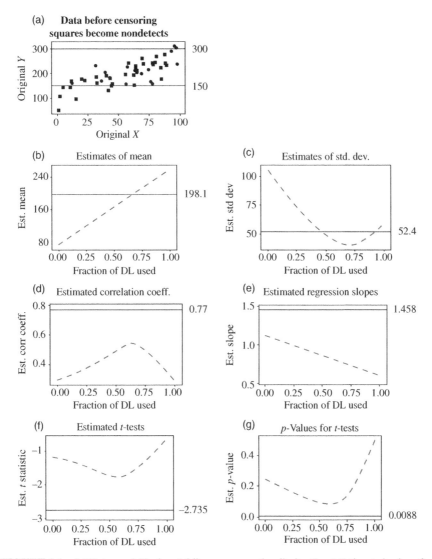

FIGURE 1.1   (a) Data used. Horizontal lines are reporting limits. (b–g) Estimated values for statistics of censored data (Y) as a function of the fraction of the detection limit (X) used to substitute values for each nondetect. As an example, 0.5 corresponds to substitution of one-half the detection limit for all censored values. Horizontal lines are at target values of each statistic obtained using uncensored values.

discussed later. Results for the median (not shown) were similar to those for the mean, and results for the interquartile range (not shown) were similar to those for the standard deviation. The arbitrary nature of the choice of fraction, combined with its large effect on the result, makes the choice of a single fraction an uncomfortable one. As shown later, it is also an unnecessary one.

TABLE 1.1   Statistics and Test Results Before and After Censoring

Procedure	Before Censoring	Range Using Substitution	Using MLE
Mean	198.1	72–258	191.3
Standard deviation	52.4	41–106	54.0
Correlation coefficient	0.77	0.29–0.54	0.55
Regression slope	1.46	0.62–1.12	1.46
t Statistic	−2.74	−1.8 to −0.68	−1.81
p-value for t	0.009	0.08–0.50	0.07

Data in the middle two columns are also shown in Figure 1.1. The right column reports the results of MLE tests expressly designed to work with censored data, without requiring substitution for censored observations.

Substitution results in poor estimates of correlation coefficients (Figure 1.1d) and regression slopes (Figure 1.1e), much further away from their respective uncensored values than was true for descriptive statistics. The closest match for the correlation coefficient appears to be near 0.7, while for the regression slope, substituting 0 would be best! With data having other characteristics, the "best" fraction will differ. Because substituted values at a given reporting limit produce a horizontal line, correlation coefficients and regression slopes are particularly suspect when values are substituted for censored observations, especially if the statistics are found to be insignificant.

The generated data were split into two groups. In the first group were data with X values of 0–40 and 60–70, while the second group contained those with X values from 40 to 60 and then 70 and above. For the most part, values in the first group plotted on the left half of Figure 1.1a, and the second group plotted primarily on the right half. Because the slope change is large relative to the noise, mean Y values for the two groups are significantly different. Before the data were censored, the two-sided t-statistic to test equality of the mean Y values was −2.74, with a p-value of 0.009. This is a small p-value, so before censoring the means for the two groups are determined to be different.

Figure 1.1f and g, and Table 1.1 report the results of two-group t-tests following substitution of values for censored observations. The t-statistics never reach as large a negative value as for the uncensored data, and the p-values are therefore never as significant. At no time do the p-values go below 0.05, the traditional cutoff for statistical significance. Results of t-tests after using substitution, if found to be insignificant, should not be relied on. Much of the power of the test has been lost, as substitution is a poor method for recovering the information contained in censored observations. Figure 1.1f and g show a strong drop-off in performance when the best choice of substituted fraction, which in practice is always unknown, is not chosen.

Clearly, no single fraction of the reporting limit, when used as substitution for a nondetect, does an adequate job of reproducing more than one of these statistics. This exercise should not be used to pick 0.7 or some other fraction as "best"; different fractions may do a better job for data with different characteristics. The process of substituting a fraction of the reporting limits has repeatedly been shown to produce poor results in simulation studies (Gilliom and Helsel, 1986; Singh and Nocerino, 2002; and many others—see Chapter 6). As demonstrated by the long list of research findings and this simple exercise, substitution of a fraction of the reporting limit for censored observations should rarely be considered acceptable in a quantitative analysis. There are better methods available.

When substitution might be acceptable? Research scientists tend to use chemical analyses with relatively high precision and low reporting limits. These chemical analyses are often performed by only one operator and piece of equipment, and reporting limits stay fairly constant. Research data sets may include hundreds of data points, and in comparison our 50 observations appears small. For large data sets with a censoring percentage below 60% censored observations, the consequences of substitution should be less severe than those presented here. In contrast, scientists collecting data for regulatory purposes rarely have as many as 50 observations in any one group; sizes near 20 are much more common. Reporting limits in monitoring studies can be relatively high compared to ambient levels, so that 60% or greater censored observations is not unusual. Multiple reporting limits arise from several common causes, all of which are generally unrelated to concentrations of the analyte(s) of interest. These include using data from multiple laboratories, varying dilutions, and varying sample characteristics such as dissolved solids concentrations or amounts of lipids present. Resulting data like that of She (1997) with 8 different reporting limits out of 11 censored observations is quite typical. In this situation, the cautions given here must be taken very seriously, and results based on substitution severely scrutinized before publication. Reviewers should suggest that the better methods available from survival analysis be used instead.

Is there a censoring percentage below which the use of substitution can be tolerated? The short answer is "who knows?" The US Environmental Protection Agency (USEPA) has recommended substitution of one-half the reporting limit when censoring percentages are below 15% (USEPA, 1998a). This appears to be based on opinion rather than any published article. Even in this case, answers obtained with substitution will have more error than those using better methods (see Chapter 6). Will the increase in error with substitution be small enough to be offset by the cost of learning to use better, widely available methods of survival analysis? Answering that question depends on the quality of result needed, but substitution methods should be considered at best "semiquantitative," to be used only when approximate answers are required. Their current frequency of use in research publications is certainly excessive, in light of the availability of methods designed expressly for analysis of censored data.

1.1.2 Statistical Methods Designed for Censored Data

Methods designed specifically for handling censored data are standard procedures in medical and industrial studies. Results for the current data using one of these methods, maximum likelihood estimation (MLE), are reported in the right-hand column of Table 1.1. The method assumes that data have a particular shape (or distribution), which in Table 1.1 was a normal distribution, the familiar bell-shaped curve.

The right-hand column of Table 1.1 shows that a method designed for censored data produces values for each statistic as good or better than the best of the estimates produced by substitution. MLE accomplishes this without substituting arbitrary values for censored observations. Instead, it fits a distribution to the data that matches both the values for uncensored observations, and the proportion of observations falling below each reporting limit. The information contained in censored observations is efficiently captured by the proportion of data falling below each reporting limit. The specific procedures used, such as the likelihood r correlation coefficient, are described in subsequent chapters. Table 1.1 shows that for two-group tests, correlation coefficients and regression slopes, true differences and nonzero slopes can be missed when substitution is used for censored observations.

1.2 WHY NOT SUBSTITUTE?—FINDING SIGNALS THAT ARE NOT THERE

Comparing two groups of data, one a possibly contaminated test group and the other a control group, is a basic design in environmental science. Trace metal concentrations in the bodies of mayflies in pristine streams could be contrasted to those in streams with industrial outfalls. Particulates in the atmosphere are compared inside and outside of a national park. Cadmium concentrations in soils are tested upwind and downwind of an old smelter site. Blood lead levels in children are contrasted between homes with old and peeling paint to those in homes with lead-free paint. Are concentrations in the test group higher than in the control group?

The classic approach for this design is the two-sample t-test. If data distributions do not follow a normal distribution, the nonparametric Mann–Whitney (also called Wilcoxon rank-sum) test is used instead. With either test, a roadblock looms in the data shown in Table 1.2—there are values below detection limits; several detection limits.

Substitution for the Table 1.2 data produces the data of Table 1.3, and a Mann–Whitney test p-value of 0.015. The equivalence of the groups is rejected, and the test group is declared higher than the control group. Expensive remediation actions might be mandated for conditions that have caused the elevated concentra-

TABLE 1.2 Contaminant Concentrations with Multiple Reporting Limits in a Test and a Control Group

Control Group		Test Group	
<1	<1	<2	<5
<1	<1	<2	<5
<1	<1	3.3	<5
<1	4.1	3.4	<5
1.0	7.0	<2	4.7
1.8	7.5	12.2	<5
2.2	15.4	<5	22.5
<2		6.6	

TABLE 1.3 Contaminant Concentrations in a Test and a Control Group After Substituting One-Half the Reporting Limit for Censored Observations

Control Group		Test Group	
0.5	0.5	1.0	2.5
0.5	0.5	1.0	2.5
0.5	0.5	3.3	2.5
0.5	4.1	3.4	2.5
1.0	7.0	1.0	4.7
1.8	7.5	12.2	2.5
2.2	15.4	2.5	22.5
1.0		6.6	

tions in the test group. Soil is removed. Industrial equipment is modified. Wells are abandoned. People are given new medications.

Now let us pull back a curtain. These data were not field data, but were computer generated. By generating data, the true situation is known. All of the data in Table 1.2 came from the same distribution—there is actually NO difference in their mean or median levels (see Figure 1.2). For the original uncensored data, the Mann–Whitney test produced a one-sided p-value of 0.43, stating that there is no evidence for difference between the two groups. Any reasonable method for analyzing the data with censored observations should also find no difference in the two groups. For example, in Chapter 9 a Wilcoxon score test is presented, a nonparametric test to compare two groups of data with multiple thresholds. No substitution is involved, and the test produces a p-value of 0.47 for the censored Table 1.2 data. No difference. No contamination. No remediation. But following substitution, a difference was declared.

The examples in these two sections have demonstrated that substitution for censored observations can lead to "finding" either false differences that are not there, or false no-differences when data are truly not equivalent. Substitution implants

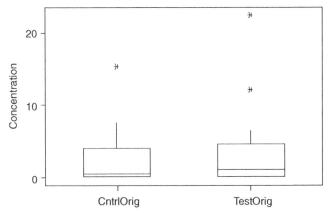

FIGURE 1.2 Boxplots for data of Table 1.2 prior to setting artificial reporting limits. Mann–Whitney test p-value (uncensored data) = 0.43.

an invasive pattern into the data that may be quite different than the pattern of the data itself. Substitution is not neutral.

1.3 SO WHY NOT SUBSTITUTE?

The only conclusion possible based on these two simulations is that substitution of values tied to the reporting limit, still the most commonly used method in environmental studies today, is NOT a reasonable method for interpreting censored data. The first simulation demonstrated that an invasive pattern not present in the original data was implanted by substitution, hiding signals that are really there. Causes of contamination are missed, and human or ecosystem health is needlessly endangered. The second simulation shows that the invasive pattern of substitution can introduce a signal that is not there in the data. Expensive cleanup measures may be implemented where none are needed. Substituting values as "real data" that are a function of the process used by the laboratory, are a function of time, or of the dilution of the samples, or of interferences in some samples but not others, or of the mass of material submitted to the laboratory, can easily impose an artificial, invasive pattern that originally was not there. The result is not just an incorrect conclusion by a hypothesis test. In the real world, contamination goes unnoticed. Remediation goes undone. Public health is unknowingly threatened.

There are better ways.

1.4 OTHER COMMON MISUSES OF CENSORED DATA

In addition to the two previous misuses of censored data:

(1) deleting/ignoring nondetects and computing the mean of what's left, or
(2) substituting a fraction of the reporting limit for censored observations, these two flawed approaches to evaluating censored data are fairly common:
  • substituting a value for the variance, standard deviation, or coefficient of variation (CV)
  • interpreting changes in the percent of detections while the reporting limit is changing.

There are methods for estimating the variability of censored data (see Chapter 6), and measures of location such as mean and median. Unknowingly, people have instead fabricated a number that seems "reasonable" to them. Fabricated values have made their way into some environmental regulations, where 0.6 for the CV (the ratio of the standard deviation to the mean) is currently popular. Douglas Adams would no doubt have chosen 0.42. These guessed values could be very far off, with unwarranted consequences either to human or ecological health, or to the cost of monitoring programs. The three methods in Chapter 6—MLE, Kaplan–Meier, and ROS—will each estimate the mean and standard deviation, and so the CV, for censored environmental data. There is little reason to guess a value.

Scientists also draw conclusions based on the percent of detected values, as that statistic changes between groups or through time. We will recommend the practice later in this book. However, this analysis is suspect when the definition of "detection" changes—the reporting limit changes—between groups or through time. Envision two sets of identical concentrations where the first was measured 10 years ago, and the second measured this year. They are exactly the same concentrations. There has been no physical or chemical change. The early data were censored with a mix of two reporting limits, at 1 and 10 µg/L:

<1 <1 <1 3 5 7 9 <10 <10 <10 <10 <10

while this year's data were measured with better instruments. Now the only reporting limit is at 1 µg/L:

<1 <1 <1 3 5 7 9 <1 2 2 3 5

The analyst then computes that there were only 33% detects 10 years ago, but now there are 67% detects of this dangerous chemical. The percentage has dramatically increased, and something must be done to correct it! As you can see, this change is entirely due to the change in the mix of reporting limits used in the two groups. Comparing percent detections between groups, over space or over time only makes sense when the mix of reporting limits is constant.

Government agencies have routinely reported percent detections of pesticides and other organics in drinking water supplies, surface waters, or ground waters by compiling existing data from multiple sources. Detection limits for each chemical usually varies by source of data and over time. Maps of percent detections purport to give a regional picture of where water quality is better or worse. Decreased detection rates are cited as evidence for improving quality. Yet with the definition of "detection" changing, a change in the proportion of data sources or amounts of recent versus early data at each site can severely skew the resulting statistics. Rather than summarizing the "percent detections," statements about "the percent of concentrations above 1 µg/L" or another well-defined threshold are much more easily interpreted. In the midst of moving detection thresholds, statements such as "Data was closely checked and it was confirmed that the detection limit changes did not affect the trend [in percent detections] significantly" (Ontario Ministry of the Environment, 2010) are hard for a reader to evaluate or believe.

Instead of computing the percent detections above a moving target, this book recommends either doing so only after recensoring all data to the highest reporting limit in the data set, a simple procedure but which may lose information, or instead using survival analysis methods that correctly account for differing reporting limits. If the metric reported and discussed is the percent of detected observations, inspect the definition of "detection" to certify that the reporting limit has not changed as in the small example above. If it has, it and not the underlying concentrations may be the cause of any shift in the percent of detections observed.

2 Three Approaches for Censored Data

Substitution is quick and easy. The lack of care it represents is evidence of a general skepticism about the information content of observations below reporting limits. In truth, a great deal of information is available in censored data. If efficient methods are used, the information extracted from them is almost equal to that for data with single known values. Their information is primarily contained in the proportion of data below the threshold value(s). Knowing that for one data set, three quarters of the data are below the detection limit, while a second data set has only 10% below the same limit, strongly indicates that the first data set contains lower values than the second. This is evident without any knowledge of values above the detection limit. Efficient procedures for censored data combine the values above the detection limit(s) with the information contained in the proportion of data below the detection limit(s) in order to reach a result.

Example—Information Content of Censored Observations:
Estimate the center (median) of the following data.

<1 <1 <1 <1 <1 <1 <1 5 12 22

If there were no information content in censored observations, the seven censored values could be discarded and the median of the three remaining values would equal 12. However, 12 is a very poor estimate of the center of a data set in which 70% of the observations are below 1. A much better estimate would be <1. There is a great deal of information in the lowest values in the data set—the issue is how to best extract that information.

There are three approaches that are far better than substitution for extracting information from data sets that include censored observations.

1. Nonparametric methods after censoring at the highest reporting limit.
2. Maximum likelihood estimation—survival analysis procedures assuming a specific distribution.
3. Nonparametric survival analysis procedures.

Statistics for Censored Environmental Data Using Minitab® and R, Second Edition.
Dennis R. Helsel.
© 2012 John Wiley & Sons, Inc. Published 2012 by John Wiley & Sons, Inc.

The first approach, nonparametric methods after censoring at the highest reporting limit, can be used as an alternate to substitution when the desire is to keep it simple. This approach does not have the power of the other two approaches, but may be all that you need without learning a new vocabulary. The other two approaches are standard survival analysis procedures. A general description of each approach follows. The methods representing all three approaches are found in the later chapters of this book.

2.1 APPROACH 1: NONPARAMETRIC METHODS AFTER CENSORING AT THE HIGHEST REPORTING LIMIT

Substitution is simple. Yet there are other simple procedures that work far better for data with censored observations. Unlike substitution, the methods represented by this approach do not add artificial, invasive data that create a pattern alien to the original observations. They do not declare that you know more than you actually do. They will not introduce and then "find" a signal such as a trend or difference among groups that is not there in the original data. So before delving into the more complex methods of survival analysis, these procedures can be applied to data containing censored observations, are relatively simple and familiar, and yet do not involve substituting a value for censored observations.

Nonparametric methods deal with ranks (or percentiles) of the data. The position each observation holds in the data set is used for analysis. Nonparametric methods are distribution-free—data are not assumed to follow any one distribution, such as the normal, in order that test results be accurate. Censored data can be directly used if there is only one reporting limit, or recensored to the highest reporting limit, to apply these relatively simple methods. The survival analysis methods described in the next two approaches better utilize the information in censored data, especially when there is more than one detection limit. In particular, survival analysis methods will incorporate the information available from uncensored observations below the highest reporting limit. The methods of this first approach treat those uncensored observations only as tied with all other observations below the highest reporting limit. However, if you wish to stay simple, this approach is familiar and may be all that you need. It will produce far more reliable results than when fabricating data, such as substituting one-half the detection limits and running a t-test or regression. Two types of simple nonparametric procedures are presented, binary methods and standard nonparametric methods.

2.1.1 Binary Methods: Above Versus Below the Reporting Limit

Methods based on the binomial distribution deal with data categorized into one of the two classes. To use binary methods, first recode censored data into two classes, either above or below the highest reporting limit (if there is just one reporting limit, it is above and below that single limit). Then compute descriptive statistics, perform hypothesis tests, and build regression models using a binary response (y) variable. Familiar methods in this category include the test of proportions (also known as contingency tables) and logistic regression.

2.1.2 Ordinal Methods, First Censoring at the Highest Reporting Limit

Binary methods do not use all of the information available for data above the reporting limit, instead categorizing all detected/quantified observations as only "equal to or above the reporting limit." This results in a loss of power, a lower ability to detect a signal if present, as compared to methods that use the relative order of uncensored observations. Standard nonparametric methods such as the Mann–Whitney test assign separate ranks to all uncensored values, preserving the information in their ordering, without either assuming a specific distribution for the data or inserting fabricated values. All data below the highest detection limit, or below a single detection limit if there is only one, are represented by tied ranks. Familiar nonparametric methods such as the Mann–Whitney (rank-sum) and Kruskal–Wallis tests can be easily applied to censored data with one reporting limit. If a signal (differences between groups, a significant regression relationship) is found, it can be believed, unlike when fabricated values have been substituted.

2.2 APPROACH 2: MAXIMUM LIKELIHOOD ESTIMATION

Maximum likelihood estimation (MLE) is increasingly used in environmental studies—Owen and DeRouen (1980) for air quality and Miesch (1967) for geochemistry are two early examples. MLE uses three pieces of information to perform computations: (a) numerical values above reporting limits, (b) the proportion of data below each reporting limit, and (c) the mathematical formula for an assumed distribution. Data both below and above the reporting limit are assumed to follow a distribution such as the lognormal. Parameters are computed that best match a fitted distribution to the observed values above each reporting limit and to the percentage of data below each limit.

The most crucial consideration for MLE is how well data fit the assumed distribution. A major problem with MLE is that for small data sets there is often insufficient information to determine whether the assumed distribution is correct or not, and so whether parameters are estimated reliably. MLE has been shown to perform poorly for data sets with less than 25–50 observations (Gleit, 1985; Shumway et al., 2002). For larger data sets, MLE is an efficient way to estimate parameters, given that the chosen distribution is correct. The term "efficient" means that the fitted parameters have relatively small variability, so that their confidence limits are as small as possible. For data sets of at least 50 observations, and where either the percent censoring is small (so that the distributional shape can be evaluated) or the distribution can be assumed from knowledge outside the data set, MLE methods are the methods of choice.

MLE methods are computed by solving a likelihood function L, where for a distribution with two parameters β_1 (mean) and β_2 (variance), $L(\beta_1, \beta_2)$ defines the likelihood of matching the observed distribution of data. The function L increases as the fit between the estimated distribution and the observed data improves. The parameters β_1 and β_2 are varied in an optimization routine, choosing values to maximize L. In practice, it is the natural logarithm $\ln(L)$ rather than L itself that is maximized, where $\ln(L)$ is the "log-likelihood", usually (though not necessarily) a negative number. Maximizing $\ln(L)$ is accomplished by setting the partial derivatives of $\ln(L)$ with respect to the two parameters equal to zero.

$$\frac{d(\ln L[\beta_1])}{d(\beta_1)} = 0 \text{ and } \frac{d(\ln L[\beta_2])}{d(\beta_2)} = 0 \tag{2.1}$$

The exact equation for L will change depending on the assumed distribution and the process under study (estimation of a mean, linear regression, etc.). However, in each case the likelihood function L is the product of two component pieces, one for censored observations and one for uncensored (detected) observations. In the uncensored piece is the probability density function $p[x]$, the equation describing the frequency of observing individual values of x. In the censored piece is the survival function $S[x]$, which is the probability of exceeding the value x. $S[x]$ equals $1 - F[x]$, where $F[x]$ is the cumulative empirical distribution function (edf) of the distribution, the probability of being less than or equal to x. Either $S[x]$ or $F[x]$ can be used when writing the likelihood function.

In the most general case, L can be considered to be the product of three pieces, where the censored data component is split into two, one for left-censored and one for right-censored data:

$$L = \prod p[x] \prod (F[x]) \prod S[x] \tag{2.2}$$

where $p[x]$ is the pdf as estimated from the uncensored observations, $(F[x])$ is the edf as determined by left-censored observations, and $S[x]$ is the survival function as determined by right-censored observations ("greater-thans"). Greater-thans are not typically found among environmental data, and so likelihood functions in environmental studies typically deal with only the first two pieces.

For censored data, two variables x and δ are required to represent each observation. The value for the measurement, or for the reporting limit, is given by x. The indicator variable δ is a 0/1 variable that designates whether an observation is censored (0) or detected (1). As one of the simpler likelihood functions, the equation for L when estimating the mean and standard deviation of a normal distribution using MLE is

$$L = \prod p[x_i]^{\delta_i} \cdot F[x_i]^{1 - \delta_i} \tag{2.3}$$

where δ is as defined above, and for a normal distribution the pdf is

$$p[x] = \frac{\exp\left[(-1/2)((x-\mu)/\sigma)^2\right]}{\sigma\sqrt{2\pi}} \tag{2.4}$$

For uncensored observations, $\delta = 1$ and the second term in equation 2.3 becomes 1 and so drops out. For censored observations, $\delta = 0$ and the first term becomes 1 and so drops out. The cumulative distribution function for a normal distribution is

$$F[x] = \Phi\left[\frac{x-\mu}{\sigma}\right] \tag{2.5}$$

where Φ is the cdf of the standard normal distribution

$$\Phi[y] = \frac{1}{\sqrt{2\pi}} \int_0^y \exp(-u^2/2)du \tag{2.6}$$

After substituting in the above and setting the partial derivatives of $\ln(L)$ equal to 0 (equation 2.1), the nonlinear equations are solved by iterative approximation using the Newton–Raphson method. The solution provides the parameters mean and standard deviation for the distribution that best matches both the pdf and cumulative distribution function (or $1 -$ survival function) estimated from the data. In other words, the estimates of mean and standard deviation will be the parameters for the assumed distributional shape that had the highest likelihood of producing the observed values for the uncensored observations and the observed proportion of data given below each of the reporting limits.

Likelihood methods can be used when performing an hypothesis test. The test is set up to determine whether $\beta = 0$, where β is the parameter of interest. This could be a slope coefficient in regression, or an estimate of the difference between two population means. The null hypothesis of $\beta = 0$ is compared to an alternative that $\beta \neq 0$ using one of two types of test procedures, either likelihood-ratio tests or Wald's tests.

Likelihood-ratio tests are based on the value for the log of the likelihood function, $\ln(L)$. The test compares the log-likelihoods for two models, one where $\beta =$ the value chosen by MLE, and the second for the "null" state, $\beta = 0$. The test statistic takes the form of $-2 \ln L(\beta) - (-2 \ln L(0))$, resulting in a positive value if $\beta \neq 0$. This difference is the likelihood-ratio test statistic, and is compared to a chi-squared distribution to produce the p value for the test. Likelihood-ratio tests are the form used by most statistical software that perform maximum likelihood.

Wald's test statistics take a form similar to t-tests in regression. The numerator of the test statistic is the MLE value for the coefficient β, and the denominator is the standard error of β. Their ratio is compared to a standard normal distribution. Wald's tests are generally not considered as accurate as are likelihood-ratio tests and the latter are preferred, though the differences in p values are often small.

2.3   APPROACH 3: NONPARAMETRIC SURVIVAL ANALYSIS METHODS

Nonparametric methods are so named because they do not involve computing "parameters," such as the mean or standard deviation, of an assumed distribution. Instead they use the relative positions (ranks) of data, a reflection of the data's percentiles. Because these methods do not require an assumption about the distribution of data, they are also called "distribution-free" methods. Nonparametric methods are especially useful for censored data because they efficiently use the available information. Censored observations are known to be lower than values above their reporting limit, and so are ranked lower. These methods do not require estimates of the unknown distances between censored observations and uncensored values, but only their relative order.

Nonparametric methods are now commonly used in the environmental sciences. There is general recognition that many variables measured in natural systems have skewed distributions, and nonparametric procedures have greater power than parametric procedures for skewed data, especially data with outliers. Normal theory tests may work well after transforming data, but a transformation that corrects nonnormality for all groups of data is often difficult to find. Textbooks such as Gilbert (1987) and Helsel and Hirsch (2002) have demonstrated nonparametric procedures and their usefulness to environmental studies. However, nonparametric score tests, developed for data with multiple thresholds, are still not familiar to most environmental scientists, and are woefully underutilized. Score tests are extensions of the more familiar rank-sum, sign, and contingency table tests to situations with multiple thresholds. They are found in statistical software along with other methods for survival analysis.

2.4   APPLICATION OF SURVIVAL ANALYSIS METHODS TO ENVIRONMENTAL DATA

Consider a typical survival analysis problem, a test of whether light bulbs with a new filament composition last longer than those with the existing filament. A group of 15 light bulbs for each filament type is connected to power, and the length of time each burns is measured. After 48 h, it is decided that this sample size is too small, and 20 additional light bulbs of each type are added to the test. After 6 weeks (1008 h) from when it was begun, the experiment is stopped. By that time, many of the bulbs have burnt out, and their burn lengths recorded. However, some of the bulbs started in the second batch are still burning when the experiment ends. Their burn lengths are recorded as "greater than 960 h," because they were still burning after 1008–48 h of use. A few of the bulbs in the original batch are also still burning, and their lengths are recorded as "greater than 1008 h."

Of interest is whether bulbs of both filament types have the same mean or median burn length. If all bulbs had burnt out, the lengths for every bulb would be known and a t-test or rank-sum test could be used to test for differences. However, for some of the bulbs the actual length is not known, but are censored as "greater-thans." Because there are two different thresholds resulting from the different times a bulb entered into the experiment, these data sets are also "multiply censored." Survival analysis methods were designed for such right-censored data sets with multiple censoring thresholds.

Environmental data are also often censored, with a number of nondetect values included in the data set. They are often multiply censored, as detection limit thresholds change over time or with varying sample characteristics or among different laboratories. The primary difference between environmental and industrial/medical data is that concentration data are dominantly left-censored, where low-level concentrations are known only to be below a laboratory reporting threshold. There are examples of right-censored environmental data: flood magnitudes that are known based on historical records to be at least a certain cubic feet per second, but probably more; or transmissivity estimates based on specific capacity measurements. Specific capacity is affected by the well, resulting in an estimated transmissivity (T) that is lower than the true value in the aquifer. T is "greater than" the estimate, but the amount greater is unknown. For these right-censored examples, survival analysis software can be used directly. The situation is a little more complicated for left-censored data with censored observations, the focus of this book.

Nonparametric survival analysis software is often hard-wired for right-censored data. Left-censored environmental data must be transformed into right-censored data before these routines can be used. Parametric maximum likelihood methods for censored data usually allow interval-censored data to be input. Interval censoring is the most flexible format for censored data entry (see Chapter 3), and data with censored observations can be entered directly with these methods.

To demonstrate the transformation from left to right censoring, consider a left-censored data set with five observations shown as a bar graph on the left side of Figure 2.1. The censored observations are shown as open white bars and the uncensored detected values as gray-shaded bars. These data have the following values:

<1, 3, <10, 12, 17

The data measured when the detection limit was 1 are the values of <1, 3, and 12. The data measured when the detection limit was 10 are the values of <10 and 17.

Figure 2.1 also shows these same data as dark bars drawn down from the upper end of the plot in addition to those from the lower end. The only difference is the datum of the base of the bar, which is now 25 and looking down rather than at 0 and looking up. The new data at 25 was chosen simply because it is larger than the largest value in the data set, so a finite bar length occurs when looking down from the new datum. The lengths of the new dark bars are 8, 13, and 22 corresponding to the uncensored values of 17, 12, and 3 (25–17, 25–12, and 25–3). The lengths of the two dark bars for the censored values are >15 and >24.

The bars drawn down from the top represent the same data, on an alternate measurement scale. These new bars, which include greater-thans, are right-censored

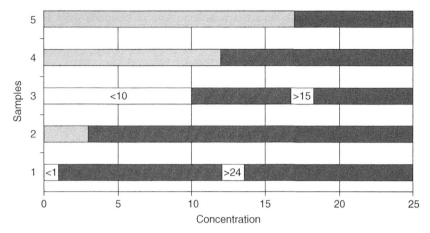

FIGURE 2.1   Five observations, including two observations recorded below one of the two detection limits. Bars from the top–down show conversion to right-censored data.

data. Left-censored data (the gray and white bars) can be transformed to right-censored data (the dark bars) by subtracting each value from the same large number (equation 2.7), in this case 25. The constant can be any value larger than the maximum in the original data. This left-to-right transformation is called "flipping" the data distribution and is a linear transformation—the transformation does not alter the shape of the data distribution other than to reverse its direction. For example, Figure 2.2 shows a boxplot of concentration on the left, and a boxplot of its flipped values on the right. They have the same shape, except one is the mirror image of the other in the vertical direction. They are "the same data"—one is a simple linear transformation of the other.

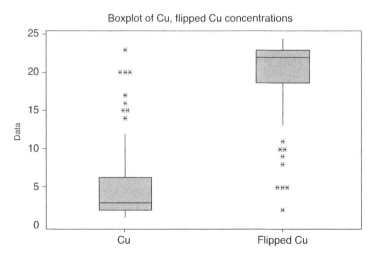

FIGURE 2.2   Copper (Cu) concentration data on the left. Flipped Cu concentrations on the right. Flipping converts left- to right-censored data.

If the survival analysis software available to you is designed only for right-censored data, environmental data can be "flipped" into a right-censored form, and the data analyzed.

$$\text{Flipped Data} = \text{Constant} - \text{Original Data} \qquad (2.7)$$

For example, SAS's PROC LIFETEST performs nonparametric tests of differences between groups of survival data. The procedure handles only right-censored "greater-thans". Data with censored observations must first be flipped in order to use these routines. Nonparametric hypothesis tests on flipped data will have the same test results for determining significance, the same p values, as if the software had allowed the original left-censored data to be used as input. The p values can be directly used to compare differences between distributions, and so on. Data are transformed by flipping only to accommodate the software input requirements. Measures of location (mean, median, and other percentiles) using flipped data must go through a reverse transformation to obtain estimates in the original units. Measures of spread or variability, such as the standard deviation and IQR, are the same for both original and flipped data, requiring no retransformation. Slopes of regression equations using flipped response (Y) variables must have their signs reversed to represent slopes of the original data. Perhaps with sufficient interest from environmental scientists, software

TABLE 2.1   Parallel Statistical Methods for Uncensored and Censored Data

Methods for Uncensored Data Sets	Methods for Censored Data
Computing summary statistics	
Descriptive statistics	Kaplan–Meier, MLE, or ROS estimates
Comparing two groups	
t-test	Censored regression with 0/1 group indicator
Wilcoxon rank-sum test	Wilcoxon rank-sum test
Paired t-test	Censored CI on differences
(Paired) sign or signed-rank test	PPWtest
Comparing three or more groups	
ANOVA	Censored regression with 0/1 group indicators
Kruskal–Wallis test	Generalized Wilcoxon test
Correlation	
Pearson's r	Likelihood r
Kendall's τ	Kendall's τ
Linear regression	
Linear regression	Censored regression
Logistic regression	Logistic regression
Theil–Sen median line	Akritas–Theil–Sen median line
Exploration of multivariate patterns	
PCA, factor analysis	PCA, factor analysis on ranks and u-scores
NMDS	NMDS on ranks and u-scores
MANOVA	ANOSIM on ranks and u-scores
Least-squares tests for trend	Test of seriation on ranks and u-scores

companies will code their routines to allow left-censored and right-censored data to be input. Alternatively, input of interval-censored data may become standard for both parametric and nonparametric routines. But neither of these expanded input formats is generally available today.

2.4.1 Power Transformations with Survival Analysis

Parametric survival analysis methods, those computed using maximum likelihood, require that data follow a specific distribution. If after looking at plots or normality tests the data do not appear to follow this distribution, prior transformation of the data is necessary. Flipping the data from left- to right-censored does nothing to alter the skewness or outliers of a data set. Power transformations such as the square root or logarithm that change the shape of a distribution must be done prior to flipping the data into a right-censored format. The order of processing is therefore, as follows:

1. Decide whether a power transformation is necessary to alter the data's shape to be closer to the assumed distribution. If so, transform the data.
2. Flip the transformed data to produce a right-censored distribution (equation 2.7).
3. Compute the survival analysis test and interpret the test results.
4. Convert estimates of mean or median back to left-censored format by subtracting from the constant used to flip the data. If necessary, retransform using the reverse power transformation to get parameter estimates in original units.

2.5 PARALLELS TO UNCENSORED METHODS

A list of maximum likelihood and nonparametric survival analysis methods that can be used to analyze censored data is given in Table 2.1. On the left are familiar methods used for uncensored data sets, data without censored observations. On the right are the equivalent methods used for censored data. Some of the censored methods are direct extensions of the uncensored procedure. Others are computationally very different, but perform the same function.

3 Reporting Limits

"Reporting limit" is an intentionally generalized term that represents a variety of thresholds used to censor analytical results. It is a limit above which data values are reported as single numbers without qualification by the analytical laboratory. Unfortunately, the terms "detection limit," "detected value," and "nondetect" are ubiquitous and used in both a specific and general sense. In this edition of the book I have attempted to be specific in meaning whenever possible. I have reserved the term "detection limit" for its more specific meaning described in this chapter, and used "reporting limit" whenever I refer to a censoring limit that might be either a detection or quantitation limit. The difference between detection limits and other types of reporting limits is an important one to understand. I have not tried to invent new terms, however, so a "nondetect" is a value below a reporting limit, and a "detected value" is a measurement above the reporting limit.

Reporting limits are set in a variety of ways, and for a variety of purposes. As stated in a report summarizing the calculation of reporting limits (USEPA, 2003): "one conclusion that can be drawn is that detection limits are somewhat variable and not easy to define." Yet there are several things each type of reporting limit has in common. Each is a threshold computed so that measured values falling below that threshold are reported differently than those falling above. Most reporting limits are based on a measure of the variability or noise inherent in the laboratory process. The two general classes of reporting limits are split between those that assume this noise is constant over different concentrations, versus those that model the noise as a function of concentration. The standard deviation of repeated measurements is used to quantify the noise of the analytical process. First we discuss reporting limits based on constant standard deviation, and then those based on varying standard deviation.

From the data users' point of view, any method that changes an observed numerical measurement into a censored value prior to reporting the data to the user is a "reporting limit." It may have been developed in a variety of ways, but all require the user to somehow interpret data labeled as a "nondetect" or "less-than." The focus of this book is to provide methods that deal with data censored at reporting limits, regardless of the type of limit employed. However, knowledge by the data user of the type of limit employed can lead to better data analysis. Chapter 4 presents methods

Statistics for Censored Environmental Data Using Minitab® and R, Second Edition.
Dennis R. Helsel.
© 2012 John Wiley & Sons, Inc. Published 2012 by John Wiley & Sons, Inc.

that distinguish data below a detection limit from those between the detection and quantitation limit.

3.1   LIMITS WHEN THE STANDARD DEVIATION IS CONSIDERED CONSTANT

Reporting limits are computed with a single value for the standard deviation when it is assumed that the noise of the measurement process is constant from concentrations of zero up to the highest reporting limit. Reporting limits for constant standard deviation can be classified into two general types, most often called detection limits and quantitation limits. The two types of limits differ in how they are computed, and in what they represent.

3.1.1   The Detection Limit

Values measured above this threshold are unlikely to result from a true concentration of zero.

A detection limit is a threshold below which measured values are not considered significantly different from a blank signal, at a specified level of probability. Measurements above the detection limit evidence a nonzero signal (at a given probability), indicating that the analyte is present in the sample. Other terms used for this type of threshold have included the "critical value" and "decision level" of Currie (1968), as well as the "method detection limit" or MDL (USEPA, 1982) and the "limit of detection" or LOD (Keith, 1992). The basic ideas were established in the seminal paper by Currie (1968), and variations since then still depend heavily on Currie's concepts. USEPA (2003) and USEPA (2007) discuss the differences among variations of what is generally recognized as thresholds having the same objective—to distinguish samples with a concentration signal from those without a signal. Brüggemann et al. (2010) review the variations used within the European Union, including those where limits are set using tolerance intervals rather than using the t-distribution.

The first step in computing a detection limit is to estimate the inherent variation to be expected at a concentration of zero, where no analyte is present. Currie (1968) envisioned repeated measurements of blank solutions, but this is difficult to do successfully. Instead, repeated measurements of a standard solution of low concentration is usually substituted for blanks. Figure 3.1 illustrates the process, using a standard at a concentration value of 2. The analyst is assuming that the measurement error at zero concentration is the same as at the low standard concentration—the standard deviation is constant between zero and the concentration standard. Measurement variation is almost always assumed to follow a normal distribution around the true value. The left-hand curve in Figure 3.1 illustrates the possible measured concentrations when the true concentration is zero. One-half of the measurements would be negative. The y-axis in the figure is the number of measurements for each value of concentration. Though the most frequently observed value is the true concentration at the center of each normal curve (assuming 100% recovery and no

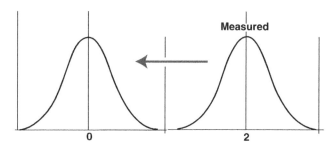

FIGURE 3.1   Error distribution of a measured standard at a concentration of 2, which is then imputed to also apply to a true zero concentration.

other bias), a variety of other measurements result from the same sample, due to random variation in the measurement process. A limit called the "critical value" (USEPA, 2007, Appendix D; IUPAC, 1997), abbreviated L_C, is set near the upper end of the distribution (Figure 3.2). When an instrument measures a value above this limit, the concentration is unlikely to have resulted from a true concentration of zero—unlikely to be a false positive.

The curve in Figure 3.2 describes the possible measured values that may result from a true concentration of zero. Due to random variability, half will be positive and half negative. Of most interest is how far away from zero in the positive direction those measurements are likely to fall. This can be described using the standard deviation of the data, relating the distance from the center to a statement of probability through the t-distribution. The mean signal has about an 18% probability of falling at least 1 standard deviation above zero when the true concentration is zero, given that the standard deviation was computed using a sample size of seven replicates. The probability of falling at least 2 standard deviations above zero, when the true concentration is zero, is about 5%. At a distance of 3 standard deviations, the

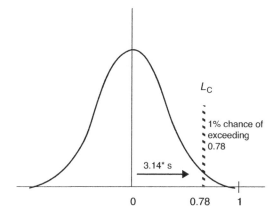

FIGURE 3.2   Setting a critical level L_C at 3.14 times the standard deviation above zero. This protects against false positives.

probability is just over 1%. The choice of a distance to represent the detection limit is made so that no more than a small percentage of the measured mean values truly originating from a zero concentration will fall above the limit. As one example of setting a detection limit, USEPA (1982) describes the procedure for computing what it calls the MDL. Seven or more replicate analyses are performed on a standard where the chemical is present at a low concentration. The standard deviation (abbreviated s) of these measurements is then multiplied by 3.14, the one-sided t-statistic for a sample size of $n = 7$ and probability of exceedance of 1% ($\alpha = 0.01$). This false exceedance rate is called the false positive or Type I error, the rate of measuring a mean further to the right than the dashed line pictured in Figure 3.2. For a standard deviation (s) equal to 0.25, the critical level (L_C, also USEPA's MDL) is set at a value of 3.14 × 0.25, or 0.78. Current USEPA guidelines are to use L_C as the detection limit (USEPA, 2007). Any measurement falling above 0.78 would be declared to have a concentration that is significantly different from zero at a 1% false-positive rate, and the analyte is declared to be present in the sample.

IUPAC guidelines differ from this (IUPAC, 1997). The critical level is just the first step in calculating their definition of the detection limit. Rather than using L_C to distinguish the measurement from a true value of 0, IUPAC recommends that the detection limit be set sufficiently high that a detected value can be distinguished from the measurement response of blanks. The distribution of measurement responses of blanks is the bell-shaped curve centered on 0. Therefore, the detection limit is raised to the point (Figure 3.3) where there is only a small probability that a concentration at the detection limit would be mistakenly measured as below L_C. This is sometimes referred to as protecting against false negatives. Using the same probability of error as when computing the critical level, the IUPAC detection limit would be twice the critical level. IUPAC guidelines calculate the distance using a probability of error of 5% for both L_C and the detection limit, while USEPA guidelines use an error rate of 1% for computing the L_C/detection limit single value.

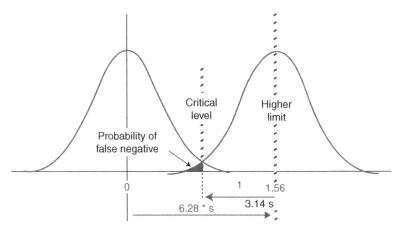

FIGURE 3.3 Establishing a higher limit at twice the critical level, or 6.28 times the standard deviation, to protect against false negatives.

These two definitions of a detection limit assume that the underlying laboratory variability can be described by a single standard deviation. In practice, measurement variability changes from day to day, analyst to analyst, instrument to instrument, and from sample to sample as a function of the sample matrix. This is particularly true in a production lab, where any given sample entering the door may be assigned to one of a number of people and instruments. One approach to account for these changes was devised by the U.S. Geological Survey and called the "Long-Term Method Detection Level" or LT-MDL (Oblinger-Childress et al., 1999). The standard deviation used to compute the LT-MDL incorporates the variability among the multiple instruments and multiple operators any given sample may be assigned to upon entering the laboratory. The LT-MDL is re-evaluated each year, as equipment and operating conditions change. This and similar processes that more correctly track the precision of production laboratories make it even more likely that reporting limits will change over time (multiply-censored data). This in turn means that end-users must understand and use interpretation methods that correctly incorporate data having multiple reporting limits.

False negatives, also called Type II errors, are one motivation to use a higher reporting limit than the critical level. The concept of a false negative provides an important example of the difference in perspective between laboratories and data users. From a laboratory's point of view, a serious error is made if an individual sample whose true value is at or above the detection limit is reported as below the critical level, and so in the range of measurement responses for blanks. The argument goes that a true concentration exactly at the critical level has a 50% chance of being recorded as below the critical level, and so erroneously reported as a nondetect (Figure 3.4). To avoid this, a higher threshold is instituted.

However, there are two counterarguments that lead to making no adjustment for false negatives, as per the USEPA definition. First, creating a higher threshold value does not in itself reduce the likelihood of false negatives. A true concentration at a higher detection

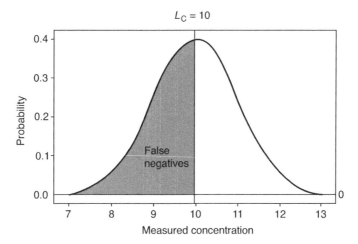

FIGURE 3.4   Probability of a "false negative" when the true concentration is at a critical level of 10. Shaded probability equals 50%.

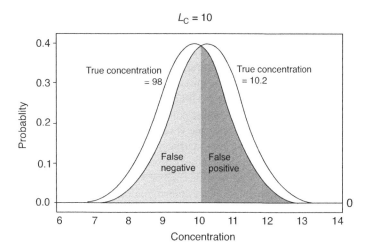

FIGURE 3.5   Balanced probabilities of false negatives and positives for true concentrations just below and above a detection limit of 10.

limit has a 50% chance of being erroneously reported as a censored value below that higher limit. The limit is higher, but the effect is the same. In fact, for any concentration there is a 50% chance that the measured value will be below the true level, assuming 100% recovery. Creating a higher threshold does not in itself solve this problem.

Second, if the true concentration were just a hairs-breadth below a detection limit, there is an almost 50% chance that the measured value will exceed the limit, and so be erroneously reported as a detected value. The same error characteristics surround each true concentration. Looking at measurement errors from a collective standpoint, as a data user would do, the positive and negative errors will tend to balance out (Figure 3.5). Without adjusting for false negatives, the proportion of values falling within each interval of concentration remains correct if each measurement has the same variability. The same percent of values near each boundary will by chance fall into a higher category as those falling into a lower category. So from a perspective of the overall data set, there is little need to censor data based on an avoidance of false negatives. They are balanced by errors in the opposite direction.

Perhaps the most helpful way to distinguish these two philosophies is that the critical level/USEPA definition provides a $(1 - \alpha)\%$ probability that a true (mean) concentration of zero will be below this limit, while the IUPAC higher definition provides a $(1 - \alpha)\%$ probability that a mean concentration at the detection limit will not overlap measurements of blanks.

3.1.2   The Quantitation Limit

Thresholds above which single numerical values (rather than an interval) are reported.

Quantitation limits specify a threshold above which reliable single numbers can be reported. Thresholds with this purpose have been given names such as the "limit of quantitation (LOQ)" (Keith, 1992), and the "practical quantitation limit (PQL)." They are not generally computed with a statistical definition, but more of an experienced feel for where the method precision warrants reporting single values. A common rule is that the quantitation limit is 10 times the standard deviation of a low standard such as the one used to define the method detection limit. The factor of 10 has been around for a number of years (USEPA, 2003), and a concentration 10 times the background variability is considered large enough by most chemists that a single number might be reliably reported. The result is a threshold that is a little over three times the value of the USEPA detection limit $((10/3.14) = 3.18)$. Another commonly used rule is that the quantitation limit is twice the detection limit. This rule is often used with the IUPAC guidelines.

Quantitation limits are used because matrix effects or other causes of variation may occur in environmental samples that are not present in the prepared standards used to establish a detection limit. With organic analyses, for example, peaks from other compounds may interfere with an exact determination of the amount of the target compound, and at low levels whether the target compound is actually present. The analyst determines how low concentrations can be before a reliable single value can be reported, given that the standard deviation of repeated measurements is still somewhat large in comparison to the signal itself. For measurements between the detection and quantitation limits the analyst believes that there is a signal (the analyte is present at a trace amount), but that the signal is small in comparison to the variability of the measurement process. To avoid reporting an unreliable single number, the analyst constructs the threshold above which a single number may be reliably reported. Measurements above this threshold may be quantified by a single value; those below are usually not.

3.1.3   Data Between the Limits

There is general agreement among laboratories that no numeric values should be placed on measurements below the detection limit because of the risk of false positives, and the confusion of reporting zero and negative readings. There is agreement that single values above the quantitation limit should be reported as measured. Disagreement remains on what to do with measurements in the region between the two thresholds. Here the chemist generally believes that the analyte is present in the sample, but at concentrations that cannot be quantified with precision. Older analyses in this region were reported as simply "trace" or "detected." The large amount of information present in these data is contained in the proportion of values below specific thresholds, so reporting the numeric value of the threshold is crucial for capturing and analysis of this information. Recent analyses (say since 1980) generally do report a threshold value, usually the higher quantitation limit, resulting in a value of $<$QL for these in-between measurements.

Data users have lobbied to get numerical values between the detection and quantitation limits (Gilliom et al., 1984). Several labs now report these values, qualified by a remark (ASTM, 1983; Oblinger-Childress, 1999). This remark indicates that the relative error for these measurements is high, and so the individual values might be somewhat smaller or larger. Most data users incorporate these values as if they were equivalent to values above the quantitation limit, effectively resetting the detection limit as the reporting limit. Most laboratories would consider this risky, as calibration standards are usually not available or accurate in this range.

If data between the limits are not handled correctly a bias may be introduced. This bias is discussed in Section 3.1.4. Three methods for handling data between the limits without introducing a bias are as follows:

1. *Use the Quantitation Limit as the Reporting Limit.* All values below the quantitation limit are considered censored observations labeled "<QL". Values between the limits are considered too unreliable to report as single numbers, and are reported as <QL, as are all values measured below the detection limit. Data analysis methods of this book may then be applied directly. However, measurements that signal presence of the analyte are lumped with measurements not distinguishable from zero. Some information is lost in comparison to the next two methods.

2. *Use the Detection Limit as the Reporting Limit.* This is the de facto result when data users take values reported between the thresholds as similar in precision to those above the quantitation limit, ignoring the qualifier. Values "<DL" are censored. Values between the limits are used as individual values. The advantage over method 1 is that measurements different than zero are recognized as higher than true censored observations. The risk is that there may be too much variability to reliably treat the data between the limits as anything other than tied with one another.

3. *Use Interval-Censoring Methods.* Data between the limits are assigned tied ranks (nonparametric methods) or assigned to an interval (parametric methods) higher in value than those assigned to data below the detection limit. The ordering of data is preserved—the <DL group is considered lower than the "DL to QL" group—without assigning single values to observations in either group. Data in the "DL to QL" group between the two limits are not distinguished from one another—all are reported as "DL to QL" and the interval-censored methods provided in this book can be used for data analysis.

The decision of which of these three methods to use is a decision that should be made by the data user in consultation with the laboratory scientist. Understanding the relative precision of the data between the limits is key for determining how best to represent them. All three methods are unbiased—the probability distribution (the percentiles) of data are not consistently shifted below or above their true values. Unfortunately, a fourth method that does introduce a bias is now sometimes used to report censored data. Called "insider censoring," it is a method that in other disciplines is called "informative censoring."

3.2   INSIDER CENSORING–BIASING INTERPRETATIONS

One of the assumptions behind all methods for interpreting censored data is that the measurement value does not influence the type of censoring process used. One does not decide to use one method of censoring for one range of concentrations, and another method for another range of concentrations. The process of censoring should be "noninformative" in regards to the concentrations themselves. When this assumption is met, the proportion of data below any threshold can be validly computed and compared to the proportion below another threshold. These proportions are another way of stating the percentiles of the data set, and represent the primary information present in data with censored values. Percentiles can be correctly computed and interpreted when the censoring mechanism is noninformative.

Insider censoring invalidates the computation of percentiles (and other summary statistics or tests). An example of informative censoring from the medical sciences may help to illustrate the problem. Informative censoring is when the survival length of patients following diagnosis of a disease, the variable being measured, influences how the data are censored (Collett, 2003). Suppose the survival lengths of two groups of patients is to be compared, one group that has received a particular medical treatment, and another which has not. The goal is to determine if receiving the treatment is beneficial and generally lengthens a patient's lifespan. Informative censoring occurs if it is known that an individual's expected survival time is short, and so the patient is refused (or is not considered for) treatment. In a feedback loop, persons who are projected to live longer are treated, and so the outcome that persons live longer with treatment cannot be ascribed just to the treatment. It may be due to the selection process itself.

A process similar to the feedback loop above is currently implemented by some laboratories in an effort to provide information their data users are requesting, while preserving a sense of protection from false-negative (Type II) errors. One such process is the Laboratory Reporting Level (LRL) of the U.S. Geological Survey (Oblinger-Childress, 1999). Another is found in the RAGS risk assessment guidelines of the Superfund program (USEPA, 1989). Such processes have not yet been widely recognized as a problem, but as insider (informative) censoring they produce biased results as interpretation methods are applied to these data. Numerical values between the detection and quantitation limits are reported, though qualified. So for values measured between the limits, the detection limit is the effective reporting threshold. For observations measured as less than the detection limit, however, values are reported as less than the quantitation limit, $<QL$. These data are reported using the (higher) quantitation limit in order to avoid false negatives. The choice of reporting limit is therefore a function of the measured concentration of the sample. The result is that all interpretations of data reported in this manner, from computation of means to hypothesis tests, will be biased. This is pictured in Figure 3.6.

Figure 3.6a shows the original measured concentrations as a bar graph. Suppose that 40% of observations are measured between 0 and the detection limit (the white bar in Figure 3.6a). Twenty-five percent are measured between the detection and quantitation limits (light gray bar), and the remaining higher measurements reported as uncensored values (dark bars). The measurements between the limits are reported along with a qualifier that these observations are "estimated," but still reside between the two limits. This bar graph applies as long as the values measured below the detection limit are reported as "$<DL$". Figure 3.6b then shows the same data after insider censoring. The difference is that values measured below the detection limit are now reported as being below the quantitation limit or "$<QL$", as if they might belong anywhere from zero up to the quantitation limit. The probability (40%) that

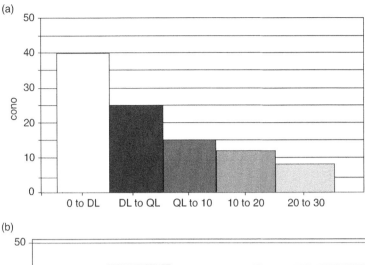

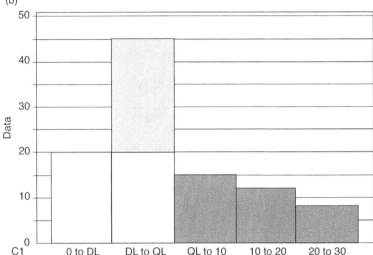

FIGURE 3.6   (a) Proportions of data within ranges of concentrations as originally measured. (b) Proportions of the (a) data within the same ranges after insider censoring. The lower end of the distribution has been shifted dramatically upward.

observations may fall below the detection limit is spread evenly along the entire range from zero to the quantitation limit. This is pictured in Figure 3.6b as white bars totaling 40% evenly split between two categories, 20% of observations in each category. The result of insider censoring is that the probability that an observation might fall between the detection and quantitation limits is exaggerated, and the probability that it would fall below the detection limit is underestimated, in comparison to the proportions actually measured. The shape of the histogram has been changed, and so too will all interpretations that follow. This upward bias is picked up by any subsequent procedure, from the simplest computation of means or percentiles to more complex methods such as maximum likelihood.

Laboratories that attempt to satisfy data users while avoiding false negatives may fall into the trap of insider censoring. Figure 3.7 (from Helsel, 2005) illustrates the bias resulting from insider censoring. Maximum likelihood estimation (see Chapter 6) was used to compute 1000 estimates of the mean of 50 observations with censored observations that had an overall true mean of 3.08. Four different censoring mechanisms

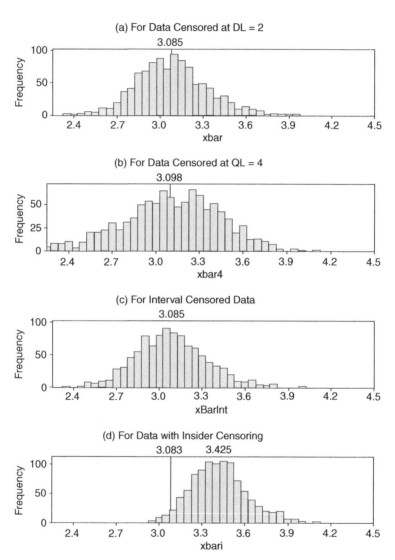

FIGURE 3.7 Histograms of 1000 estimates of the mean (from Helsel, 2005). (a) Censoring at the detection limit. Unbiased estimates around the true mean of 3.08. (b) Censoring at the quantitation limit. Unbiased estimates with a bit higher variability. (c) Interval censoring "0–2" and "2–4." Unbiased estimates with errors similar to (a). (d) Insider censoring. Data measured below 2 reported as <4. Biased estimates almost 0.4 units above the true mean.

are shown in graphs (a)–(d). The first three graphs show the results of the three recommended methods of the last section. All three resulted in unbiased estimates of the mean. Graph (d) shows the effect of insider censoring, where all values measured below the detection limit were reported as <QL while data between the limits were reported as single (qualified, "J-value") numbers. An approximate 10% upward bias results from how the data were censored. This bias may not be the largest source of error in the sampling and analytical process, but it is totally avoidable. Any of the three valid options of the previous section would avoid this bias. For labs that report data in this way, it is imperative that data users recensor their data using one of the three unbiased methods listed in the previous section prior to performing statistical analysis.

3.3   REPORTING THE MACHINE READINGS OF ALL MEASUREMENTS

Gilbert (1987) and Porter et al. (1988) were early advocates that laboratories report the original machine readings instead of censoring data in any way. Accompanying the original readings would also be a measure of their analytical error (i.e., 0.3 ± 0.8). If the machine readings are negatives or zeros, so be it. More recently, Davis and Grams (2006) also recommend reporting uncensored values regardless of whether the signal is being swamped out by the measurement noise. This is at first glance an attractive idea, especially given the opportunity for bias and inconsistency between laboratories in the censoring process. There are at least two issues with reporting uncensored machine readings to the user. They are of concern not in estimating a mean, the situation usually being discussed when the suggestion to not censor arises, but in performing hypothesis tests or other situations where the relative ordering of individual observations is important.

First, reporting machine readings gives a false sense of precision. When the quantitation limit equals 1, for example, readings of 0.3 and 0.4 appear to be distinct with one higher than the other. Any usual statistical routine will consider the 0.4 as reliably larger than the 0.3. Yet the actual concentrations might easily be the same, or even reversed, with the concentration resulting in the 0.4 in fact lower than the one resulting in 0.3. The "precision" indicated by the machine readings may in fact be misleading information. It is more than you actually can know about the data. Over time and many samples the random nature of which measurement is higher than another will even out. For computations of the mean or other statistics with large data sets where the ordering of individual values is not important, the errors even out. Yet for the one small data set provided to you, the user looking at whether concentrations in one group are higher than another, it certainly may not. Antweiler and Taylor (2008) demonstrated the problems of depending on this misleading information. They analyzed a series of trace constituents using two methods, a research-grade ultralow detection limit laboratory analysis, and a more typically available method resulting in censored observations. Various censored data techniques were used to compute a mean for the typical method's data (see Chapter 6). One of those was to use the machine readings from the typical method. Estimated statistics were compared to the "true mean" of the research-grade data. Machines readings produced estimates markedly less reliable than using Kaplan–Meier and other censored techniques on data that had been censored. Machine readings followed by a standard computation were not as good as considering the data tied within low-level intervals.

Second, machine readings must be combined with estimates of the precision in order to correctly interpret data in this low signal to noise range. A mean, for example, could be computed by weighting each observation by the inverse of the variance of each measurement. Less reliable data would have less influence on the estimate. Using both the measurement values and their variability is necessary for a statistical analysis to recognize that two low-level observations are essentially indistinguishable, even though one is reported as 0.3 and the next as 0.4. These more complicated, weighted analyses are rarely if ever performed by the typical user. A study comparing variance-weighted methods to the established censored data methods of the remainder of this book needs to be done before the use of uncensored machine readings can be recommended with assurance.

3.4 LIMITS WHEN THE STANDARD DEVIATION CHANGES WITH CONCENTRATION

Many papers in analytical chemistry have over the years modeled the variation of concentration as a function of concentration itself. Variation takes on a form something like plus or minus a percentage of the measurement, rather than plus or minus a constant number across the range of concentrations. Higher concentrations have higher variability. Logarithm or square root transformations of concentration are often used to convert measured concentrations to data approaching a more constant variance. Figure 3.8 is an example of data where the variability of measurements of

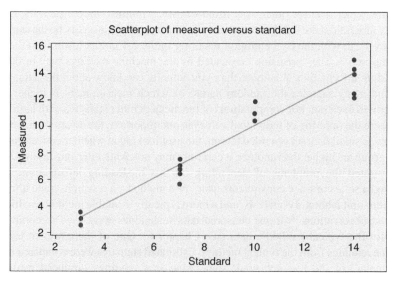

FIGURE 3.8   Variance of measurements increases with concentration.

prepared standards increases as the concentration of those standards increase. There are five measurements for each standard solution. The five replicates for a standard concentration of 3 overlap to the extent that not all five can be distinguished on the plot, while the five at higher concentrations show much more scatter.

Gibbons and Coleman (2001) state that this situation is common—variation increases as concentration increases. If this is true, detection and quantitation limits calculated assuming a constant standard deviation will often be inaccurate. Those computed using a higher concentration standard will obtain a higher estimate for the standard deviation. Projecting this higher standard deviation all the way down to zero, higher detection and quantitation limits are estimated than would be obtained with a lower concentration standard. Two laboratories that have the same underlying method precision will get different estimates for detection and quantitation limits when they use different concentration standards to set those limits.

This last statement should be of great interest to data users. Labs with the same precision characteristics should have the same detection and quantitation limits, but may set different limits due to the use of different standard concentrations. Censored observations from identical samples sent to these labs will be assigned different numbers if substitution is used! The variety of reporting limits resulting from causes unrelated to characteristics of a submitted sample is a strong reason to avoid substitution methods. The substituted number may be more strongly influenced by the concentration standard used 2 months prior to set the detection limit than it is by the concentration of the analyte present in the sample.

Gibbons and Coleman (2001) propose five steps for calculating censoring limits when the standard deviation varies with concentration (calling these "calibration-based limits"):

1. Measure the concentrations of replicates for several concentration standards ranging from very low to above the level expected for the quantitation limit.
2. Perform a weighted least-squares regression of measured concentration (Y) versus concentration of the standards (X). Weights may be computed as the inverse of the observed variance of the replicates at each concentration standard. Or they may be modeled using a regression of the variance versus the concentration standard values.
3. Guess a concentration for the detection and quantitation limits, and compute initial estimates of the standard deviation for those concentrations.
4. Use these initial estimates to compute initial values for the detection and quantitation limits.
5. Iterate between steps 3 and 4, alternatively computing estimates of standard deviation and subsequent censoring levels, until the estimates converge.

Using weighted least-squares, measurements with higher variability further from a concentration of zero have less influence on the final estimate of standard deviation than do those with lower variability closer to zero concentration. So if variability indeed increases with concentration, weighting produces a better estimate for the standard deviation near zero concentration, and therefore a better estimate of the detection limit, than does an assumption that the measurement variation is constant. Recently, this approach for determining detection limits was used by the USEPA to define the LCMRL, the lowest concentration minimum reporting level (Martin et al., 2007; Winslow et al., 2006). It is unclear as of late 2010 whether the LCMRL will replace USEPA's recommendation to use the single-standard MRL, or whether it will go alongside the MRL with the choice to use one or the other in differing circumstances.

3.5   FOR FURTHER STUDY

A summary of the development and history of detection and quantitation limits from the perspective of the USEPA is found in USEPA (2003). Ideas to date on how to set analytical censoring limits are reviewed, though limits computed with a single standard deviation are emphasized. They also discuss both analytical and statistical issues raised by a variety of people involved with Clean Water Act determinations. This is primarily a review of how definitions of censoring thresholds have changed over time.

Currie's seminal 1968 article first called for standardization of methods for computing censoring limits of radiochemical data. His concepts have since been adapted for use in many subdisciplines of environmental science. Currie's subsequent articles over the years have clarified and amplified elements of the topic. See Lindstrom (2001) for a summary of Currie's contributions.

Gibbons and Coleman (2001) and Gibbons et al. (1997) present a detailed discussion of the process for setting detection and quantitation limits when the standard deviation is judged to be a function of the concentration value, rather than assuming that the standard deviation is constant between zero and the concentration standard. These concepts have been adopted by the American Society of Testing Materials (ASTM,1991, 2000) and given the acronyms "interlaboratory detection estimate" or IDE and "interlaboratory quantitation estimate" or IQE. Several comments on the 1997 article have also appeared; see Kahn et al. (1998) and Rigo (1999) as well as the responses from the original authors that follow those comments.

Another paper by Gibbons (1995) on the deficiencies of then-current practices of setting detection and quantitation limits is followed by several discussion papers debating the points which Gibbons makes. Of special note is the discussion by White and Kahn (1995) who defend USEPA practices. The original and discussion papers together provide a history of the current debate over the process of setting reporting limits. The more recent direction of using changing variance at multiple concentration standards for setting a LCMRL is discussed in detail, with many references, by Martin et al. (2007) and Winslow et al. (2006).

Other papers of note include Hyslop and White's (2008) simultaneous determination of detection limits for multiple analytes using X-ray fluorescence, the Kaus (1998) review of practices for setting reporting limits within commercial laboratories in the European Union, and the tutorial on detection and quantitation limits by Thomsen et al. (2003).

4 Reporting, Storing, and Using Censored Data

What is the best way to report data that include censored values to users, and store them in a database? Can they be stored in such a way as to make analysis easier? What kind of analyses will be available when the project interpretation is ready to be performed?

4.1 REPORTING AND STORING CENSORED DATA

Goals for database storage of censored data sets include

- detections and censored observations are both clearly identified and distinguished;
- left censored (nondetect) data can be distinguished from right-censored (greater-than) values;
- data are easily incorporated and used by statistical software;
- censoring by detection versus quantitation limits can be distinguished and recorded.

4.1.1 The Key: Storing the Reporting Limit Values

All of the methods discussed in this text require a numerical value for the reporting limit. The knowledge that an observation is below a specific threshold (the reporting limit) is the primary information contained in censored observations. Without the threshold value, no information is available. Older data sets reporting censored observations as "0" or "trace," or newer recommendations to report data between the detection and quantitation limits as "DNQ"—detected but not quantified—do not have the necessary information to be used in numerical computations. To use nondetect data when the thresholds have not been reported, either a record of the reporting limits used by the laboratory at that time of analysis must be obtained,

Statistics for Censored Environmental Data Using Minitab® and R, Second Edition.
Dennis R. Helsel.
© 2012 John Wiley & Sons, Inc. Published 2012 by John Wiley & Sons, Inc.

or the trace values can be designated as $<\min(d)$, where $\min(d)$ is the minimum detected value recorded for that location and time. These attempts to later recover a threshold value are inferior to reporting the threshold values as the data are stored. In short, the numerical values of the censoring thresholds must be reported in order to fully utilize nondetect data.

On paper, designating a value as a nondetect is trivial—a text character is used, the less-than sign ($<$). However if the text character "$<$" is placed into a database or spreadsheet cell, the entire entry is considered text and therefore not available for computations. Another method of data storage is required. Two easily used methods for storing data with censored observations provide the necessary information:

1. Indicator variables
2. Interval endpoints

Both methods require two columns of data for each data entry. The indicator variable approach is consistent with representing values as left-censored. The approach is commonly used by nonparametric procedures in statistical software, particularly the Kaplan–Meier and related methods. The interval endpoints format represents each observation as interval-censored data. This is the more flexible and intuitive format of the two. It is most often used by parametric maximum likelihood methods, though some nonparametric procedures (Turnbull) will use this format as well. The data analyst may have to use both the indicator variable and interval endpoint formats for the same data set in order to compute both parametric and nonparametric analyses.

4.1.2   Method 1: Indicator Variables for Left-Censored Data

Using an indicator variable, two columns are required to represent each observation. In the first column is the numerical value for a detected observation, or the numerical value of the reporting limit for that observation. In the second column is an indicator of whether the value in the first column is a detected concentration or the reporting limit for a censored observation. The indicator may be either a number or text. In this book, a value of 0 will generally designate a detected observation, while a 1 designates a nondetect. Examples of using text as an indicator include phrases such as "censored" and "uncensored," "above" and "below," or letters (qualifiers) such as "E" or "J". If text phrases are not allowed in the database or statistical software, the variable name for the numerical indicator should remind the user which state refers to which number. A column name such as "BDLeq1" reminds the user that below detection or reporting limit, data have values equal to 1. Zeros must then be uncensored values. If the indicator variable is ignored, summary statistics will be biased high, because the reporting limit will be erroneously considered as the measured value for all censored observations.

Differing text indicators have been used to designate different types of censored data. One letter may indicate that values are below a reporting limit, while another used for values between the detection and quantitation limits. A "DNQ" is a valid indicator of values between the detection and quantitation limits, as long as the numerical values of both limits are correspondingly stored as well. A different letter could indicate a right-censored "greater than" value. Within statistics software these letters are used to translate the data into the interval censoring format, such as "0 to 1" (below a detection limit of 1 µg/L) or "1–3" (between the detection and quantitation limits) in order to use the data in a statistical routine.

The indicator variable method is able to meet all four of the listed goals for storage of censored observations if multiple indicators are used to distinguish data between the limits from true censored observations, and greater-thans from less-thans. The simpler binary 0/1 indicator is only able to meet the first and third goals for data storage. The indicator method's primary disadvantage is that it can sometimes be difficult to remember which state each numerical indicator value refers to.

A Simple Example   Three values of <1, <5, and a detected 10 are represented by the two columns below. A 1 in column 2 indicates a censored observation. If a fourth observation that was between the detection limit of 1 and the quantitation limit of 3 were also present, the indicator variable format would not easily handle it.

Column 1: Concentration	Column 2: BDLeq1
1	1
5	1
10	0

4.1.3   Method 2: Interval Endpoints for Interval-Censored Data

The interval endpoints format is the easiest and most flexible way to store censored data. It is also the format that most closely represents how software uses censored data. All values in the data set are represented by the interval those data fall within. Endpoints for the interval, high and low, are stored in separate columns. For uncensored observations the values in both columns are identical. For censored data the values differ.

Interval endpoint storage is a flexible storage system that meets all four of the goals for data storage listed previously. Left-censored data (censored observations) can be distinguished from right-censored (greater-thans) data, so that both may be possible within the same data set. The format is illustrated below.

A Simple Example, Cont.   For the data set consisting of a <1, <5, and 10, two variables represent each observation, one at the lower end (Start) of an interval within which the measured values lie, and a second at the upper end (End) of the interval. For left-censored concentration data, a value of 0 is entered in the Start variable to denote that the nondetect is no lower than zero, and the detection limit is entered as the largest possible value in the End variable. The detected observation of 10 has the same value for both variables.

Start	End
0	1
0	5
10	10

Nondetects have a lower bound of 0 because concentrations do not go negative. If it is possible for data to extend below zero, a missing value indicator is used by statistical software to represent minus infinity. Common missing value indicators include an asterisk (*) or a period (.), but check the indicator your software uses. The missing value states that there is no known boundary. A right-censored value of >100 would have the missing value in the upper End column, and so be designated as

Start	End
100	*

while a "DNQ" value that is detected above the detection limit of 1, but below the quantitation limit of 3, would be listed as

Start	End
1	3

Where there is a boundary, such as with a lower bound of 0, that boundary should be used. Maximum likelihood methods (discussed later) will produce different results depending on whether 0 or a missing value is found in the Start variable. Entering a missing value indicator for the Start variable produces estimates that are too low because values below zero are considered possible by the procedure when they actually are not possible for bounded environmental data.

Measurements where negative values are truly possible can be represented using the interval endpoints format.

Start	End
−10	5

"DNQ" observations between the detection limit of 1 and the quantitation limit of 3 (second entry below) will be recognized as higher than true nondetects below the detection limit of 1 (first entry below).

Start	End
0	1
1	3

The interval endpoints format is entirely numeric—no text is needed to designate censoring. Reading a direct printout of the data set is clear—it is perhaps the least confusing and most efficient data storage format for censored data.

4.2 USING INTERVAL-CENSORED DATA

Data between the detection and quantitation limits contain sufficient noise that chemists are hesitant to place a single number on them, even though users have been requesting this for years. Using the interval endpoints notation may be the best solution for reporting these data. All values in the interval DL to QL are given the same value, the same interval—they are not distinguished from one another. This recognizes the noise of data in this region—an estimated or nonquantified 1.8 cannot really be distinguished from a nonquantified 1.5 or 1. Yet they are recognized as being higher than measurements below the detection limit of 1, and lower than quantified values.

For an example of how interval-censored data can be used in calculations, consider a data set with five measurements below the detection limit of 1 µg/L, seven measurements above 1 but below the quantitation limit of 2, and seven values measured and reported as individual numbers above the quantitation limit. There are 19 observations in all. The top line shows the data as they were originally reported, with a "*J*" value for data between the detection and quantitation limits. The bottom lines are the same data in interval-censored format. The (1,2) format is an interval-censored value falling between 1 and 2 µg/L.

Original format:

```
<1  <1  <1  <1  <1  1.1J  1.2J  1.3J  1.5J  1.8J  1.9J
1.3J  2.2  2.4  2.7  3.0  3.3  3.9  4.8
```

Interval-censored format:

```
(0,1) (0,1) (0,1) (0,1) (0,1) (1,2) (1,2) (1,2) (1,2) (1,2) (1,2)
(1,2) 2.2  2.4  2.7  3.0  3.3  3.9  4.8
```

The nonparametric approach to interval censoring assigns the average rank of all values below the detection limit to those measurements. All five (0,1) values below the detection limit are assigned the average of ranks 1 through 5, or 3. Values between the two thresholds are similarly assigned the average of their ranks, recognizing that they are higher than values below the detection limit. The seven values are assigned the average of ranks 6 through 12, or a rank of 9. Values above the quantitation limit are represented by the same ranks as those they would have received if there had been no censoring, starting with the rank of 13 and going up to a rank of 19. Ranks of the observations are shown in italics below the interval-censored data themselves:

(0,1)	(0,1)	(0,1)	(0,1)	(0,1)	(1,2)	(1,2)	(1,2)	(1,2)	(1,2)	(1,2)
3	*3*	*3*	*3*	*3*	*9*	*9*	*9*	*9*	*9*	*9*

(1,2)	2.2	2.4	2.7	3.0	3.3	3.9	4.8
9	*13*	*14*	*15*	*16*	*17*	*18*	*19*

These ranks would be used in a nonparametric test, say if some of these data belonged to one group and some to a second group. See Chapters 9 and 10 for more on testing among groups. The median is the central value in the data set. With 19 observations, the median is the 10th ranked observation from the bottom. Counting up to the 10th ranked observation, a (1,2), the median of these data is a value between the detection and quantitation limits. The ranks of interval-censored data will be computed and used in a variety of nonparametric procedures presented later in this book.

Parametric methods solve for parameters using maximum likelihood estimation (MLE). MLE methods easily incorporate interval-censored observations. They do not need individual quantified values for every observation in order to compute a mean or test a hypothesis. The mean of the above interval-censored data is 1.89 by maximum likelihood assuming a normal distribution. Below is the output using Minitab®:

```
Characteristics of Distribution
Standard 95.0% Normal CI
                            Estimate    Error      Lower      Upper
Mean(MTTF)                  1.88636     0.279905   1.33776    2.43497
Standard Deviation          1.19779     0.198577   0.865493   1.65767
Median                      1.88636     0.279905   1.33776    2.43497
First Quartile(Q1)          1.07847     0.314310   0.462431   1.69450
Third Quartile(Q3)          2.69426     0.306238   2.09405    3.29448
Interquartile Range(IQR)    1.61579     0.267876   1.16753    2.23616
```

Chapter 6 will provide more details on MLE, and later chapters expand the idea to testing for differences between group means. Whether using parametric or nonparametric methods, computations can be performed without substituting a single value for interval-censored observations.

EXERCISES

All data sets for the exercises in this book are found on the web site http://www.practicalstats.com/nada in both Minitab (*.mtw) and Excel (*.xls) file formats. Also found there are all Minitab macros (*.mac) used throughout the book for computing the in-text examples and exercises.

4-1 Millard and Deverel (1988) measured copper and zinc concentrations in shallow groundwaters from two geological zones underneath the San Joaquin Valley of California. One zone was named the Alluvial Fan, the other the Basin Trough. Their data are found in the data set CuZn (use CuZn.xls if using software other than Minitab). In addition to the two columns of concentrations, there are paired columns in the Indicator Variable format designating which of the observations represent detected concentrations, and which are "less-thans." The indicator variable names (CuLT = 1 and ZnLT = 1) show that "less-than" observations have a value of 1, while uncensored observations are indicated by a 0.

Create two new variables in the Interval Endpoints format, StartCu and EndCu, that will contain the same information given by the current variables Cu and CuLT = 1.

4-2 What problem may have occurred with the following censored data set? What characteristics lead to that conclusion?

0.55   0.6   0.8   0.85   0.9

<1   <1   <1   <1   <1 1.0   1.2   1.7   1.8   2.2   2.6   3.5

4-3 Flip the copper concentrations for the Alluvial Fan zone to a right-censored format and store in a new variable named something like "FlipCu." Plot both Cu and FlipCu with either a boxplot or a histogram (ignoring the less-than indicators at this point). How do the plots of the two variables compare? Given that the variable Cu is skewed, take logarithms and repeat the process.

5 Plotting Censored Data

Plots that display percentage information can be extended to censored data, as that information is still available even though individual values are not. Boxplots can illustrate the distribution (shape, typical values, outliers) of censored data. An empirical distribution function (edf) depicts individual observations with more precision, but perhaps less familiarity, than does a boxplot. A survival function plot is an edf specifically developed for censored data. Probability plots provide a visual check of conformance to a specific distribution such as the normal, and associated tests quantify the goodness of fit. Scatterplots for X–Y data can be adapted for displaying interval-censored values. Commercial statistics software will compute some of these for you, while macros for Minitab® and R are available on this book's web site for all of them. Become familiar with the survival analysis section of the software you use, and if these plots are not available there, they are not difficult to construct using a few sequential commands.

5.1 BOXPLOTS

Boxplots (Helsel and Hirsch, 2002, Chapter 2) are one of the most intuitive ways to visualize a data set. They employ three percentiles (25th, 50th, and 75th) that define the central box. The relative positions of the percentiles show the center, spread, and skewness of the data. Boxplots also represent outliers as individual points. Censored observations can be incorporated into boxplots by using the information in the proportion of data below the highest reporting limit. Boxplots should never be drawn by deleting censored observations, drawing the graphic using only uncensored values. Deleting censored observations destroys all meaning of the percentiles of the data set, which is what the box of a boxplot represents. There is a better way.

To draw a boxplot for data with a single reporting limit, all censored observations are set to any single value lower than the limit. A horizontal line is drawn at the reporting limit. All uncensored values will be represented correctly, but the distribution below the reporting limit is unknown and should not be represented in the same way as the portions above the limit. The portions of the box below the reporting limit line are blanked out, usually by covering with a rectangle of the same color as background. This "boxplot at sunrise" (Figure 5.1) is an accurate representation of

Statistics for Censored Environmental Data Using Minitab® and R, Second Edition.
Dennis R. Helsel.
© 2012 John Wiley & Sons, Inc. Published 2012 by John Wiley & Sons, Inc.

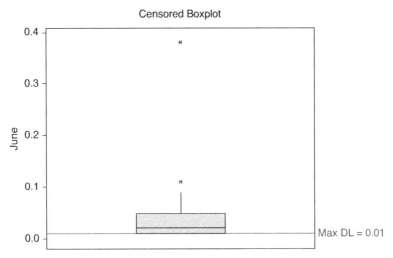

FIGURE 5.1   Boxplot of censored atrazine data. The proportion of censored data is between 25 and 50%, as shown by the presence of a line for the 50th, but not the 25th, percentile.

the information contained in the data set. The proportion of data censored is indicated by how much of the graphic is below the horizon. In Figure 5.1, the lack of a lower line for the box, but presence of the central median line, shows that between 25 and 50% of the data are censored observations (there are 9 of 24 values or 38% below 0.01 for the atrazine data set).

For multiple reporting limits, only data above the maximum reporting limit is known exactly. Portions of the box above this limit are drawn with solid lines. In the most conservative approach, everything below the maximum reporting limit would not be shown (Figure 5.2). However, if too much of the box is invisible below the

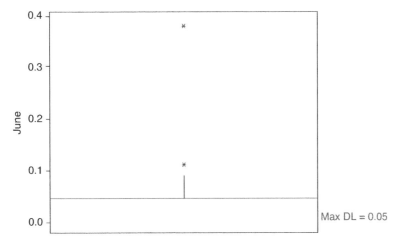

FIGURE 5.2   Censored boxplot for altered atrazine data with reporting limits at 0.01 and 0.05. The 25th, 50th, and 75th percentiles are all below the higher reporting limit and not shown.

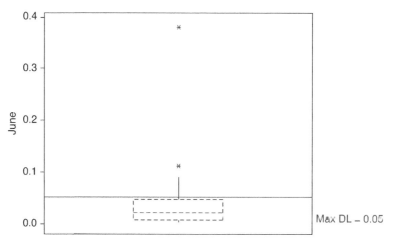

FIGURE 5.3 Boxplot for altered atrazine data with reporting limits at 0.01 and 0.05. The 25th, 50th, and 75th percentiles have been estimated using the robust ROS method of Chapter 6.

horizon, the 25th, 50th, and 75th percentiles can be estimated if necessary and drawn with dashed lines (Figure 5.3). Percentiles are estimated using either Kaplan–Meier (KM) or robust ROS (see Chapter 6). These methods incorporate the proportion of observations occurring below each reporting limit when calculating percentiles.

The cenboxplot command in the NADA for R package will draw censored boxplots. Here the ShePyrene data set is used to illustrate the procedure. There are two arguments to the command, first the name of the column containing the detected values plus reporting limits, in this case Pyrene. The second argument is the column of censoring indicators, PyreneCen. A value of 1 (TRUE) for the censoring column indicates a censored nondetect value, so that the corresponding reporting limit is located in that row of the Pyrene column. To draw the boxplot, the command is

```
> cenboxplot(Pyrene, PyreneCen)
```

resulting in Figure 5.4. Note that the estimated lower portions of the box are shown below the maximum reporting limit line. ROS is used to estimate the lower portions.

5.2 HISTOGRAMS

Histograms are not particularly useful plots for depicting censored data. This is partly because there is not one histogram that is unique to a data set—many equally valid histograms might be drawn from the same data. In Figure 5.5, the censored observations are drawn with their own (dark) bar to show that 38% of the data set is below the single reporting limit. For data with multiple reporting limits the bar would include all data below the highest limit. Above that limit, uncensored observations are categorized into ranges and the percentage in each category shown

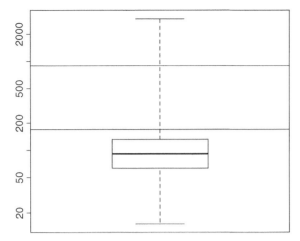

FIGURE 5.4   Boxplot for the ShePyrene data set with a maximum reporting limit of 174. The 25th, 50th, and 75th percentiles have been estimated using the robust ROS method of Chapter 6.

(gray bars in Figure 5.5). The reporting limit (it would need to be the maximum reporting limit for multiply-censored data) is represented by a vertical line in Figure 5.5.

5.3   EMPIRICAL DISTRIBUTION FUNCTION

A plot of the empirical distribution function, also called a quantile plot, shows the sample percentiles (quantiles) of each observation in the data set. Edfs are sample approximations of the true cumulative distribution function (cdf) of a continuous

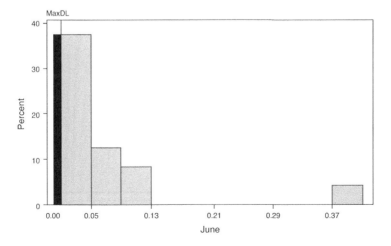

FIGURE 5.5   A censored histogram of the June atrazine data.

random variable. The vertical axis lists the quantiles (median = 0.5 or 50%, etc.) ranging between 0 and 1 (or 100%). The horizontal axis covers the range of numerical values of the data. Data points are plotted in sequence from low to high, and connected by straight lines to form the graph. By selecting one percentile value (say 50%) and reading across to the curve, the percent of observations below that value in the data set is obtained.

To construct an edf, the data are ranked from smallest to largest. The smallest value is assigned a rank $i = 1$, and the largest a rank $i = n$, where n is the sample size of the data set. Ranks are converted to percentiles using a "plotting position" p. For the cdf of a population where all possible data are known, the plotting position is $p = i/n$. With a sample of data instead of the entire population, the edf should use a plotting position where the largest value is slightly less than i/n. When data are tied, as for censored observations, each is assigned a separate plotting position (the plotting positions are not averaged). Tied values are seen as a vertical "cliff" on the plot, like the one in Figure 5.6 for atrazine data at the reporting limit of 0.01.

The plotting position for an edf is an estimated percentile, the probability of being less than or equal to that observation. With commercial software, the largest observation is usually assigned the plotting position $i/n = 1$ (the 100th percentile), having a zero probability of being exceeded. For the cdf of the total population this is appropriate, but with a sample of only part of the total population it would be wise to recognize that there is a likelihood of exceeding the largest value observed to date. This can be represented on the graph by using a plotting position less than i/n on the vertical axis, although unfortunately, commercial statistics software almost always uses i/n as the plotting position.

Different plotting positions can be used depending on the purpose and the tradition of a procedure. Numerous plotting position formulae have been suggested that will assign the highest observation a percentile below 100% (Helsel and Hirsch, 2002), many having the general formula

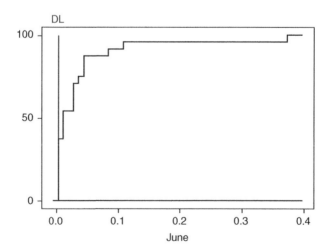

FIGURE 5.6   Empirical distribution function of the June atrazine data.

$$p = (i - a)/(n + 1 - 2a)$$

where a varies from 0 to 0.5. Each differs in the probability of exceedance above the largest observation. Five of the most common formulae are

Name	a	Formula
Weibull	0	$i/(n + 1)$
Blom	0.375	$(i - 0.375)/(n + 0.25)$
Cunnane	0.4	$(i - 0.4)/(n + 0.2)$
Gringorten	0.44	$(i - 0.44)/(n + 0.12)$
Hazen	0.5	$(i - 0.5)/n$

The Weibull formula is used in many areas of science. The Blom plotting position is used most often on probability plots to compare data to a theoretical normal distribution.

Commercial survival analysis software will draw edfs when there are one or multiple censoring thresholds. Percentiles are estimated using Kaplan–Meier methods and plotted on a "survival function plot", which is in essence an edf. However, the i/n plotting position is usually employed for these plots.

5.4   SURVIVAL FUNCTION PLOTS

Survival function plots (Figure 5.7) are an edf plot flipped side to side, plotting $(1 - p)$, the probabilities of exceedance of an observation. For left-censored data flipped and plotted using software for right-censored survival analysis, the largest observations are represented on the left side of the plot. In Figure 5.7, the detected atrazine observations are plotted as open circles. The atrazine scale was added to the bottom of Figure 5.7 for reference, and shows the largest values at the left side of the plot. Survival plots will typically label the x-axis as "Time" or "Survival time."

The survival function presents the probabilities of exceeding a value of " Survival time." Figure 5.7 shows a 97% probability of exceeding a "time" of 0.62. Time was created by subtracting atrazine from a value of 1.0, so that this point is also an atrazine concentration of $(1 - 0.62) = 0.38\,\mu g/L$. The concentration of 0.38 represents the 97th percentile of the atrazine distribution, and so has a 97% probability that atrazine will be less than or equal to 0.38. Percentiles plotted on the y-axis are calculated with the Kaplan–Meier method described in Chapter 6.

Left-censored environmental data can be converted to right-censored "Time" data by subtracting each observation from a number larger than the maximum in the data set (see Chapter 2). This changes the "Atrazine Concentration" scale into the "Survival Time" scale of flipped data. In Figure 5.7, a survival time of 0.99 corresponds to an atrazine concentration of 0.01, the reporting limit. The jagged survival-function probability line intersects the vertical reporting limit at a "Cumulative proportion surviving" or probability of 0.38. Survival probabilities,

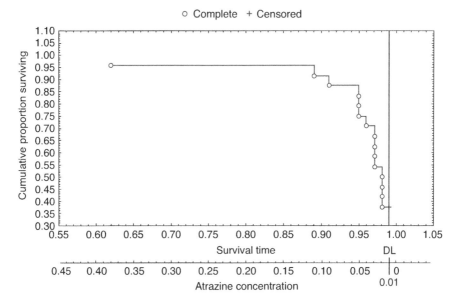

FIGURE 5.7   Survival function plot of the June atrazine data, with an additional atrazine concentration scale added for reference at the bottom.

equal to percentiles of the original observations, are found on the y-axis. There is a 38% probability of "surviving" to a value greater than 0.99, as well as a 38% probability of being less than the reporting limit, just as previously calculated. Only the uncensored data are plotted on a survival function, but their positions are influenced by the censored as well as the uncensored observations. Further detail on Kaplan–Meier computations is given in Chapter 6.

Some survival analysis software will also compute the "cumulative failure time" probability p as well as the survival time $(1 - p)$. "Failures" or "deaths" in survival analysis are the "detects" of environmental science—the observations with a single known value. Representing the data as interval censored and plotting a cumulative failure time plot results in Figure 5.8, again an edf where the atrazine data are plotted in the usual "low values on the left" format. This format is more familiar to environmental scientists than the reversed survival plot format.

In Figure 5.9, the edf for groups of lead concentrations in the blood of herons is plotted by dosage groups, one group using a dashed line and one using a solid line. This is accomplished with the plot command in NADA for R, operating on the results of the cenfit command.

```
> bloodPb = cenfit(Blood, BloodCen, DosageGroup)
> plot(bloodPb)
```

Cenfit computes Kaplan–Meier percentiles for left-censored data. See Chapter 6 for more on the cenfit command.

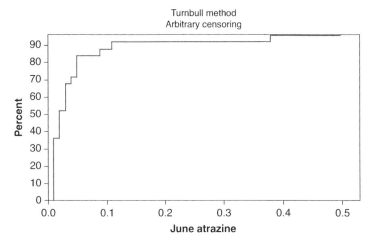

FIGURE 5.8   An edf for the atrazine data produced as a "cumulative failure time" plot for interval (or arbitrary) censored data.

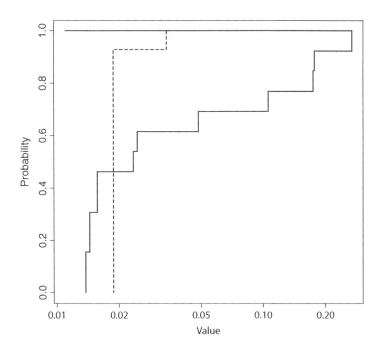

FIGURE 5.9   Empirical distribution functions for lead concentrations of the Golden data set produced by plotting the results of the cenfit command in NADA for R.

5.5 PROBABILITY PLOT

Probability plots check the similarity of data to a normal or other specified distribution. The linear percentile scale of an edf is altered to match the percentiles of the assumed distribution. If the data follow that distribution, then the plotted points will fall along a straight line. For a normal distribution, the percentile scale is stretched out at both ends (see Figure 5.10). Each observation is then plotted individually.

Waller and Turnbull (1992) surveyed several types of plots including probability (or Q–Q) plots for censored data. For censored data, only uncensored observations are plotted because the location at which to place a censored value is unknown. Substitution of numbers such as one-half the reporting limit is invalid, as the shape of the plot will change depending on the values substituted. Worse yet, some casual readers will then miss the fact that no numerical values are actually available for these data. However, the proportion of data below each reporting limit is computed in order to determine the placement of the uncensored data, including those that fall between reporting limits. So censored data affect the positions of the uncensored data on the plot. All uncensored values above the highest reporting limit have probabilities (percentiles or "normal quantiles") that are identical to what they would have been if all the data had been uncensored. Positions of uncensored data below the highest reporting limit will be affected by the censored values. They should be. Probability plots should not be drawn by simply deleting the censored values, plotting only the uncensored observations. Doing this will result in incorrect calculated percentiles for the plotted, uncensored observations, distorting the shape of the distribution. A probability plot computed only using uncensored observations will not evaluate whether the entire data set fits that distribution.

A probability plot for the June atrazine data is shown in Figure 5.10. Note that no individual points are plotted for the lowest 35% or so of data. That is the effect of the censored values. Had the censored data been deleted and a probability plot drawn, the

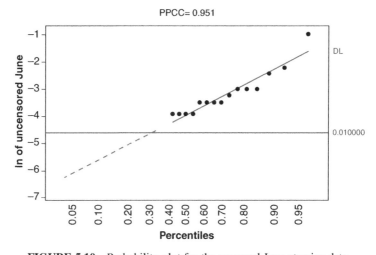

FIGURE 5.10 Probability plot for the censored June atrazine data.

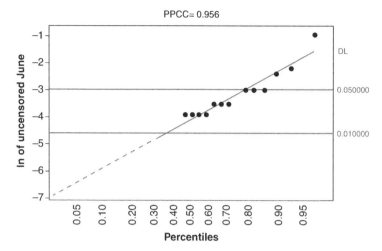

FIGURE 5.11 Probability plot for logarithms of censored data with two reporting limits.

lowest uncensored value at 0.02 (natural log of approx. −4) would have plotted at the 5th percentile. This is obviously incorrect—33% of the original data are known to be below 0.02. The lowest uncensored observation plots above the 33rd percentile when the censored observations are correctly accounted for. Based on Figure 5.10, the atrazine data appear to reasonably fit a lognormal distribution (the straight line).

For multiple reporting limits, statistical software will again assign percentile values to uncensored observations while taking into account the proportion of censored data below each reporting limit. All uncensored observations can be plotted, even those located between reporting limits. Either Kaplan–Meier or robust ROS (see Chapter 6) is used to estimate percentiles for the uncensored observations. A probability plot for the atrazine data, altered to add a higher reporting limit at 0.05, is shown in Figure 5.11. Note that seven uncensored observations are shown between the two reporting limits.

Figures 5.8 and 5.10 were drawn using the %cros macro. This uses the ROS estimates of percentiles with the Weibull plotting position. These are lognormal probability plots, as the logarithms of data are compared to percentiles of the normal distribution. Variations on probability plots include transforming the nonlinear percentile scale to a linear "normal quantile" scale of a standard normal distribution, as in Figure 5.12. Figure 5.12 is identical to Figure 5.10 except for the scale used on the horizontal axis representing the normal distribution. In Figure 5.12, the median and mean are at a standardized value of 0, and the slope of the line is in standard deviation units. A point plotting at a normal quantile of 1 is one standard deviation above the mean, or at about the 84th percentile. The linear scale is not as intuitive for the user as a percentile scale, but is helpful in procedures where regression is computed on a probability plot, as with ROS (see Chapter 6). The Weibull plotting position results in the largest value of 0.38 plotting approximately at the 96th percentile, and it is shown to be somewhat of an outlier, lying off the center line representing the (log)normal distribution.

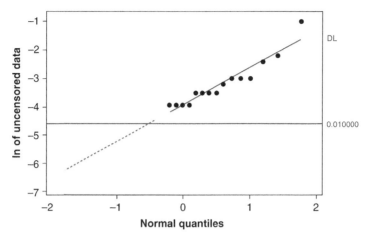

FIGURE 5.12   Probability plot of the atrazine data using quantiles of the standard normal distribution on the horizontal axis instead of a nonlinear percentile scale.

A probability plot for the June atrazine data is drawn using Minitab's Survival Analysis software in Figure 5.13 using the command.

```
Stat>Reliability/Survival>Distribution Analysis
(arbitrary censoring)>Distribution ID Plot
```

There are four differences between this plot and the ROS probability plot of Figure 5.10, two important and two a matter of preference. The most important

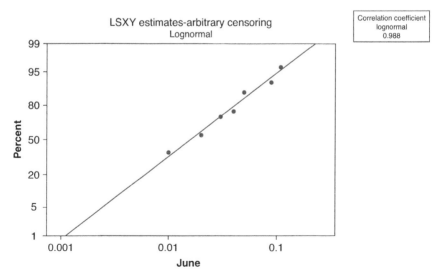

FIGURE 5.13   Probability plot of the atrazine data using Minitab's survival analysis software. A plotting position of i/n is used, so the largest observation is not shown.

difference is that a plotting position of i/n was used, resulting in a percent value of 1.0 that is off the scale of the plot. Therefore the largest value is not shown! This is not good practice for empirical sample data, but is standard practice with commercial software. The second important difference is that tied values are overplotted, assigning all of them the same median percentile value rather than individual percentiles. There is no reason not to assign individual percentiles to tied data as long as it is obvious that there is no way to determine which percentile is associated with which of the tied observations. Using the median percentile and representing multiple points as one point tends to increase the correlation coefficient shown on the plot (see next section). Third, Kaplan–Meier estimates of percentiles are used, which are slightly different than the ROS estimates. Kaplan–Meier has much more theoretical support. KM percentiles result in a slightly different value for the correlation coefficient than the similar (but not identical) ROS percentiles. Finally, the axes are reversed from Figure 5.10, with the data values on the horizontal scale and the percentiles on the vertical scale. This is just a matter of preference.

A fourth version of a probability plot is shown for the atrazine data in Figure 5.14. Here the fit to the distribution is performed by maximum likelihood (see Chapter 6) rather than computing a least-squares line on the probability plot. Rather than using the correlation coefficient of plotted data as the test for similarity to the lognormal distribution, the Anderson–Darling (AD) test is used. Again note that due to the i/n plotting position, the largest observation is missing.

5.5.1   Know the Procedures Used by Your Software

To allow plotting of the highest point in the data set, Waller and Turnbull (1992) recommended using the Weibull $i/(n + 1)$ plotting position, and provided a similar modification to Kaplan–Meier percentiles for data with censoring. One way to

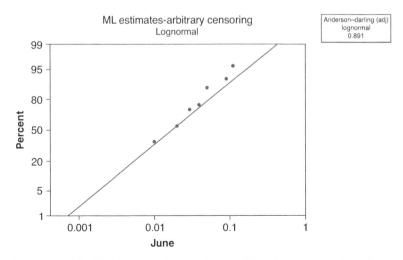

FIGURE 5.14   Probability plot of the atrazine data using maximum likelihood.

accomplish this using commercial software is to add one artificial value higher than the highest data value to the data set, and replotting. The artificial observation is then the one not shown, and the actual data are plotted using the Weibull plotting position of $i/(n + 1)$. The highest true observation will have a percentile less than 1 and be shown on the plot. This fix-up may work, or not, depending on how and whether the software is using that artificial, highest point. Make sure you understand what your software is doing before taking this step! Many attributes of the plot may not be correct if the numerical value of the artificial observation is used. Estimates of mean and standard deviation, for example, will change if the added observation is included in the computations. The test statistic for distributional fitting may no longer be valid. But the plot will be correct.

Using Minitab as an example of one commercial software package, the mean, standard deviation, and distribution-fitting test statistic will change based on the value of the added artificial point IF the analysis is performed using maximum likelihood. It will not matter as long as least-squares computations are performed. Minitab's least-squares software does not use the highest, 100th percentile observation at all! It was not using the highest of your original data values when estimating the mean, standard deviation, or correlation coefficient if you did not add the artificial point! Without adding an additional point, the estimates given by Minitab's least-squares procedure are incorrect because an inappropriate plotting position is being used. Therefore to obtain a plot and test equivalent to Figure 5.8, but computed with Minitab's built-in software for interval-censored data with a least-squares fit (allowing the highest true atrazine observation to be at the 96th percentile), add an artificial observation higher in value than all of your data and run the procedure. The value of the artificial point will not matter, because the 100th percentile point is not used in the least-squares calculations. Figure 5.15 shows the resulting probability plot, again with axes switched from those in Figure 5.8, and with one point and percentile representing all tied observations. The resulting PPCC of 0.973 is the equivalent to one of 0.951 in Figure 5.8, with the difference being the different percentiles used (Turnbull/Kaplan–Meier in Figure 5.15; ROS in Figure 5.8) and the single point representing tied observations. The latter is the more important effect, tending to produce a higher correlation coefficient as seen in Figure 5.15.

For the maximum likelihood fitting procedure in Minitab, however, DO NOT add an artificial, high point. The highest point is used in MLE calculations, resulting in incorrect estimates. With an added point, the line estimate of the fitted distribution will change depending on the value of the artificial point. Instead, the mean, standard deviation, and Anderson–Darling coefficient using the original input data set is correct. Unfortunately, the highest point in the data set will have a plotting position of i/n and not be shown on the plot (Figure 5.14).

Using NADA for R, a probability plot for the censored TCE data (Figure 5.16) is drawn by plotting the results of an ROS analysis performed with the censos command (see Chapter 6 for more on the censos command):

```
> tceros=cenros(TCEConc, TCECen)
> plot(tceros)
```

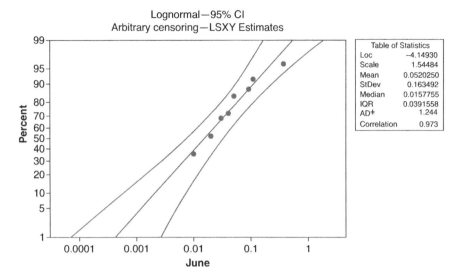

FIGURE 5.15   Minitab's probability plot for the interval-censored June atrazine data following the addition of one artificial point higher than all of the original data. That artificial 100th percentile point is not used in the least-squares calculations. Adding the point ensures that the highest actual observation is used in the calculations.

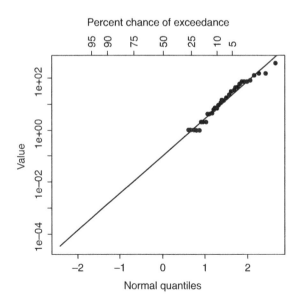

FIGURE 5.16   Probability plot for the TCE data set using NADA for R.

Note that a large proportion of the data are censored below reporting limits of 1, 2, and 5 µg/L. Therefore, the first detected value (a detected 1) is plotted as a point at a normal quantile of about 0.5, or near the 70th percentile. Seventy percent of observations are less than 1 µg/L.

5.5.2 Testing Adherence of Censored Data to a Distribution Such as the Lognormal

In addition to plots, hypothesis tests can provide a numeric measure of whether or not data follow a specific distribution. Tests based on probability plots, including tests applied to censored data, have been used for decades.

The Kolmorgorov–Smirnov (K–S) test for adherence to a distribution was applied to censored data by Barr and Davidson (1973). The K–S test is considered lower in power than most other tests for distributional adherence today, but a modification of it, the Anderson–Darling test, was applied to censored data in 1976 (Pettitt, 1976). The Anderson–Darling test is one of several tests commonly used today for this purpose. Another popular test is the probability-plot correlation coefficient (PPCC) test. This test is essentially the same as the Shapiro–Wilk test, the standard procedure for testing distributional shape hypotheses over the past several decades. The PPCC test is easy to compute. It is Pearson's correlation coefficient between the data and their quantiles of a distribution such as the normal—the data on a probability plot. If the data fall exactly on a straight line, the correlation coefficient equals 1. As the correlation coefficient decreases below 1, evidence builds against the hypothesis that data follow the assumed distribution. PPCC is today one of the most commonly used tests for distributional shape. It was applied to censored data by Verrill and Johnson (1988). Royston (1993, 1995) showed that the PPCC test can adapt to provide a p-value when there is one censoring limit. Unfortunately, commercial software has not often taken advantage of this fact. Hawkins and Oehlert (2000) provided tables of critical values for PPCC with two censoring thresholds, plotting the Kaplan–Meier percentiles for uncensored values on the probability plot.

Figure 5.17 shows the fit of the June atrazine data to four distributions. An artificial point was added to the top so that the $i/(n + 1)$ Weibull plotting position is used, and the largest actual point is visible on the plot. In addition to the visual evaluation of which plot is most straight, the value of PPCC closest to 1.0 provides a numerical measure. The lognormal distribution appears to be the best choice among these four distributions. In particular, a comparison of the lognormal to the normal distribution is a common decision to be made in environmental studies. The higher PPCC here indicates that the lognormal is a better choice for the atrazine data than is the normal distribution.

A major difficulty with these procedures is that even though test statistics can be calculated, with censoring there are currently no closed-form solutions to provide a calculated p-value for the test. When the tests were developed, their authors conducted Monte Carlo experiments to tabulate the p-values for a given test statistic, sample size, and percent of censoring. The resulting tables are awkward to place into statistical software, and so are not; p-values for censored data procedures are generally not

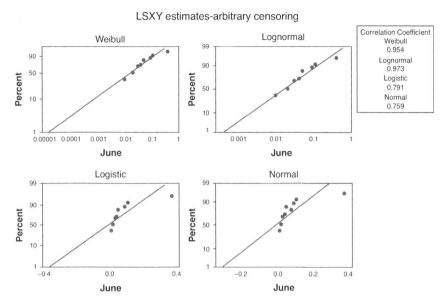

FIGURE 5.17   Probability plots for four possible distributions of the June atrazine data. The lognormal distribution fits best. Its PPCC is closest to 1.0.

calculated in commercial software. The user may take test statistics such as AD or PPCC and go to a published table for tests on censored data. But *p*-values are not generally reported in commercial software when testing the distribution of censored data. As one newer approach, Ren (2003) used a bootstrapped approach to compute *p*-values for tests on data that include interval-censored values.

5.6   *X–Y* SCATTERPLOTS

Scatterplots compare the values of two continuous variables, usually denoted X for the variable plotted along the horizontal axis and Y for the variable on the vertical axis. The paired *X–Y* values are visually inspected for patterns of correlation or trend—are values of Y predictably high or low for given values of X? A dilemma comes when either variable is censored—what numeric value should be used to place that observation on the plot?

Unfortunately, the most common practice is to substitute one-half the reporting limit and plot the fabricated data as if it were measured. The result is a false impression that these values are actually known, and that they are in all cases the same number. Neither is true. For multiple reporting limits, plotting a fabricated value gives a false impression of the comparison between observations. A value of <10 plotted as a 5 is shown as if it were larger than a <3, when in fact the reverse might be true. All of the disadvantages of substitution in numeric procedures carry over into scatterplots. A signal that is present may be obscured. A signal may be shown that in reality does not exist.

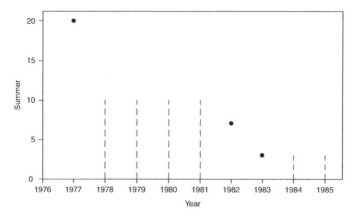

FIGURE 5.18   Scatterplot of dissolved iron concentrations over time. Censored observations shown as dashed lines. Data from Hughes and Millard (1988).

Since censored values are known to be within an interval between zero and the reporting limit, representing these values by an interval, such as with a line segment, provides a visual picture of what is actually known about the data. Figure 5.18 shows a scatterplot of dissolved iron concentrations in summer samples from the Brazos River, TX, reported by Hughes and Millard (1988). They investigated whether trends occurred in iron concentrations over a 10-year period. Iron concentrations were censored at two reporting limits during the study, at 10 µg/L in the earlier years and at 3 µg/L in later years. Uncensored observations are shown as single points, as usual.

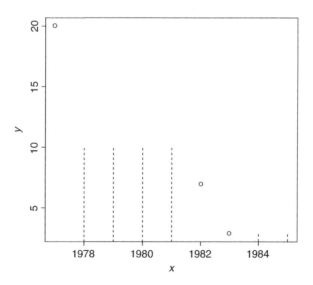

FIGURE 5.19   Censored scatterplot of dissolved iron concentrations over time using cenxyplot in NADA for R. Data from Hughes and Millard (1988).

Censored observations are shown as dashed gray lines between zero and the reporting limit.

Censored scatterplots clearly illustrate the interval-censoring of the data. Lines are best shown as grayed-out or dashed rather than fully dark to emphasize that any one location within the interval is less likely than the single location shown for each detected observation. In NADA for R, the cenxyplot command produces a similar plot (Figure 5.19), where the arguments for the command line are (x, x censoring indicator, y, y censoring indicator).

```
> cenxyplot(Year, YearCen, Summer, SummerCen)
```

EXERCISES

5-1 Plot a censored boxplot and censored histogram for Millard and Deverel's (1988) zinc concentration data found in the data set CuZn. Use the Minitab macros chist. mac and cbox.mac. The zinc concentrations will need to be split into two columns, one for each zone, to plot the censored histograms. The censored boxplots can be plotted with one command

%cbox c3 c4 c5 (or abbreviating, %cbox c3-c5)

and a box will be drawn for each group listed in column c5. Describe the results— what characteristics of the data will likely be important for further data analysis?

5-2 The atrazine data used in this chapter are found in the data set Atra. Draw an empirical distribution function plot (also called a cdf) for the June atrazine data. In Minitab this is done using the Graph > Empirical cdf command. Using the Distribution dialog box, select "lognormal" as the best-fitting distribution. A lognormal distribution will be plotted as a blue line, and the edf with a red step function. Compare the resulting plot to the survival function plot (black solid line) of Figure 5.7. How are the two plots related?

6 Computing Summary Statistics and Totals

More articles have been written comparing and recommending methods to compute summary statistics than for any other type of analysis of censored environmental data. As early as 1967, Miesch recommended use of an approximate maximum likelihood estimation (MLE) for computing estimates of mean abundances (mass) of metals with censored measurements in rock samples (Miesch, 1967). Nehls and Akland (1973) recommended using one-half the reporting limit (RL) to compute summary statistics for air quality data. Gilbert and Kinnison (1981) applied several methods, including probability plotting procedures, to censored radiochemical data. Chung and Spirito (1989) found that MLE provided better estimates of summary statistics than did substitution. Helsel (1990) recommended the use of survival analysis methods for censored data, though the idea of transforming left-censored data to right-censored values had already been published by Ware and DeMets (1976). Use of each of these methods continues today. A review of several papers testing these methods and variations thereof for estimation of means, medians, variances, and other parameters is found at the end of this chapter. First, however, the primary methods available for computing summary statistics of censored data are introduced. Methods are classed into the three approaches of Chapter 2—simple nonparametric methods after censoring at the highest reporting limit, maximum likelihood methods for survival analysis, and nonparametric survival analysis methods. Imputation procedures (robust ROS, robust MLE) are also included. Following a detailed description of each method, articles comparing their performance under a variety of conditions are reviewed and summarized.

6.1 NONPARAMETRIC METHODS AFTER CENSORING AT THE HIGHEST REPORTING LIMIT

6.1.1 Binary Methods

Consider a small example data set of 11 observations:

< 1   < 1   3   < 5   7   8   8   8   12   15   22

Statistics for Censored Environmental Data Using Minitab® and R, Second Edition.
Dennis R. Helsel.
© 2012 John Wiley & Sons, Inc. Published 2012 by John Wiley & Sons, Inc.

To perform binary methods the data are recoded as either less than the highest reporting limit (LT), or greater than or equal to the highest limit (GE). For the 11 observations with a highest reporting limit of 5, there are four LTs and seven GEs:

<1	<1	3	<5	7	8	8	8	12	15	22
LT	LT	LT	LT	GE	GE	GE	GE	GE	GE	GE

The binary summary of data is stated as follows: there are 7 of 11 or 64% of the data equal to or exceeding a value of 5. A mean, a median, a standard deviation are not computed. There are 64% "detects" at a reporting limit of 5.

6.1.2 Ordinal Methods

Ordinal methods are those that use the ordering (or ranking) of data to perform a procedure. Ordinal methods have the advantage that they can be used on data without a numeric value, but which can be ranked as smaller or larger than other data. Examples of ordinal data include qualitative measurements on a scale such as low/ medium/high. Censored observations below a threshold are ranked as smaller than uncensored observations above the threshold. With censoring there will be many ties. Again consider the small set of 11 observations, along with their ranks:

Data:	<1	<1	3	<5	7	8	8	8	12	15	22
Ranks:	2.5	2.5	2.5	2.5	5	7	7	7	9	10	11

The ranks for data tied at the same value are themselves tied at a value equal to the average or median of the ranks they would have had if there had been no ties. So for the three tied uncensored values at 8, which would have had ranks 6, 7, and 8 if they could be distinguished, all three are assigned a rank of 7, the median of the three ranks. This preserves the sum of the ranks, a statistic used in many nonparametric tests. Similarly, all four of the lowest values are recensored as below the highest reporting limit of 5, and assigned a rank of 2.5, the median/mean of ranks 1–4. Note that we know that the detected 3 is higher than the <1s, but we do not know which of these are higher or lower than a <5, so we cannot uniquely assign ranks to them. We do know that all are below 5, and must consider them all tied below 5 until we understand the concept of scores in nonparametric survival analysis procedures. If we erroneously assigned half the reporting limit to the <5 it becomes a 2.5 and would be considered lower than the detected 3. That is a possibly false ordering that is not known from the data; it is fabricated and is more than what we truly know. This fabricated ordering could contribute to a false result in a subsequent test if substitution were used. Instead, identical ranks below the highest reporting limit state no more than what is known, that these are the four lowest values in the data set, below the remaining detected data. Following this with a nonparametric test using the ranks gives a result that will not give a false positive (false rejection), as substitution could. It is simple and reliable. To obtain more power, and therefore fewer false negatives (not seeing a signal that is there), use the later survival analysis methods.

Percentiles can be computed using the ordinal approach. The 50th percentile (median) is the observation whose position is one-half the way between 0 and $n + 1 = 12$, or the 6th observation from the bottom. Half of the 11 observations are equal to or above the 6th ranked value, and half are equal to or below. The 50th percentile is therefore an 8, one of the three observations tied at that level. Note that if we knew the numerical value of the <5, the median would still be 8. If we knew the numerical values of the <1s, the median would still be 8. Therefore, we can confidently state that the sample median of this small data set is 8, and the 75th percentile is 12. The 25th percentile is <5. Percentiles for all values above the highest reporting limit are known. The percentiles for these data can be listed as follows, with a range of percentages representing the censored and tied observations.

```
Data:          <1    <1    3    <5     7    8     8     8    12    15    22
Ranks:        2.5   2.5  2.5  2.5     5    7     7     7     9    10    11
Percentiles:       [0.08 to 0.33]        0.42  [0.50 to 0.67]  0.75  0.83  0.92
```

6.2   MAXIMUM LIKELIHOOD ESTIMATION

Maximum likelihood estimation requires the assumption that a distribution (normal, lognormal, or some other distribution) will closely fit the shape of the observed data. A mean and standard deviation for the distribution are computed based on the observed uncensored values, and the observed proportions of data below one or more censoring thresholds. Optimization of the mean and standard deviation produces the specific distribution that best fits the observed data.

In the late 1950s and early 1960s, several papers by A.C. Cohen introduced MLE for determination of the mean and variance of censored data. The method was fairly computer intensive, beyond the computing power available to most people at that time. So Cohen developed a version that uses a lookup table to estimate the mean and variance of a singly censored (one reporting limit) normal distribution by adjusting downward the statistics for uncensored (detected) observations in response to the amount of censoring (Cohen, 1959). He presented an expanded lookup table with more detail in a subsequent article (Cohen, 1961). Though used by Miesch (1967) to estimate statistics of geochemical data, the method was not popularized for environmental sciences until Gilbert's 1987 book on environmental pollution monitoring.

Cohen's table-adjustment method is unnecessary today, since more accurate and versatile solutions of the likelihood equations are possible with modern statistical software. It has one serious drawback—the tables are restricted to the case of one reporting limit. Most environmental data today contain multiple limits. These are easily handled by MLE methods available in commercial statistical software, but not by Cohen's method. However, the table-adjustment method is still sometimes recommended (USEPA, 2002a), and was considered "new" in some fields as late as 1990 (Perkins et al., 1990).

Environmental data are more often similar to a lognormal than to a normal distribution, so the mean and variance of the logarithms are more typically estimated by MLE and subsequently reconverted to estimates in original units. The traditional formulae for reconversion are derived from the mathematics of the lognormal distribution, and are found in many textbooks, including Gilbert (1987) and Aitchison and Brown (1957):

$$\hat{\mu} = \exp\left(\hat{\mu}_{\ln} + \frac{\hat{\sigma}_{\ln}^2}{2}\right) \tag{6.1}$$

$$\hat{\sigma}^2 = \hat{\mu}[\exp(\hat{\sigma}_{\ln}^2) - 1] \tag{6.2}$$

$$\text{C.V.} = [\exp(\hat{\sigma}_{\ln}^2) - 1]^{1/2} \tag{6.3}$$

where $\hat{\mu}_{\ln}$ and $\hat{\sigma}_{\ln}^2$ are estimates of the mean and variance, respectively, of the natural logarithms of the data. These equations will work reasonably well if the data are close to lognormal in shape, and if the estimates in log units ($\hat{\mu}_{\ln}$ and $\hat{\sigma}_{\ln}^2$) are close to their true values. However, for small samples the estimates are typically poor enough to bias estimates in original units (Cohn, 1988), leading to overestimation of the mean and variance. MLE methods have not been found to work well for estimating the mean or variance of small ($n < 30$; 50–70 for skewed populations) samples in the papers reviewed later, particularly those assuming a lognormal distribution.

Estimates for percentiles are obtained by computing the percentiles in log units, assuming that the logarithms follow a normal distribution, and then retransforming. The kth percentile is therefore computed as

$$p_k = \exp(\mu_{\ln} + z_k \sigma_{\ln}) \tag{6.4}$$

where p_k is the kth percentile value in original units, and z_k is the kth percentile of a standard normal distribution. For the median, $k = 0.5$ and $z_k = 0$, so that $p_{0.5} = \exp(\mu_{\ln})$. The exponentiated mean of the logarithms is sometimes given a special name, the geometric mean. When the logarithms of data follow a normal distribution, the geometric mean estimates the median of the data's original units (and not the mean).

6.2.1   An Example

Arsenic concentrations were measured in urban streams on the island of Oahu by Tomlinson (2003). The 24 arsenic concentrations (in µg/L) making up the Oahu data set are

```
0.5   0.5   0.5    0.6    0.7    0.7    <0.9  0.9    <1.0  <1.0  <1.0  <1.0
1.5   1.7   <2.0   <2.0   <2.0   <2.0   <2.0  <2.0   <2.0  <2.0  2.8   3.2
```

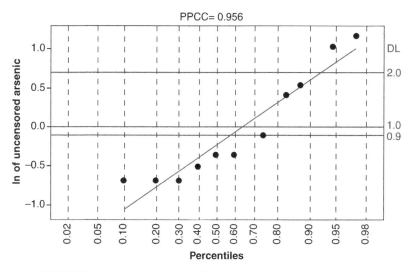

FIGURE 6.1   Lognormal probability plot for the Oahu arsenic data.

Three reporting limits are listed, along with uncensored observations below the lowest reporting limit. It is likely that these data have been subjected to insider censoring (the <0.9 was actually a nondetect of <0.45, etc.), but that is not dealt with here. MLE estimates can be thought of as the statistics of the distribution most likely to have produced the observed data, both censored and uncensored, given that the underlying process follows the assumed distribution. Checking this distribution assumption before computing estimates is a crucial first step. One method for checking the shape of a distribution is the probability plot (see Chapter 5). Figure 6.1 is a probability plot for the Oahu arsenic data. Uncensored values are plotted as solid circles, and a lognormal distribution with the same mean and standard deviation as the data is represented by the straight line. The data points follow a straight line pattern reasonably well for all but the lowest concentration. Therefore, an assumption of a lognormal distribution for these data should produce reasonable estimates for summary statistics.

6.2.2   Cohen's Table Adjustment Method—Example

To illustrate how Cohen's table adjustment method is computed, the procedure of Gilbert (1987) is applied to the Oahu data. In order to do this, all data below the highest reporting limit must be considered censored, as the method works only for one reporting limit. In other words, all values below 2 become <2, resulting in a data set of

<2.0 <2.0 <2.0 <2.0 <2.0 <2.0 <2.0 <2.0 <2.0 <2.0 <2.0 <2.0
<2.0 <2.0 <2.0 <2.0 <2.0 <2.0 <2.0 <2.0 <2.0 <2.0 2.8   3.2

1. Compute h, the proportion of measurements censored. For the above data, $h = 22/24$ or 0.917. 91.7% of the observations are "censored" below the highest limit.

2. Compute the mean and variance from the uncensored observations. Given that most concentration data more closely follow a lognormal than a normal distribution, first convert the data to natural logarithms and then compute the mean_u and variance_u of the uncensored observations: $\text{mean}_u = 1.096$ and $\text{var}_u = 0.0089$ (log units).

3. Compute $\gamma = ((\text{var}_u)/(\text{mean}_u - DL)^2)$, where DL is the reporting limit (in log units).

$$= \frac{0.0089}{(1.096 - 0.693)^2} = 0.0549$$

4. Estimate λ from the table in either Cohen (1961) or Gilbert (1987). For $h = 0.90$ and $\gamma = 0.05$, $\hat{\lambda} = 3.314$.

5. Estimate the mean and variance of the log-transformed data.

$$\mu_{\text{ln}} = \text{mean}_u - \lambda(\text{mean}_u - DL) \quad = 1.096 - 3.314(0.403) \quad = -0.24$$
$$\sigma_{\text{ln}}^2 = \text{var}_u + \lambda(\text{mean}_u - DL)^2 \quad = 0.0089 + 3.314(0.162) \quad = 0.547$$

6. Estimate the mean and variance in original units using equations 6.1 and 6.2:

$$\hat{\mu} = \exp(-0.24 + 0.547/2) = 1.034$$
$$\hat{\sigma}^2 = (1.034)^2 \cdot [\exp(0.547) - 1] = 0.778$$

For a given data set and assumed distribution, differences in summary statistics between Cohen's table-lookup estimates and MLE estimates using statistical software can be attributed both to the approximations built into the lookup table, and to the additional censoring required by table lookup for data sets with more than one reporting limit.

6.2.3   MLE Using Statistical Software—Example

MLE methods are available in commercial statistics packages for data in the interval endpoints format (see Chapter 3). For true nondetects or values reported with a less-than indicator (such as <1), the lower bound is considered to be zero. Minitab®'s Stat > Reliability/Survival menu selection includes routines for interval-censored data in its MLE procedures. The appropriate option is Parametric Distribution Analysis – Arbitrary Censoring. This is a parametric analysis because a distribution will be assumed. It is arbitrary censoring because the data are not right-censored—"arbitrary" censoring covers the other options, including left- and interval-censored data. A lower and upper endpoint must be specified for each observation. For censored observations the lower endpoint ("Start") is often set at zero, and the upper endpoint ("End") is set to the reporting limit. This states that censored observations are located between zero and the reporting limit (the "censoring interval"). Uncensored observations have the same value in both the Start and End variables. The resulting parameter estimates describe the distribution

TABLE 6.1   Summary Statistics for the Oahu Data Using the Lognormal MLE Method in Minitab

	Estimate	Standard Error	95.0% Normal CI Lower	Upper
Mean (MTTF)	0.9453	0.1511	0.6910	1.2931
Standard deviation	0.6559	0.1979	0.3631	1.1849
Median	0.7766	0.1325	0.5559	1.0850
First quartile (Q1)	0.5088	0.1098	0.3334	0.7766
Third quartile (Q3)	1.1854	0.1903	0.8653	1.6238
Interquartile range (IQR)	0.6766	0.1472	0.4417	1.0363

with the maximum likelihood of having produced a data set with the observed uncensored values and the proportions of censored data below each reporting limit. Applying MLE to the Oahu data results in the output shown in Table 6.1.

6.2.4   MLE Methods Using R

The R statistical software package (see Chapter 14) contains functions for computing distributional characteristics and performing regressions with interval-censored data. The contributed package NADA simplifies and standardizes the data input structure for censored data routines in R. Using the NADA package, compute the MLE estimates by typing

```
data(Oahu)
attach(Oahu)
LowAs=As*(1-AsCen)
AsStats=cenmle(As,AsCen, dist="gaussian")
AsStats
            n       n.cen      median        mean          sd
24.0000000  13.0000000   1.0200176   1.0200176   0.7451676
```

and if assuming a lognormal distribution,

```
AsLog=cenmle(As,AsCen, dist="lognormal")
AsLog
            n       n.cen      median        mean          sd
24.0000000  13.0000000   0.7766007   0.9452585   0.6559261
```

6.2.5   Interval Censoring with Detection and Quantitation Limits

Both detection and quantitation limits can be incorporated into a parametric framework by differentiating between the interval endpoints for the two types of data being input to maximum likelihood. In the data set below, a quantitation limit of 1 was used to censor results. Values measured between the detection limit of 0.5 and the quantitation limit were reported as qualified with a remark code signaling a warning. The remark code was dropped by the user, as is often the case, and numbers between 0.5 and 1 used as though they were measured as precisely as values above the quantitation limit of 1. Values measured below 0.5 were reported as <1, an insider censoring procedure. The resulting data set is given in the left-most column named "Values from the lab."

Value from the lab	StartInt	EndInt	Start	End
<1	0	0.5	0	0.5
<1	0	0.5	0	0.5
<1	0	0.5	0	0.5
<1	0	0.5	0	0.5
<1	0	0.5	0	0.5
<1	0	0.5	0	0.5
<1	0	0.5	0	0.5
<1	0	0.5	0	0.5
<1	0	0.5	0	0.5
<1	0	0.5	0	0.5
0.5	0.5	1.00	0.5	0.5
0.55	0.5	1.00	0.55	0.55
0.6	0.5	1.00	0.6	0.6
0.6	0.5	1.00	0.6	0.6
0.7	0.5	1.00	0.7	0.7
0.7	0.5	1.00	0.7	0.7
0.9	0.5	1.00	0.9	0.9
1.5	1.5	1.5	1.5	1.5
1.7	1.7	1.7	1.7	1.7
2.8	2.8	2.8	2.8	2.8
3.2	3.2	3.2	3.2	3.2
5.7	5.7	5.7	5.7	5.7
8.1	8.1	8.1	8.1	8.1

Biased (high) estimates of the mean and percentiles will result from insider censoring if the data are used as reported in the left-hand column. To more correctly compute a mean for these data, maximum likelihood will be performed after recoding censored values into intervals (the StartInt and EndInt columns). All values reported as <1 (actually measured as below 0.5) are considered to be within the interval 0 and 0.5, showing that they were true nondetects. This restores the original measured range of values and avoids insider censoring. To address uncertainty in the values between the two limits, all values measured between 0.5 and 1 are recoded to be within the interval 0.5–1—between the limits. Quantified data, values measured above 1, are input with the same value in both the StartInt and EndInt columns. The software reads this as a quantified single value rather than an interval. Minitab produces the summary statistics of Table 6.2 using MLE and arbitrary censoring, assuming a lognormal distribution:

TABLE 6.2 Summary Statistics from the Lognormal MLE Method for Interval-Censored Data

	Estimate	Standard Error	95.0% Normal CI Lower	Upper
Mean (MTTF)	1.35667	0.517576	0.642303	2.86557
Standard deviation	2.81139	2.09520	0.652472	12.1138
Median	0.589618	0.184836	0.318957	1.08995
First quartile (Q1)	0.246835	0.103307	0.108682	0.560605
Third quartile (Q3)	1.40842	0.425117	0.779488	2.54482
Interquartile range (IQR)	1.16158	0.383806	0.607862	2.21973

Data below the reporting limit and data between the limits were input as different intervals.

If insider censoring had been incorrectly used so that the 10 censored observations were assigned an interval between 0 and 1 instead of 0–0.5, a biased-high estimate of the mean would have been produced. Interval censoring methods for data analysis may alleviate the perceived need for insider censoring, resolving some of the conflict between a user's request for "numbers" and a laboratory analyst's protective reporting measures. Of course if the values measured between the limits of 0.5 and 1.0 are considered sufficiently reliable, those numbers can be considered individual values (Start and End columns at the right-hand side of the data set) rather than using an interval for data between the limits. The reporting limit is then the detection limit. MLE estimates become those of Table 6.3. Note there is little difference for parameter estimates using the two coding schemes. This should provide confidence that interval censoring captures most of the information present in the data between the limits.

6.3 THE NONPARAMETRIC KAPLAN–MEIER AND TURNBULL METHODS

The standard method for estimating summary statistics of censored survival data is the nonparametric Kaplan–Meier (KM) method. Yet as Ware and DeMets stated in 1976,

TABLE 6.3 Summary Statistics from the Lognormal MLE Method Where Only Values Below the Detection Limit are Censored

	Estimate	Standard Error	95.0% Normal CI Lower	Upper
Mean (MTTF)	1.31499	0.502001	0.622270	2.77888
Standard deviation	2.74088	2.04200	0.636402	11.8045
Median	0.568820	0.178298	0.307726	1.05143
First quartile (Q1)	0.237544	0.0993348	0.104663	0.539132
Third quartile (Q3)	1.36208	0.411049	0.753931	2.46081
Interquartile range (IQR)	1.12454	0.371099	0.588954	2.14719

"Although the Kaplan–Meier estimate is fundamental to survival data analysis, it is often overlooked when a left or right censored data [sic] arises in other settings" (Ware and DeMets, 1976). This has certainly been true in the setting of environmental sciences.

Kaplan–Meier is implemented in commercial statistics packages offering routines for survival analysis. However, it only accepts right-censored data. Interval-censored data such as the Oahu arsenic data must either be treated as left-censored and flipped, or the Turnbull method for interval-censored data (a variation of Kaplan–Meier) is used to compute summary statistics. Minitab (version 16) has the KM method in its `Stat > Reliability/Survival` menu, under a submenu selection named `Distribution Analysis – Right Censoring > Nonparametric Distribution Analysis`. If all interval-censored data have a lower bound of 0, nonparametric analyses can incorporate them as flipped left-censored values, ignoring the magnitude of the lower bound. The lowest values are simply the lowest values with the lowest ranks, and when flipped become the highest flipped values with the mirror image ordering of ranks. Information that a lower bound of zero exists is not used. However, when some of the interval-censored values have a nonzero lower bound as with data between the detection and quantitation limits, interval-censored methods such as the Turnbull estimator should be used.

To illustrate the use of Kaplan–Meier, the maximum value in the Oahu data is 3.2, so 5 is arbitrarily chosen as a value larger than the maximum. Let M equal this flipping constant of 5. Right-censored data are constructed by subtracting all observations from M

$$\text{Flip}_i = M_i - x_i \qquad (6.5)$$

for all observations x_i. The result is stored in a column labeled Flip in Table 6.4.

The KM method produces estimates of the survival probability function S for right-censored data. The survival function S is the probability that a data value

TABLE 6.4   Computation of Kaplan–Meier Survival Probabilities for the Oahu Data ($n = 24$)

As (μg/L) (detects)	Flip	Rank r	Number at Risk b, $b = (n - r + 1)$	Number of Detects d at that Value	Incremental Survival $p = (b - d)/b$	S, Survival Probabilities
3.2	1.8	1	24	1	23/24	0.9583
2.8	2.2	2	23	1	22/23	0.9167
1.7	3.3	11	14	1	13/14	0.8512
1.5	3.5	12	13	1	12/13	0.7857
0.9	4.1	17	8	1	7/8	0.6875
0.7	4.3	19	6	2	4/6	0.4583
0.6	4.4	21	4	1	3/4	0.3438
0.5	4.5	22	3	3	0/3	0.0000

Censored observations are accounted for through the rank statistic r.

$T > y$ for any specific values y. If computed using flipped data, $S = \text{Prob}(\text{Flip} > y)$, or $\text{Prob}(M - x > y)$, or $\text{Prob}(x < M - y)$. The latter expression shows that survival probabilities are also the cumulative distribution function of the original x data. Computation of the survival function after flipping the Oahu data is illustrated in Table 6.4.

Table 6.4 has one row for each unique value of uncensored observations in the data set (tied data get one row). The survival probabilities S are computed for each unique value. Using the flipped values, the uncensored observations ("failures" or "deaths" in survival analysis terminology) are ranked from small to large, accounting for the number of censored data in between each detected observation. For example, there are eight censored values for the Oahu data at <2.0, between the flipped values of 2.2 and 3.3 (see Table 6.4). The rank of the surrounding flipped observations therefore jumps from 2 to 11; KM places each nondetect at its reporting limit prior to ranking. The "number at risk" b equals the number of observations, both detected and censored, at and below each detected concentration. The number of uncensored observations at that concentration is d, where d is greater than 1 for tied values. The incremental survival probability is the probability of "surviving" to the next lowest uncensored concentration, given the number of data at and below that concentration, or $(b - d)/b$. The survival function probability is the product of the $j = 1$ to k incremental probabilities to that point, going from high to low concentration for the k uncensored observations.

$$S = \prod_{j=1}^{k} \frac{b_j - d_j}{b_j} \tag{6.6}$$

For example, the survival function probability of 0.6875 for the concentration at 0.9 equals 0.7857(7/8). Note that for the case of ties, KM assigns the smallest rank possible to each observation, rather than the average rank as is done for most nonparametric tests. KM will assign a probability of 0 to the smallest observation (largest flipped value), if there are no censored observations below this value in the data set. This represents a plotting position of i/n for the empirical distribution function (edf) of flipped values, so that the probability of exceeding the last value is 0. If the smallest concentration is a censored value, as is usually the case, the smallest detected observation will have a nonzero exceedance probability, while probabilities are indeterminate for all censored observations below the lowest detected observation.

A plot of the survival function for the Oahu data is shown in Figure 6.2. The KM analysis of Flip produces the following estimates of summary statistics:

	Standard	95.0% Normal CI	
Mean (MTTF)	Error	Lower	Upper
4.0510	0.1647	3.7283	4.3738
Median =	4.3000		
IQR =	0.4000	Q1 = 4.1000	Q3 = 4.5000

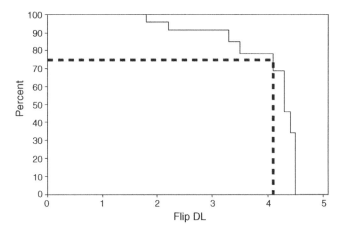

FIGURE 6.2   Survival function plot (Kaplan–Meier method) for the flipped Oahu data.

Location estimates for flipped data (mean, median, other percentiles) must be retransformed back into the original scale by subtraction from the constant M used to flip the data. When the above estimates for mean, median, and Q1 and Q3 are subtracted from the flipping constant of 5, the results are those printed in Table 6.5. Estimates of variability (variance, standard deviation, standard error, IQR) are the same for both the flipped and original units; no retransformation is needed.

How were these summary statistics estimated by KM? For percentiles, the estimate is the minimum X-value on the survival function graph that is intersected by the line drawn at the probability value from the Y-axis. It is the smallest flipped observation having a survival probability equal to or less than the stated probability of the percentile. The 25th percentile (Q1) has a survival probability (probability of exceedance) of 0.75. A horizontal line drawn from 0.75 on the Y-axis intersects the vertical line at an X-value of 4.1. Looking at Table 6.4, the flipped observation at 4.1 is the smallest flipped value for which the survival probability is 0.75 or less. Subtracting this from the flipping constant of 5, the 75th percentile of the original data is 0.9. The process is similar for other percentiles. When more than 50% of data are below the lowest reporting limit, and the smallest observation (largest flipped value) is censored, the median cannot be estimated as a single number. It is simply <RL, where RL is the lowest reporting limit. The KM software will report this as a missing value. A method that assumes some sort of model for the data distribution must be employed if a single number estimate for

TABLE 6.5   Summary Statistics Using Kaplan–Meier for the Oahu Arsenic Data

Mean	Standard Deviation	Standard Error	Q1	Median	Q3
0.949	0.807	0.1647	0.500	0.700	0.900

the median is required. Two possible methods for doing so include parametric MLE (previous section) and robust ROS (next section).

KM methods include an estimate for the standard error of the survival function. Like the function S itself, the standard error is a step function that changes for each detected (uncensored) observation. Standard errors are computed most often to estimate confidence intervals around the estimated percentiles, describing the certainty with which that percentile value is known (see Chapter 7). Plots of survival functions often include the interval boundaries based on the standard error. The standard error formula, known as Greenwood's formula, is derived in many books on survival analysis, including Collett (2003),

$$\text{Std. error of } S = \text{s.e.}[S] = S\sqrt{\sum_{j=1}^{k} \frac{d_j}{b_j(b_j - d_j)}} \tag{6.7}$$

where b_j is the "number at risk" and d_j is the number of uncensored observations (see Table 6.4) at each of the k values for uncensored observations.

6.3.1   KM Estimates of Mean, Variance, and Standard Error

The mean is generally considered less useful than the median in survival analysis, as distributions of medical or other "lifetime" data are sufficiently skewed that the mean is not a typical value, but is strongly influenced by a few unusual values. The mean is often not reported by the software. However, the Kaplan–Meier estimate of the mean is unbiased and as efficient or more so than parametric methods for estimation (Meier et al., 2004). It may be computed by integrating the area under the KM survival curve. To see why this is so, consider the usual equation for the mean of n observations, $\mu = \sum x/n$. The equation can be stated as $\mu = \sum (f_i/n)x_i$, where f_i is the number of observations at each of the i unique values of x, so that f_i/n is the proportion of the data set at that value. The mean is the sum of the products of the proportion of data for each value times the magnitude of the observation's value. This is just what is accomplished when integrating under the KM survival curve (Figure 6.3). The curve is divided by drawing horizontal lines at the value of each detected observation. The resulting multiple rectangles have as their height the estimated proportion of data at that value, with the proportions summing to 1. The width of the rectangle is the magnitude of the observation, x. The mean is estimated by multiplying the width of each rectangle by its height to get the area, and then summing the areas over all rectangles.

When the smallest observation (largest flipped value) is censored, the end of the edf/survival function is unknown. This is not an issue with establishing percentiles, but is when integrating areas to compute the mean because the width of the final bar is unknown. To estimate the mean, the convention in survival analysis is to use the censoring threshold to represent the censored value. A <1 stops the width of the bar at 1, as it is unknown how much farther down toward zero the concentration actually goes. This convention is called "Efron's bias correction" (Klein and Moeschberger,

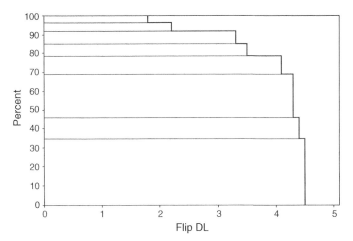

FIGURE 6.3 Computing the mean by integrating under the Kaplan–Meier survival curve. The total area is the KM estimate of the mean.

2003, p. 100) because it is less biased than stopping the final bar at the lowest detected value somewhere above 1. It produces an estimate of the mean in original units that is biased high (the mean of the flipped data is biased low). If there is only one reporting limit, the result is that the mean is identical to a substitution of the reporting limit for censored observations. As this occurs only at the smallest reporting limit when there are multiple limits, the positive bias in the KM mean is much less than with simplistic substitution of the reporting limit when there are multiple reporting limits. However, the "stubbornness" of a nonparametric procedure in not extrapolating below the lowest reporting limit reduces the usefulness of the KM method for data when estimating the mean (not the percentiles) with one reporting limit.

Estimates of the standard error of the mean are often produced by survival analysis software to establish a confidence limit around the mean. These limits assume either that the data follow a normal distribution, or that there are sufficient numbers of observations to invoke the Central Limit Theorem (CLT). The CLT states that even if the data do not follow a normal distribution, with sufficient data the distribution of the sample mean itself follows a normal distribution. The theorem is used with parametric approaches to data analysis, and is becoming less important as bootstrap procedures replace purely parametric approaches. Lee (2003) shows that for m uncensored values, n total observations, and the cumulative areas A_r, $r = 1, \ldots, m$ under the Kaplan–Meier curve, the variance of the mean is computed using the formula inside the square root sign of equation (6.8). Note the multiplication by the ratio of $m/(m-1)$ to obtain an unbiased sample estimate of the variance. Finally, the standard error is the square root of the variance of the mean.

$$\text{Std. error} = \sqrt{\left(\frac{m}{m-1}\right) \sum_{r=1}^{m} \frac{A_r^2}{(n-r)(n-r+1)}} \tag{6.8}$$

Estimates of standard deviation of observations are even of less interest than the mean in traditional survival analysis due to the skewness found in most survival data. It provides a poor measure of the variability of data when those data are strongly skewed. Environmental data are similarly skewed, but if an estimate for the standard deviation is justified, it can be computed "through the back door" using equation (6.9). For uncensored data, the standard error of the mean equals the standard deviation (s.d.) divided by the square root of the sample size n.

$$\text{Std. error} = \frac{\text{s.d.}}{\sqrt{n}} \tag{6.9}$$

Therefore, multiplying the standard error by $\sqrt{n}$ will estimate the standard deviation. The standard deviation for the Oahu data estimated by this method is shown in Table 6.5.

6.3.2   Minitab Calculation of Kaplan–Meier Procedures

The Minitab macro KMStats (available on the Practical Stats web site practicalstats.com/nada) performs all of the flipping, computation, and retransformation of results back to the original units. If your censored data all have a lower bound of 0, and so can be considered left-censored, they can be flipped to right-censored data and Kaplan–Meier computations performed in one easy step. If some of your censored data have a nonzero lower bound, instead use the interval-censored methods of the next section.

For the Oahu data the command

```
%kmstats c1 c2;
   cens 0.
```

produces the output

```
Statistics using Kaplan–Meier, with Efron bias correction
   Left-Censored data
Mean Arsenic                        0.948958
Standard error                      0.164689
Standard Deviation                  0.806807
95th Percentile                     2.80000
90th Percentile                     1.70000
75th Percentile                     0.900000
Median                              0.700000
25th Percentile                            *
10th Percentile                            *
```

6.3.3   Calculation of Kaplan–Meier Procedures Using R

The NADA for R function cenfit will compute nonparametric Kaplan–Meier estimates of summary statistics. For the Oahu data,

```
Askm=cenfit(As,AsCen)
Askm
          n       n.cen      median         mean           sd
24.0000000  13.0000000   0.7000000    0.9489583    0.8068068
```

and typing summary of a cenfit object will give the table of percentiles (prob)

```
summary(Askm)
     obs  n.risk  n.event   prob         std.err       0.95LCL      0.95UCL
1    0.5  3       3         0.0000000    NaN           NaN          NaN
2    0.6  4       1         0.3437500    0.15381284    0.04228238   0.6452176
3    0.7  6       2         0.4583333    0.15669579    0.15121522   0.7654514
4    0.9  8       1         0.6875000    0.12592662    0.44068835   0.9343116
5    1.0  12      0         0.7857143    0.09842664    0.59280161   0.9786270
6    1.5  13      1         0.7857143    0.09842664    0.59280161   0.9786270
7    1.7  14      1         0.8511905    0.08200806    0.69045764   1.0000000
8    2.0  22      0         0.9166667    0.05641693    0.80609151   1.0000000
9    2.8  23      1         0.9166667    0.05641693    0.80609151   1.0000000
10   3.2  24      1         0.9583333    0.04078938    0.87838763   1.0000000
```

Figure 6.4 shows that plotting a cenfit object in NADA for R results in an edf of
the data.

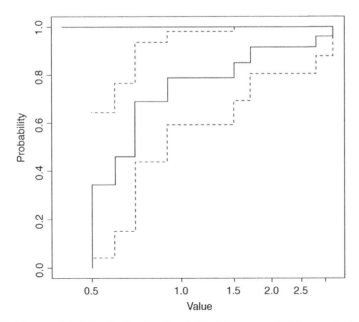

FIGURE 6.4   Empirical distribution function (edf) of the censored Oahu arsenic data, using
NADA for R's cenfit command.

6.3.4   Turnbull Interval-Censored Method: Statistics for Data that Include (DL to RL) Values

Censored environmental data are most easily represented as interval-censored, with a lower and upper boundary of concentration. Data that are true nondetects have a lower boundary of zero—concentrations might truly be zero—and the data can be considered left-censored as well as interval-censored. Censored data above the detection limit, such as those between the detection and quantitation limits, have a nonzero lower boundary. These data are interval—rather than left-censored. In an important paper, Turnbull (1976) demonstrated how to define a survival function (edf) for interval-censored data and thereby establish percentiles. Mathematical details of the procedure may be found in Turnbull's original article or in the book by Sun (2006).

For n observations within the intervals $[a_i, b_i]$, $i = 1$ to n, where a_i are the lower bounds and b_i are the upper bounds, the survival function $S[a_i, b_i]$ can be computed where

$$S[a_i, b_i] = S[a_i] - S[b_i] \tag{6.10}$$

In other words, the probability of being between a_i and b_i equals the probability of exceeding the lower bound a_i minus the probability of exceeding the upper bound b_i.

Percentiles for the Oahu data are estimated using the Turnbull method in Minitab's

```
Stat > Reliability/Survival > Distribution analysis (arbitrary
censoring) > Nonparametric distribution analysis
```

procedure. Data are input as intervals, and no flipping is required. The lower end of each interval is 0 for this data set, so results should agree with those using Kaplan–Meier after flipping. Shown below from the output are the values for uncensored observations, and the "cumulative failure probability" or $1 - S$, which estimates the probability of being at or below each value, in essence their percentiles:

```
Value           Probability
0.5             0.343749
0.6             0.458333
0.7             0.687500
0.9             0.785714
1.5             0.851190
1.7             0.916667
2.8             0.958333
```

A concentration of 0.5 is at the 34th percentile, of 0.6 at the 46th percentile, and so on. For values of specific percentiles such as the median with $p = 0.5$, the convention is to use the observation whose cumulative failure probability is at or above the desired percentile. The median would then equal 0.7. The 75th percentile would similarly equal 0.9 and the 90th percentile equal 1.7. Percentiles below the 34th are less than the lowest value, or <0.5. Note that only percentiles are calculated by the method—it does not estimate a mean or standard deviation. Comparing the results to the Kaplan–Meier procedure used earlier after flipping the data, we get

	Left-Censored (KM)	Interval-Censored (Turnbull)
Mean Arsenic	0.949	
Standard error	0.165	
Standard Deviation	0.807	
95th Percentile	2.8	2.8
90th Percentile	1.7	1.7
75th Percentile	0.90	0.90
Median	0.70	0.70
25th Percentile	*	<0.5
10th Percentile	*	<0.5

Although not demonstrated here, right-censored observations may also be input to the Turnbull procedure, where a "> 100" is represented as [100, *]. The missing value indicator * represents infinity. The nonparametric Turnbull method, like the parametric maximum likelihood methods discussed elsewhere, may be used when both less-thans and greater-thans are present in the same data set. This is because the Turnbull procedure is the nonparametric maximum likelihood solution, where the likelihood function is (Collett, 2003):

$$\prod_{i=1}^{lc}\{1 - S_i(b_i)\} \prod_{i=lc+1}^{lc+rc} S_i(a_i) \prod_{i=lc+rc+1}^{n} \{S_i(a_i) - S_i(b_i)\} \qquad (6.11)$$

Here there are lc left-censored observations, rc right-censored observations, and n observations overall. The first term $1 - S_i(b_i)$ is the product of cumulative failure rates for the upper bound (probability of being at or below the reporting limit) for left-censored observations. The second term $S_i(a_i)$ is the probability of exceeding the lower bound of right-censored observations. The final term is the probability of being within the interval of interval-censored observations. The probabilities that best match the observed percentages of data are the Turnbull solution to this equation.

The Turnbull procedure and parametric MLE methods are perhaps the best solutions to incorporating values between the detection and quantification limits, the "J" values over which there is so much controversy as to their reliability. To use interval-censored procedures, report "J" values as an interval where a is the lower (detection) limit and b is the higher (quantification) limit.

6.4   ROS: A "ROBUST" IMPUTATION METHOD

Methods that calculate summary statistics with least-squares regression on a probability plot are called "regression on order statistics," (ROS). One version is fully parametric, the method found in Minitab when "least-squares" is selected for probability plot routines, and is generally less efficient than maximum likelihood. In this fully parametric mode (described in the next paragraph) ROS has little advantage over MLE for computing numerical statistics, though the plot itself is a useful guide as long as the procedure plots the highest observation (see Chapter 5). A second "robust" implementation of ROS uses sample data whenever possible, assuming a distribution to impute values for the censored portion of the distribution. Uncensored observations are used in their own right. In the robust form, abbreviated MR by Helsel and Cohn (1988) and LPR by Hewitt and Ganser (2007), ROS is an attractive alternative to the more restrictive parametric assumptions of maximum likelihood. Its most common application is to small data sets ($n < 30$), where MLE estimation of parameters becomes inaccurate.

ROS computes a linear regression for data or logarithms of data versus their normal scores, the coordinates found on a normal probability plot (e.g., see Figure 5.12). The regression parameters (slope and intercept) are computed using the uncensored observations. Due to the definition of normal scores, fitting this line is fitting a normal distribution (a lognormal distribution if logs are used on the y-axis) to the observed data. For the fully parametric version of this method, the intercept and slope of the regression line estimate the mean and standard deviation, respectively, of the data or their logarithms. The intercept, the y value associated with a normal score of 0 (50th percentile) at the center of the plot, estimates the mean of the distribution. The slope of the line equals the standard deviation, as normal scores are scaled to units of standard deviation. The parametric and robust forms of ROS have often been confused in the literature. Travis and Land (1990) recommended the fully parametric method, though for justification they reference the results of Gilliom and Helsel (1986), who actually used the robust ROS. The two forms of ROS have different performance characteristics, and care should be taken to treat them separately.

If logarithms are used for the y-axis, the intercept and slope of a probability-plot regression estimate the mean and standard deviation of the logarithms. These summary statistics must be retransformed to provide estimates of statistics in original units. Transforming moment statistics (mean and standard deviation) across scales with power transformations such as the logarithm results in transformation bias. The mean in one set of units will not provide an accurate estimate of the mean when converted to a second set of units using a power transformation. A simple example of transformation bias is given in Table 6.6. The mean logarithm of 2, retransformed back to 100 in the original units, is biased low when compared to the mean of over 2000 for data in the original units.

Transformation bias is just as much of concern for parametric ROS as it was for MLE. If the estimates in log units are simply exponentiated, the resulting geometric mean estimates the median, just as the 100 in Table 6.6 is biased low as an estimate of the mean. If equations 6.1 and 6.2 are used for retransformation to avoid transformation bias, the resulting estimates are biased high (Cohn, 1988) for small samples due to the inaccuracy in estimating the standard deviation of the logarithms.

To avoid transformation bias, summary statistics can be computed by imputing numbers for the censored observations based on a parametric model. Imputation is the

TABLE 6.6   An Illustration of Transformation Bias

	Original Units	Logarithms (base 10)
	1	0
	10	1
	100	2
	1000	3
	10000	4
Mean	2222.2	2
$10^{\text{mean log}}$		100

process of guessing values for observations based on a statistical model. Combining imputed estimates with uncensored observations to compute summary statistics avoids transforming calculated means and standard deviations across scales, from logarithms to original units. Avoiding the resulting transformation bias is why the imputation procedure was labeled as "robust." Using this approach, a more limited assumption of normality (or lognormality) is used—only values below reporting limits follow a specified distribution (Helsel and Cohn, 1988).

After fitting a regression equation using the uncensored observations on a probability plot, values for individual censored observations are predicted from the regression model based on their normal scores (the explanatory or x variable in the regression equation). Predicted values from the equation are imputed (or logs imputed and exponentiated if y is in log units) and combined with uncensored observations to compute summary statistics as if no censoring had occurred. Retransforming individual values instead of the fitted parameters, and calculating summary statistics only after returning to the original units, avoids the problem of transformation bias. Computations by robust ROS for the Oahu arsenic data are given in Table 6.7 as an example.

Multiple reporting limits are accounted for in the following way. First the probability of exceeding each reporting limit is computed using the proportion of values in the data set that are at or exceed each limit. For the Oahu data there are 2 out of 24 observations at or above the highest reporting limit of 2 µg/L. So the probability of detection at 2 µg/L is 2/24 or 0.083. Then the probability of detection at the next highest limit (1 µg/L) is computed. There are 14 observations below 2 µg/L that can be compared to a reporting limit of 1 µg/L. Of these, 2 observations are at or exceed 1 µg/L and 12 observations are below 1 µg/L. So an estimate of the probability of detection at a limit of 1 µg/L is (2/14) × 0.917 + 0.083 = 0.214, where 0.917 is the probability of being at or below 2 µg/L (or 1 − 0.083). Finally, to determine the probability of detection at the lowest reporting limit of 0.9 µg/L, of the eight observations below 1 µg/L there is one at or exceeding 0.9 µg/L. So an estimate of the probability of detection at a limit of 0.9 µg/L is (1/8) × 0.786 + 0.214, or 0.313.

In general, the probability of exceeding the jth reporting limit is

$$pe_j = pe_{j+1} + \frac{A_j}{A_j + B_j}\left[1 - pe_{j+1}\right] \tag{6.12}$$

TABLE 6.7 Computation of Summary Statistics Using Robust ROS for the Oahu Data ($n = 24$)

As (μg/L) (Detects)	log e Conc	Prob of Detection	Plot pos Percentile	Rank r	Predicted logs	Observed + Estimated Concentration
3.2	1.163		0.972	24		3.2
2.8	1.030		0.945	23		2.8
<2		0.083	0.815		0.349	1.42
<2			0.713		0.134	1.14
<2			0.611		−0.047	0.95
<2			0.509		−0.215	0.81
<2			0.407		−0.381	0.68
<2			0.306		−0.559	0.57
<2			0.204		−0.766	0.46
<2			0.102		−1.05	0.35
1.7	0.531		0.873	14		1.7
1.5	0.405		0.829	13		1.5
<1		0.214	0.629		−0.018	0.98
<1			0.471		−0.276	0.76
<1			0.314		−0.543	0.58
<1			0.157		−0.881	0.41
0.9	−0.105		0.737	8		0.9
<0.9		0.313	0.344		−0.49	0.61
0.7	−0.357		0.589	6		0.7
0.7	−0.357		0.491	6		0.7
0.6	−0.511		0.393	4		0.6
0.5	−0.693		0.295	3		0.5
0.5	−0.693		0.196	3		0.5
0.5	−0.693		0.098	3		0.5

where A_j is the number of observations detected between the jth and $(j + 1)$th reporting limits, and B_j is the number of observations, censored and uncensored, below the jth reporting limit.

When j is the highest reporting limit, $pe_{j+1} = 0$ and $A_j + B_j = $ n. The number of censored observations below the jth reporting limit is defined as C_j:

$$C_j = B_j - B_{j-1} - A_{j-1} \qquad (6.13)$$

Plotting positions are then calculated in order to compute a normal score for each observation (see Table 6.7). Normal scores for uncensored observations are used to construct the regression equation relating the log of concentration to normal scores. Normal scores for censored observations are input to that regression equation to predict a log concentration, which is then retransformed to estimate concentrations for the set of censored observations. So plotting positions and normal scores are needed for both censored and uncensored observations. Plotting positions are at values spread equally between exceedance probabilities, and are computed separately for uncensored and censored observations. For the two uncensored observations above the highest reporting limit of $2\,\mu g/L$, plotting positions are two values equispaced between $(1 - 0.083 = 0.917)$ and 1.0, or at $0.917 + (1/3) \times 0.083$ and $0.917 + (2/3) \times 0.083 = 0.972$. The $C_3 = 8$ censored observations known to be $<2\,\mu g/L$ are spread evenly between probabilities of 0 and 0.917, or $(i/9) \times 0.917$, where $9 = C_3 + 1$ and $i = 1$ to 8. The two uncensored values between the reporting limits of 1 and $2\,\mu g/L$ are spread evenly at 1/3 and 2/3 the distance between probabilities of $(1 - 0.214) = 0.786$ and 0.917. The $C_2 = 4$ censored observations known to be $<1\,\mu g/L$ are spread evenly between probabilities of 0 and 0.786, or at $(i/5) \times 0.786$, where $5 = C_2 + 1$ and $i = 1$ to 4. The one detected observation between a <0.9 and $1\,\mu g/L$ plots at a position halfway between detection probabilities of $(1 - 0.313) = 0.687$ and 0.786. The $C_1 = 1$ nondetect at <0.9 plots halfway between probabilities of 0 and 0.687. Finally, the six uncensored values below 0.9 plot at probabilities of $0 + (i/7) \times 0.687$, where $i = 1$ to 6.

In general, plotting positions for uncensored observations are

$$pd_i = (1 - pe_j) + \left[\frac{i}{A_j + 1}\right] \cdot \left[pe_j - pe_{j+1}\right] \quad \text{for } i = 1 \text{ to } A_j \qquad (6.14)$$

and for censored observations are

$$pc_i = \left[\frac{i}{C_j + 1}\right] \cdot \left[1 - pe_j\right] \quad \text{for } i = 1 \text{ to } C_j \qquad (6.15)$$

These equations follow the pattern of Hirsch and Stedinger (1987), who extended the traditional use of probability plotting in flood hydrology to the case of censored records of historical floods. After considering Bayesian and other methods for assigning plotting positions, their Appendix C provides equations 6.12 and 6.14 for determining plotting positions with multiple censoring levels. Helsel and Cohn (1988) extended these to equation 6.15 for censored data.

Estimated values produced for censored observations (the right-most column of Table 6.7) should not be assigned to any individual sample. For data sets with multiple observations below the same reporting limit, there is no valid way to do so. Which estimate belongs to which sample is unknown. The corporate collection of estimates below each reporting limit is sufficient to compute overall statistics, and yet does not allow the scientist to fall into the trap of indicating that the value for an individual censored observation is known, as is implied with simple substitution methods. Even when there is only one nondetect below a reporting limit, declaring that a value is known for it is untrue. The value is known only to be within the interval from zero to the reporting limit.

For the Oahu arsenic data, ROS estimates assuming a lognormal distribution produced by the %cros macro in NADA for Minitab are given in Table 6.8.

TABLE 6.8 Summary Statistics Using ROS for the Oahu Arsenic Data

MEAN	STD DEV	Pct25	MEDIAN	Pct75
0.972	0.718	0.518	0.700	1.103

Use the cenros function in NADA for R to get the same result, also assuming a lognormal distribution:

```
> Asros=cenros(As,AsCen)
> Asros
        n         n.cen      median        mean          sd
24.0000000 13.0000000  0.7000000  0.9698429  0.7185456
```

And the summary function provides details on the regression equation versus normal quantiles on the x-axis:

```
> summary(Asros)

Call:
lm(formula = obs.transformed ~ pp.nq)

Residuals:
     Min         1Q    Median        3Q       Max
-0.28975   -0.11613   0.02519   0.12463   0.37624

Coefficients:
               Estimate   Std. Error   t value   Pr(>|t|)
 (Intercept) -0.23712      0.06971     -3.402    0.00785 **
 pp.nq        0.64876      0.06790      9.555    5.22e-06 ***
 —
Signif. codes: 0 '***' 0.001 '**' 0.01 '*' 0.05 '.' 0.1 ' ' 1
Residual standard error:   0.2192 on 9 degrees of freedom

Multiple R-squared: 0.9103,   Adjusted R-squared: 0.9003
F-statistic:    91.3 on 1 and 9 DF, p-value: 5.22e-06
```

while the plot function provides the probability plot (Figure 6.5) and fitted regression equation:

```
> plot(Asros)
```

Shumway et al. (2002) improved the robust ROS method by determining whether data best fit a lognormal, normal, or square root–normal distribution prior to performing ROS. This was done by choosing the units that produced the largest log-likelihood statistics when fit by MLE. They state that one of these three distributions generally matches the observed shape of environmental data. With this prior evaluation of distributional shape, they found that robust ROS produced estimates of the same quality as did MLE for moderate ($n = 50$) sized data sets, and of better quality than MLE for small ($n = 20$) data sets.

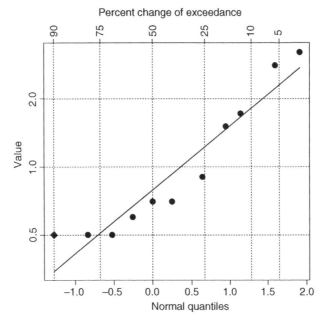

FIGURE 6.5   Probability plot and ROS line using the cenros function in NADA for R.

A side-by-side comparison of the estimates computed in this chapter for the Oahu arsenic data is given in Table 6.9. The simple ordinal method is not that helpful for this data set, as the highest reporting limit was quite high as compared to uncensored measurements, so that most statistics would be <2. The Kaplan–Meier method, independent of any distributional assumption, is the most generally applicable of the methods. If data appear to follow a lognormal distribution (the Oahu data are close except for the highest "outlier" observation) and there is a "large" set of uncensored data to adequately estimate parameters for the distribution (there is not here), then MLE should provide the best estimates. The value for "large" increases with increased skewness, but is generally 30–50 uncensored observations. These

TABLE 6.9   Summary Statistics Using Several Estimation Methods in Minitab—Oahu Arsenic Data

METHOD	MEAN	STD DEV	Pct25	MEDIAN	Pct75
Simple ordinal	---	---	<2	<2	<2
Cohen's (ln)[a]	1.034	0.882			
MLE (ln)	0.945	0.656	0.509	0.777	1.185
robust ROS (ln)	0.972	0.718	0.518	0.700	1.103
KM	0.949	0.807	0.500	0.700	0.900

[a] Method is approximate and so not recommended for use.

TABLE 6.10   Summary Statistics Using Three Estimation Methods in NADA for R—Oahu Arsenic Data

	Median	Mean	sd
KM	0.7000000	0.9489583	0.8068068
ROS	0.7000000	0.9698429	0.7185456
MLE	0.7766007	0.9452585	0.6559261

conditions are not met in many environmental studies, where small samples sizes are the norm. The ROS method, less dependent on a distributional assumption than MLE, is most useful for smaller data sets. It can be an alternative to Kaplan–Meier when more than 50% of the data are censored and an estimate of the median is desired. Cohen's MLE method is not recommended as it never is better than a true MLE estimate.

A similar summary table (Table 6.10) is available in NADA for R using the censtats command:

```
> censtats (As,AsCen)
```

6.5   METHODS IN EXCEL

An Excel worksheet is available for computing the Kaplan–Meier estimates for summary statistics, at www.practicalstats.com/nada. Worksheets for ROS may also be found on the web. An approximate MLE solution can be performed using Excel's SOLVER function (Flynn, 2010). However, Excel does not always perform computations in the way true statistical software should. For example, Excel does not calculate percentiles as a linear interpolation between two bracketing points, the typical definition in statistics programs. Instead, it uses the observation value that is closest to but below the desired percentile. If the 90th percentile were to be calculated, Excel would select and report the value of the next lowest observation at the 88th percentile, as one example, instead of interpolating between observations located at the 88th and 94th percentiles. Excel ranks tied observations differently than what was described in this chapter. If using Excel, investigate and understand the procedures that it is following—they will likely differ from standard procedures available in commercial statistics software.

6.6   HANDLING DATA WITH HIGH REPORTING LIMITS

A "high nondetect" value, a censored observation whose reporting limit is higher than all uncensored values in the data set, has no information content. "High nondetects" can be dropped from the data set without penalty. For example, if a value of <100 is included in data where the highest value is a detected 55, the <100 is of no use. KM and any other statistics-based (as opposed to substitution) procedure will ignore high nondetects when computing percentiles or the mean. There is no way to determine the probability of the <100 being above or below the detected 55 or any other detected observation, and so it adds nothing to the determination of the proportion of data below each detected observation. A <20 is known to be below the detected 55, and so is used when computing the percentile for the detected 55. The location of a <100 is indeterminate in relation to the observed detections of that data set. MLE, ROS or KM software will give identical estimates of mean and percentiles whether indeterminate high nondetects are included or not. This is one great advantage of these methods over substitution. Just one high nondetect, when one-half of that reporting limit is substituted, can radically (and incorrectly) alter estimates of the mean and standard deviation.

High nondetects do not meet the quality control standards of the rest of the data set. Because they contain no information, they may be dropped from the analysis. Or methods such as ROS, MLE and KM that appropriately discount them should be used.

6.7 A REVIEW OF COMPARISON STUDIES

Summary statistics for censored data is the most studied topic in the treatment of censored environmental data. The confusing element is that each article seems to find a different method to be "best." Why do conclusions differ so much on the choice of methods? Four important characteristics that strongly influence findings have varied among these studies. They are as follows:

1. *Sample Size.* MLE methods work far better for larger sample sizes (at or above 50) than smaller sizes. Some studies have used small samples, others larger.
2. *Transformation Bias.* Some studies have computed estimates assuming a normal distribution, and simply stated that the results apply to lognormal or other distributions after transformation. This ignores transformation bias, the additional error resulting from moment statistics (mean and standard deviation) not being invariant to scale changes. The robust ROS and robust MLE methods attempt to overcome transformation bias.
3. *Robustness.* Some studies generate data only from the same distribution that will be assumed in computing parametric methods. Method errors then reflect only the best-case scenario for parametric methods, ignoring the real-life errors involved when the underlying distributions of data are mis-specified.
4. *Details of Method Computation and Terminology.* For ROS, some studies use the fully parametric approach, others the robust approach, and both are called by the same name. For MLE, older studies used Cohen's table lookup, while most now use the computer solution to likelihood equations. Some studies have incorrectly named substitution as "imputation," while others appropriately use that name for results from statistical modeling.

The 15 papers reviewed below have directly compared methods for computing summary statistics using simulated data, to evaluate their performances. Root mean squared error (RMSE) and bias are usually computed as measures of the inadequacy of each method, with the best methods having low values for both. While this list of papers is not exhaustive, it does provide a summary of the major findings on this topic in environmental statistics to date. Many of the studies generated data from a single distribution, usually the lognormal distribution— Gleit (1985) used a normal distribution—and so do not evaluate errors due to mis-specifying the distribution of the data. This gives a great advantage to the parametric MLE and ROS methods, at the expense of robust and nonparametric methods. Gilliom and Helsel (1986), Helsel and Cohn (1988), Kroll and Stedinger (1996), and She (1997) compared methods where the data distribution was not the same as the distribution assumed by maximum likelihood or ROS, to evaluate the robustness of each method.

1. Owen and DeRouen (1980) estimated means for air contaminant data, finding that MLE methods had high errors, especially for data of small sample size and a large proportion of censored values. They generated data of $n = 5$ to 50 having some true zeros, and found that the delta estimator, which assigns zeros to all censored values while modeling uncensored data as a lognormal distribution, had less error than MLE. This success might be attributed (though they did not do so) to the negative bias of assuming zeros counteracting the positive transformation bias produced when using formula 6.1 to retransform the estimate of the mean, which each of their methods used. The delta estimator was also favored when the data representing "truth" were generated to include true zeros.

2. Gilbert and Kinnison (1981) evaluated deleting censored values, substitution, Cohen's table lookup and the fully parametric ROS method to estimate statistics of radionuclide data. They assumed lognormal data, and recom-mended Cohen's method and parametric ROS for larger data sets with less than 50% censoring. With more censoring, they gave up trying to produce reasonable estimates, instead substituting the reporting limit and reporting the mean as an upper limit. Their study showed that even by 1981 there were methods known to be better than substitution.

3. Gleit (1985) generated small ($n = 5$ to 15) data sets from normal distributions censored at one reporting limit and found that MLE methods did not fare well for such small sample sizes. Even though the assumed distribution was correct, MLE had difficulty estimating parameters with such little information. Sub-stitution methods also had high errors—substituting the reporting limit performed better than one-half or zero substitution, though no reasons are evident. The consistently best-performing method of the ones tested was a "fill-in with expected values" imputation, where an initial estimate of the mean and standard deviation are guessed, and using order statistics as with the robust ROS method in this chapter, estimates for individual observations imputed.

Using the filled-in data set, the mean and standard deviation are recomputed, and the process repeated until convergence of the mean and standard deviation is achieved. Unfortunately, no evaluation was made of how to retransform estimates if logarithms of the data were used. No evaluation of robustness was made if the data were not normally distributed. Yet Gleit's study sounded themes that echo through later simulations—"fill-in" or imputation methods can work well, substitution methods work poorly, and MLE methods work poorly for small sample sizes.

4. Gilliom and Helsel (1986) compared substitution, MLE, and the robust ROS procedures for a variety of generating data shapes censored at one censoring level. Substitution methods worked poorly. MLE methods worked well when the distribution assumed by the method reasonably matched that of the data. MLE methods assuming lognormal or normal distributions did not work well when the generating distributions were gamma with high standard deviation and skew. The robust ROS method performed better on these high-skew distributions than did MLE, and performed similarly on the other distributions, and so was judged best overall.

5. Helsel and Cohn (1988) extended the results of Gilliom and Helsel (1986) to more than one reporting limit. Whenever data did not follow the distribution assumed by maximum likelihood, and particularly with small sample sizes, the robust ROS generally produced better estimates for the mean and standard deviation than did MLE. Percentile estimates generally had smaller errors using a bias-corrected MLE than any other method. They introduced multiply censored plotting positions to the ROS method that previously had been used only for flood frequency analysis. They also corrected MLE for transformation bias using Cohn's (1988) method.

6. Shumway et al. (1989) compared variations in computing maximum likelihood estimates of the mean and confidence interval on the mean. Originating data were from normal, lognormal, and square-root normal distributions of sample sizes 20 and 50, all with small variance and skew compared to some environmental data. Estimates improved when first determining from sample data to use either untransformed data, or log or square-root transformations based on selecting the maximum log-likelihood of the three MLE equations, rather than always assuming a lognormal or normal distribution. Confidence intervals for the mean were smaller (better) when an asymptotic optimization procedure was used instead of bootstrapping.

7. Haas and Scheff (1990) applied a Type II bias correction to estimates from Cohen's table lookup (even though censored environmental data are Type I) and found the result to be better than the parametric ROS method for small sample sizes ($n = 12$) generated from normal distributions. As the percent of censored data increased, a "restricted maximum likelihood" adjustment performed slightly better than Cohen's method. When applied to a lognormal data set, none of the methods worked well, presumably due to transformation bias.

8. Rao et al. (1991) applied Cohen's table method to data generated from a normal distribution, comparing it to several substitution methods such as $DL/\sqrt{2}$ and fill-in methods from normal and uniform distributions. Their goal was to estimate the mean and confidence interval on the mean for skewed air-quality data. A bootstrap method was also employed to produce confidence bounds for the uniform fill-in. The bootstrap and MLE methods consistently produced better estimates of mean and confidence intervals. They then applied MLE to the logs of air data, and retransformed using equation 6.1, finding that the MLE no longer produced acceptable values. As they may not have recognized this to be due to transformation bias and not the MLE itself, it kept them from recommending MLE for regular use.

9. El-Shaarawi and Esterby (1992) computed exact biases for common substitution methods, stating that there is no further need to use Monte Carlo studies to evaluate these methods (though many continue to do so). They used an example to illustrate the fully parametric ROS and MLE methods, though retransformation bias was ignored. Two problems were highlighted:

   (a) The value of the reporting limit is not used in computing estimates by parametric ROS, and they stated that ROS is therefore more directly related to Type II than Type I censoring.

   (b) No estimates of standard error are computed with ROS and therefore confidence limits cannot be directly constructed [this was later solved by Shumway et al. (2002)].

10. Kroll and Stedinger (1996) implemented a robust imputation procedure not only for ROS, but also for MLE and a probability-weighted moments estimator. They showed that the procedure used for circumventing transformation bias with robust ROS—retransforming single estimates from log space and computing summary statistics in the original units—could be done just as easily with MLE. Their "robust MLE" performed somewhat better than robust ROS (the third method was not as good as these two). The advantages cited by Helsel and Cohn (1988) for robust ROS over MLE were shown to be due to the "robust" imputation of data, avoiding the transformation bias inherent in equation 6.1 for highly skewed and/or smaller sample sizes. Their work clearly shows the advantages of a "robust" imputation, and that the order of choice for MLE and ROS methods should be robust MLE > robust ROS >> lognormal MLE > (fully parametric) lognormal ROS.

11. She (1997) compared the lognormal MLE, the (fully parametric) lognormal ROS, Kaplan–Meier and one-half substitution methods on both lognormal data and data from a gamma distribution. The nonparametric Kaplan–Meier was consistently the best or close to the best method for data from both distributions. The MLE performed well for data from the lognormal distribution when the skew was low. For highly skewed distributions, the moderate sample size ($n = 21$) resulted in poor parameter estimates using MLE, even when the assumed distribution matched that of the data generated. The fully parametric ROS performed no better, and usually worse, than MLE.

12. Shumway et al. (2002) found that the robust ROS method ("robust regression ROS") had smaller errors than standard MLE for data from lognormal distributions for sample sizes of 25 and 50. ROS had similar errors to MLE when the MLE was followed by a jackknife method to compensate for retransformation bias, rather than using equations 6.1 and 6.2. When the original data were from a gamma distribution, however, they found that MLE estimates had less error than ROS when first transformed to approximate normality, especially for sample sizes of 50. They found, as did Cohn (1988), that the theoretical lognormal retransformation using equation 6.1 over-compensates for transformation bias and itself produces biased estimates. A jackknife correction for bias performed far better.

13. Lubin et al. (2004) compared substitution to imputation methods, both when estimating a mean and for computing regression equations. They found that substitution was ". . . ill-advised unless there are relatively few measurements below DLs." Single imputation, performing the imputation step just once as in the robust MLE method of Kroll and Stedinger (1996), produced excellent estimates of the mean but estimates of variance or standard deviation were too low. They instead recommended multiple imputation, a bootstrap approach resulting in many imputed sets of numbers for each censored observation, for the case when an imputed value is required for censored data. When a single number is not required, but only parameter estimates for the mean or regression slope, standard MLE procedures worked well. They noted that their epidemiologic applications had an ample numbers of observations so that MLE performed well.

14. Hewitt and Ganser (2007) evaluated methods to estimate the mean and the 95th percentile using data generated from lognormal and contaminated-lognormal distributions with one and three reporting limits. They found that MLE estimates performed best, with ROS a close second. Robust MLE and ROS were less biased but had greater RMSE, so that the regular methods outperformed their robust counterparts. Substitution results were strongly biased, and KM did not perform well overall. This study is an interesting contrast to the next one, which obtained almost the opposite results. Perhaps the data generated here were of lower skew and more similar to a lognormal distribution even after a mild contamination than those of Antweiler and Taylor, explaining why MLE methods optimized for the lognormal distribution performed best?

15. Antweiler and Taylor (2008) analyzed a series of trace constituents using two chemical methods, a research-grade ultralow detection limit analysis, and a more typically available laboratory method resulting in censored observations. Estimated statistics for analyses by the typical procedure with censored values, computed using a variety of methods, were then compared to the "true mean" of the research-grade data. Since the "true" parameter estimates were chemical measurements rather than computer-generated values, they did not necessarily follow any specific distribution. Kaplan–Meier performed best for

estimating a mean, while substituting half the reporting limit, imputing a uniform random number between zero and the reporting limit, and robust ROS all provided reasonably good estimates. MLE did not perform well, presumably because the data did not consistently follow the assumed distributions. In an interesting and unique result, using machine readings (the raw result from the chemical instrument, including negative numbers) from the typical laboratory method followed by a standard computation did not perform as well as other methods. This point is an important one, as reporting all machine readings is one of the suggested solutions to the censoring issue today. Antweiler and Taylor noted that using one-half RL in their study may have given better results than in a typical study because their reporting limits were single-machine method detection limits rather than quantitation limits from multiple instruments and laboratories, and were not multiplied by 2 or another factor in an insider-censoring mode.

Additional papers to those above could certainly be cited:

- Baccarelli et al. (2005) reviewed a variety of methods for handling censored observations in a study of dioxin exposure, finding that imputation methods designed for censored data far outperformed substitution of values such as one-half the reporting limit.
- Ganser and Hewitt (2010) developed a new substitution method that they state provides results with similar accuracy and precision to MLE. However, they used smaller sample sizes so that MLE would not be expected to perform well, and did not compare their new method to Kaplan–Meier, which is its most logical competitor as both can be considered "simple to compute."
- El-Makari and Aboueissa (2009) developed a new, modified MLE method.
- Jain et al. (2008) compared a variety of imputation methods, including those of Lubin (2004).

6.7.1   A Recommended Course of Action

A recommended course of action that takes these fifteen articles into account (though undoubtedly it would not be endorsed by all of the above researchers) is given below, both in text and as Table 6.11. The recommendation of the Kaplan–Meier or Turnbull method for multiply censored data with up to 50% censoring follows its predominant use in other disciplines as well as its well-developed theory. KM is the nonparametric maximum likelihood estimator for constructing the survival function (Klein and Moeschberger, 2003). It requires no assumption of a particular distributional shape. If there were no censoring, KM produces the familiar sample estimates for mean and percentiles. The Turnbull procedure allows interval-censored data with a nonzero lower bound to be incorporated.

The cutoff at 50% censoring in Table 6.11 reflects the fact that Kaplan–Meier or Turnbull does not provide an estimate for the median (other than <RL) when more than 50% of observations are below the lowest, or single, reporting limit. Some

TABLE 6.11   Recommended Methods for Estimation of Summary Statistics

	Amount of Available Data	
Percent Censored	<50 Observations	>50 Observations
< 50% censored observations	Imputation or KM/Turnbull	Imputation or KM/ Turnbull
50–80% censored observations	Robust MLE, robust ROS or multiple imputation	Maximum likelihood or multiple imputation
>80% censored observations	Report only % above a meaningful threshold	May report high sample percentiles (90th, 95th)

distribution must be assumed in that case, at least for the censored portion of the distribution, to obtain a value other than <RL. The recommendation of other methods than KM with one reporting limit is a recognition of the "stubborn" nonparametric refusal to identify any values outside the range of known data. Consequently, KM in essence substitutes the reporting limit for data below the lowest, or single, reporting limit. The cutoff at a sample size of 50 reflects the inability of MLE to accurately estimate parameters with small, skewed data sets. Several of the above studies found that estimation errors increase dramatically between 60 and 80% censoring, and that above 80% censoring any estimates are merely guesses. Therefore at 80% censoring and above, methods that dichotomize the data into proportions of detect/nondetect should replace attempts to estimate the central location or spread of a censored data set. Note that for other purposes such as hypothesis testing and regression, it is shown later that a simple percent censoring cutoff for methods is not helpful, and (relatively large) signals can still be distinguished when censoring levels are near 80%.

For Less Than 50% Censoring. If there are multiple reporting limits, compute Kaplan–Meier or Turnbull estimates, the standard procedure in other disci- plines and one that does not depend on the assumption of a distributional shape. For single reporting limits use either the multiple imputation of Lubin et al. (2004) or the single imputation of robust ROS or robust MLE.

For Large Sample Sizes ($\geq$50) and 50–80% Censoring. Use MLE or multiple imputation.

For Smaller Sample Sizes (<50) and 50–80% Censoring. Use either multiple imputation, robust MLE, or robust ROS so that estimates of the median and other useful percentiles may be made. If considering distributions other than the lognormal, check goodness of fit to the distribution either by maximizing the probability plot correlation coefficient (see Chapter 5), or by maximizing the (negative) log-likelihood statistic produced by maximum likelihood (Shumway et al., 2002).

Above 80% Censoring. Report the proportions of data below or above the maximum reporting limit, rather than estimating statistics that are unreliable. Sample estimates of high percentiles such as the 90th or 95th may be available for large data sets, even with this much censoring. Any other estimates will be highly dependent on whichever distribution the data are assumed to follow.

6.8   SUMMING DATA WITH CENSORED OBSERVATIONS

One of the common tasks in data analysis is to sum a series of numbers, as when estimating the yearly total mass of a contaminant entering a water body. Twelve monthly measured values are summed to produce the total. A more complicated summation is when individual values are weighted unequally, and the weighted values summed to a total. The latter is what is involved in performing ecological risk assessments (USEPA, 1998b).

In risk assessments an overall numerical measure of the effects on organisms of chemicals such as PCBs, dioxins, and furans is needed. These chemicals are actually each a class of compounds, with each compound (congener) having a different toxicity to organisms. Toxicity equivalent concentrations (TECs) are calculated to summarize the general toxicity from all congeners in the class by assuming that toxicities of individual congeners are additive (USEPA, 2001). TECs are a critical component in issuing fish consumption advisories to protect human health, and so their computation may have significant environmental and economic consequences. Chemical congeners have differing toxicities to organisms, and so each dioxin or furan congener is "normalized" to the toxicity level of the most toxic congener, 2,3,7,8-tetrachlorodibenzo-p-dioxin or TCDD, using a toxic equivalent weighting factor (TEF) of relative toxicity (USEPA, 2001). TEFs were developed by consensus by panels of scientists for each class of organism (Van den Berg et al., 1998). TCDD has a TEF of 1 while less toxic congeners have TEFs closer to 0. Measured concentrations are multiplied by the TEF to obtain the TEC for that congener. The total TEC is the sum of the individual congener TEC values in the sediment or soil. At times, congener concentrations are below their respective reporting limits, and the issue at hand is how to use these censored observations in the summing process when computing a total TEC.

Current USEPA draft guidance for computing toxicity equivalents is silent on how to incorporate censored data, other than that 0 and the reporting limit can be substituted and the range of possible TEC values reported (USEPA, 2008). When the range of possible values is wide this method is not very helpful, tempting scientists to substitute one-half of each reporting limit to obtain a single total TEC. An example of using substitution when computing a TEC is shown in Table 6.12. One-half the reporting limit is multiplied by the TEF to estimate the toxic equivalent concentration. TECs are then summed to produce the total TEC for this location. Substitution of one-half the reporting limit before computing a sum produces a serious problem—the least precise measurements, data with high reporting limits, often have a strong influence on the resulting total TEC. For example, suppose a less precise method had been used for analysis of 1,2,3,7,8-PeCDF and instead of a <0.8 the lab had reported a value of <5. One-half of this or 2.5 would have been used to compute the TEC for this (toxic) congener, and the total TEC would have increased by 0.63 or by 19% over the current TEC. This increase is caused only by falsely translating a loss in precision (higher reporting limit) into a higher concentration by using substitution. The KM estimate in the same situation increases only by 1.5%.

TABLE 6.12   TEC Calculations Using Substitution of One-Half the Reporting Limit and Using Kaplan–Meier

Compound	Concentration	One-Half RL	TEF	TEC $^1/_2$ RL	TEC KM
1,2,3,4,6,7,8-HpCDD	25		0.01	0.25	0.25
1,2,3,4,6,7,8-HpCDF	1.8		0.01	0.018	0.018
1,2,3,4,7,8,9-HpCDF	<0.56	0.28	0.01	0.003	<0.006
1,2,3,4,7,8-HxCDD	0.26		0.1	0.026	0.026
1,2,3,4,7,8-HxCDF	<0.6	0.3	0.1	0.03	<0.06
1,2,3,6,7,8-HxCDD	2.1		0.1	0.21	0.021
1,2,3,6,7,8-HxCDF	0.33		0.1	0.033	0.033
1,2,3,7,8,9-HxCDD	0.77		0.1	0.077	0.077
1,2,3,7,8,9-HxCDF	0.37		0.1	0.037	0.037
1,2,3,7,8-PeCDD	0.18		1	0.18	0.18
1,2,3,7,8-PeCDF	0.24		0.03	0.007	0.007
2,3,4,6,7,8-HxCDF	<0.14	0.07	0.1	0.007	<0.014
2,3,4,7,8-PeCDF	<0.8(<5.0)	0.4(2.5)	0.3	0.12(0.75)	<0.24(<1.5)
2,3,7,8-TCDD	1.7		1	1.7	1.7
2,3,7,8-TCDF	5.1		0.1	0.51	0.51
OCDD	220		0.0003	0.066	0.066
OCDF	44		0.0003	0.013	0.013
[0,1-6]					
Sum				3.29(3.92)	3.21(3.26)

Values in parenthesis are the result of increasing the reporting limit for one congener from 0.8 to 5.

6.8.1   An Alternate Method for Summing Data with Censored Observations

A sum and a sample mean are the same phenomenon—the mean is the sum, standardized by the number of values summed. Reversing the equation, the sum equals the mean multiplied by n. For data with censored observations, the mean can be estimated using a reliable method that does not involve substitution, and the total is then computed by multiplying by the mean by n. The reliable method used here for censored data is the Kaplan–Meier (KM) procedure. Table 6.13 provides an example data set of six congener TEC values, two of which are censored.

TABLE 6.13   Quantiles for Six Observations when Censoring Is and Is Not Ignored, as Computed by Kaplan–Meier

Concentration	TEF	TEC	Quantile Ignoring the < Symbol	Quantile Accounting for Nondetects
2.10	0.10	0.2100	0.833333	0.833333
0.77	0.10	0.0770	0.666667	0.666667
<0.60	0.10	0.0600	0.500000	.
0.33	0.10	0.0330	0.333333	0.444444
<0.14	0.10	0.0140	0.166667	.
0.24	0.03	0.0072	0.000000	0.000000

These data are illustrated in Figures 6.6 and 6.7. (* = percentiles not computed)

First ignoring the less-than symbol and using the reporting limit, each observation is assigned quantiles $1/n$ apart from each other, and so have quantiles at 5/6, 4/6, 3/6, 2/6, 1/6, and 0 to form the edf (column titled "Quantile ignoring the $<$ symbol"). Looking at the rectangles in Figure 6.6 that make up this area, the height of each rectangle is $1/n$, or 0.16667 for $n = 6$. The area of each rectangle is 1/6 times the data value, and so the area under the cdf curve equals the mean, 0.067. The data are plotted from right to left, seemingly backward from typical plots, because the left-censored data have been flipped to perform the KM analysis. If the height of each rectangle were set to be 1 rather than $1/n$, the area equals the sum of the six numbers, 0.4012 and the histogram is a picture of the sum itself. The mean is simply a scaled version of the sum.

Now recognize that two observations are actually censored. KM computes quantiles only for uncensored observations, but the number and position of censored observations influences the quantile calculated for uncensored observations (column titled "Quantile accounting for nondetects"). For the highest observation of 0.21 there are still six observations at and below it, with five below, so its quantile is 5/6, just as it was when the nondetect designation was ignored. It is clear that the two censored observations at <0.06 and <0.014 are both below a detected 0.21. The second highest detected observation is also as before, and so has the same quantile at 0.667. The third highest value is a < 0.06. Its position relative to all values below 0.06 cannot be known, so a quantile is not calculated for it. However its influence shows in the calculation for the next lower value, a detected 0.033. This observation has three values that are known to be at or below it, including a lower nondetect. Its quantile is therefore calculated as 2/3 the previous quantile of 0.667 or 0.444. This is higher than the quantile assigned to the same observation when the two censored observations were treated as uncensored values because there is some chance that the <0.06 lies below this detected 0.033. The lowest detected observation lies at a quantile of 0, as before. Figure 6.7 shows the resulting histogram and area (i.e., KM mean) when the

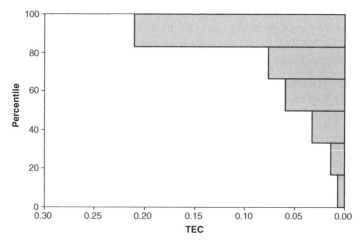

FIGURE 6.6   Kaplan–Meier method for estimating the mean without nondetects. The mean equals the total area inside the bars. Percentile equals the quantile *100.

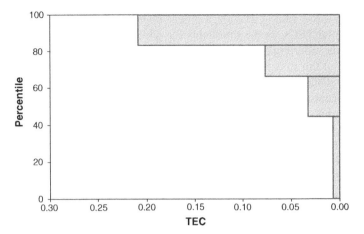

FIGURE 6.7   Kaplan–Meier method for estimating the mean, with two nondetects. The mean again equals the total area inside the bars. Percentile equals the quantile * 100.

censored observations are accounted for. With these KM percentiles the rectangles in Figure 6.7 have unequal heights corresponding to the unequal difference in percentile values between the uncensored observations. These unequal differences reflect the information available in the censored observations—their relative positions in regards to uncensored observations. The Kaplan–Meier mean for the four uncensored and two censored observations is 0.058, which when multiplied by $n = 6$ observations results in a sum of 0.35 for this data set. This KM estimate of 0.35 lies between the estimates that result when 0 (0.327) and the reporting limit (0.401) are substituted for all censored observations. It was obtained without substituting any values for the censored observations, and without assuming that the six observations follow any specific distributional shape.

6.8.2   When NOT to Use Kaplan–Meier for Summing Data with Censored Observations

Other methods than KM are available for summing censored observations in specific cases. If strong correlations exist between congeners in a series of samples so that the concentrations of one congener can be reliably predicted from others, the correlation can be used to predict values for concentrations measured as below the reporting limit. This procedure has been used when insufficient amounts of sample prevented concentrations from being measured for some congeners (Cook et al., 2003). The resulting sum will be more accurate than by using Kaplan–Meier if the estimated concentrations are close to the unknown, true concentrations for the respective congeners. Criteria for how strong a correlation should be to produce estimates with this method versus Kaplan–Meier are not known.

   Two other situations exist in which KM should not be used to sum censored data. The first is when there is only one censoring threshold. As shown in a previous section, with one limit the KM estimate of the mean will equal that of substituting the reporting limit for censored values. Having a single threshold is unlikely in the case of computing total TECs, as the thresholds for different congeners are computed by multiplying the reporting limit by the TEF weighting factor, which differs for different congeners. But this may happen with other applications.

The second situation is when a high nondetect value, higher than all TECs from detected concentrations, occurs for one of the highest toxicity congeners with TEFs close to 1. In this situation no calculation procedure can give a reliable estimate of the total TEC. KM and any other statistics-based (as opposed to substitution) procedure will ignore this high nondetect, as it has no information content. A lower reporting limit must be implemented before reliable estimates of the total TEC can be made using any calculation method. All that can be done in this situation is to substitute the reporting limit in order to provide a worst-case value for the total TEC, realizing that the true total may be far lower.

EXERCISES

6-1 The copper data from the Alluvial Fan zone of Millard and Deverel (1988) is found in the data set MDCu + (use either MDCu + .mtw or MDCu + .xls). One observation has been changed from the data in their article. The largest reporting limit of <20 was altered to become a <21, larger than all of the uncensored observations reported (the largest detected observation is a 20). Compute Kaplan–Meier estimates of the mean and median for two situations, one with the <21 in the data set and a second with the <21 removed from the data set. Demonstrate from the results that a censored observation whose threshold is above the largest detected observation has zero information content and can always be discarded. Also demonstrate why this is so by computing plotting positions by the robust ROS method for these data.

6-2 The silver.mtw data set contains analyses from 56 laboratories for a quality control standard silver solution (Helsel and Cohn, 1988). There are 12 reporting limits reported by the different labs. Produce a survival function plot for the silver data using Kaplan–Meier software. Also produce a censored probability plot using robust ROS and with lognormal MLE. The ROS method can be computed using the Minitab macro Cros.mac. Compare and contrast the three plots. Which better illustrates how well the data are fit by the assumed distribution? Which would you use to get a rough estimate for the percentiles of the distribution?

6-3 Estimate the mean, standard deviation, median, 25th, and 75th percentiles of the silver data using (lognormal) maximum likelihood estimation, Kaplan–Meier, and the robust ROS methods. Compare and contrast the three results. How must the KM percentiles be rescaled in order to compare them with those from the other methods? Based on the percent of data censored, the sample size and the fit to the distribution, which method would you choose to use?

7 Computing Interval Estimates

Confidence, prediction, and tolerance intervals are often needed for data that include censored observations. Two-sided confidence intervals bracket the likely values for a parameter such as the mean. The likelihood is represented by a confidence coefficient, a statement about how likely it is that the process used resulted in an interval that contains the true mean. A "95% confidence interval around the mean" is a statement of belief that the unknown true mean, a target of the investigation, is contained with a 95% probability between the lower and upper ends of the interval. The truth of that probability will rest on some assumptions about the distribution of the data if the method used is a parametric interval. When the data do not fit the assumed distribution well, the truth of the probability associated with parametric intervals will be in doubt. A "95% confidence interval" for an ill-fitting data set may in fact have a much higher probability than 5% of not including the true mean within the interval. Or it may be so wide that it is of little use.

One-sided confidence bounds are of interest when the concern is whether values are too large, or too small, but not both. An upper 95% confidence bound on the mean (sometimes abbreviated UCL95) is a statement that the true mean should be below the UCL95 with a 95% probability. If the UCL95 is below the relevant regulatory limit, there is more than a 95% probability that the true mean of the data lies below that limit. If the UCL95 is greater than the limit, there is more than a 5% probability that the true mean exceeds the limit, even though the observed sample mean may be below the limit. Confidence bounds such as the UCL95 have at times been written into environmental regulations. Of course, our interest is how to compute these bounds when some of the data are censored observations.

Prediction intervals provide a range bracketing the likely values for one or more individual observations not currently in the data set. Prediction intervals are wider than confidence intervals for the same set of data and same confidence coefficient. Intervals can be computed to enclose the range likely to hold one new observation, or many new observations, with a specified confidence. As with confidence intervals, parametric prediction intervals rely on the assumption that data come from a specific distribution, often the normal distribution. For parametric prediction intervals both

Statistics for Censored Environmental Data Using Minitab® and R, Second Edition.
Dennis R. Helsel.
© 2012 John Wiley & Sons, Inc. Published 2012 by John Wiley & Sons, Inc.

the data used to construct the interval and the new data that are to fall within the interval are assumed to originate from the same distribution. If data do not fit the assumed distribution, they should be transformed prior to constructing the interval. Otherwise the intervals may be too large, or the stated confidence of inclusion may overstate the probability actually attained by the interval.

Tolerance intervals bracket possible values for percentiles of the distribution, and so include within their range a specified proportion of observations. Sample percentiles can be computed for the observed data (e.g., the sample 90th percentile equals or exceeds 90% of the measured values), but as with all other sample statistics, these only estimate the true underlying population statistic. A tolerance interval puts boundaries on the location of the true population percentile. An upper 95% tolerance bound on the 90th percentile, for example, provides a limit beyond which there is 95% confidence that no more than 10% of all population values fall.

Hahn and Meeker (1991) provide detailed descriptions of these three types of intervals. In the following sections, calculation of each type of interval is illustrated when some proportion of the data are recorded as censored observations. Each of the three types of intervals can be computed using one of the general approaches for censored data discussed so far—substitution, maximum likelihood, or nonparametric methods. Use of the first approach, substitution, is strongly discouraged.

7.1 PARAMETRIC INTERVALS

Parametric confidence, prediction, and tolerance intervals are built using estimates of the mean and standard deviation, along with an assumption that data follow a normal distribution. If the distribution of data does not follow this shape, estimates of the interval endpoints can be severely in error. Parametric two-sided intervals follow the general equation of

$$\overline{x} - ks, \ \overline{x} + ks$$

where $\overline{x}$ is the mean, s is the standard deviation, and k is a constant that is a function of the sample size n, the two-sided confidence coefficient $(1 - \alpha/2)$, and the type of interval desired (confidence, prediction, or tolerance interval). One-sided bounds concentrate the possible error rate α onto one side of the mean, following the general equation of $\overline{x} + ks$ for an upper confidence bound, and $\overline{x} - ks$ for the lower bound. For these one-sided bounds, k is a function of the sample size n, the one-sided confidence coefficient $(1 - \alpha)$, and the type of interval desired.

Estimates for the mean and standard deviation may be computed using any of the techniques discussed in Chapter 6. Better parameter estimates will produce better interval estimates. Therefore, estimates of mean and standard deviation from maximum likelihood (for large samples) or from the robust ROS or robust MLE estimators should produce the best interval estimates when the shape of data follows a normal distribution.

7.1.1 Validity of Assuming a Normal Distribution

Parametric intervals discussed in this book require that the data (or transformed data) used to construct the interval were randomly sampled from a population possessing the shape of a normal distribution, the familiar "bell-shaped curve." In that case most of the data will be in the center, with outlying data symmetrically departing from the center to more and more infrequent values. The center is defined by the mean and median, both of which are at the same value. The probability of being more than two standard deviations above the mean is identical to the probability of being more than two standard deviations below it.

Most investigators have found air, water, soils, and tissue concentrations to be somewhat skewed, unlike a normal distribution. A lower bound of zero concentration prevents the distribution from looking symmetric—concentrations can only go down so far, and no further. Infrequently occurring observations–outliers—occur primarily on the high (right) side, and so data are right-skewed. Due to both right-skewness and variability that is often proportional to the concentration, logarithms (and less frequently, square roots) are more often nearly symmetric, and a normal distribution for these transformed data can be more reasonably invoked than for the original data prior to transformation.

Parametric prediction intervals for one or more observations, and tolerance intervals around a specified proportion of data, are highly sensitive to the assumption of normality. If the data are skewed and these intervals computed directly using untransformed data, the endpoints of the intervals will not reflect the desired confidence level. The interval lengths will be unrealistic, perhaps going negative in the lower direction, an impossible result. The process of attempting to summarize a skewed distribution with symmetric intervals may produce an interval that is noticeably too large. A more insidious problem is that the interval may not contain the desired result as frequently as indicated by the confidence level. A supposed 95% confidence level for a prediction interval may in fact have only a 60% probability of containing the next observation, as one example.

However if prediction and tolerance intervals are cursed with strict dependence on the normality assumption, they also are blessed with an easy solution. Once the data are transformed to approximate normality and interval endpoints constructed on transformed data, those endpoints can be directly retransformed back to original units while preserving their meaning and purpose. If natural logarithms of concentration have a nearly normal distribution, a 95% upper tolerance bound computed on the 90th percentile of logarithms may be retransformed by exponentiating its value. The result is a 95% tolerance bound on the 90th percentile of concentration. Prior to constructing prediction or tolerance intervals the Box–Cox transformation series, also called the Ladder of Powers (Velleman and Hoaglin, 1981), should be used to find a suitable transformation to near-normality. Then construct the desired interval, apply the reverse transformation and the result is the desired interval in original units.

Confidence intervals whose target is estimation of the mean have a different blessing and a different curse. Their curse is that an interval around the mean of data transformed using a power or other nonlinear transformation cannot be directly retransformed to produce a confidence interval around the mean in original units. This is because the mean of transformed data once retransformed does not estimate the mean of the original data. The mean and median of transformed data are identical when the transformed data are symmetric. Once retransformed the resulting value remains an estimate of the median in original units, but not of the mean. So the geometric mean, the mean of logarithms retransformed back to original units, estimates the median (if the log-transformed data were symmetric) rather than the mean. Therefore transformation does not help in constructing confidence intervals on the mean, as it does for prediction and tolerance intervals. Confidence intervals constructed around the mean of the logs when retransformed become confidence intervals for the median and not the mean, assuming the logs were symmetric.

How then can a confidence interval on the mean of skewed data be reliably estimated? An often-invoked blessing is the Central Limit Theorem, the property of a mean that states that the variability in estimates of the mean follows a normal distribution under certain conditions, even when the underlying data do not. When data sets are "large" and data "not too skewed," a normal theory confidence interval can be directly computed without transformation (Hahn and Meeker, 1991). But how large must "large" be, and how much skewness is allowed? Boos and Hughes-Oliver (2000) show that the sample size required to invoke the theorem is a function of the type of interval and the severity of skewness. For two-sided confidence intervals based on the t-statistic built from data of moderate skewness (skewness coefficient = 1), somewhere around 30 observations is large enough. However for a one-sided interval such as the upper 95% confidence bound on the mean (UCL95), a skewness coefficient of 1 results in a requirement of about 126 observations! With smaller sample sizes and right-skewed data, an upper confidence bound computed using the t-statistic will most often be too small (Boos and Hughes-Oliver, 2000), undershooting the true value ("miss on the left"). Environmental data generally have more than a moderate amount of skew (with a skew coefficient > 1), so that a sample size of 50 or more is not an unreasonable requirement to invoke the Central Limit Theorem for two-sided intervals. For one-sided confidence bounds, sample size requirements are quite large, greater than 126. If the skewness coefficient is known or can be estimated from past data, approximate sample size requirements can be determined from equations in Boos and Hughes-Oliver (2000). Even when sample sizes are sufficient, the resulting t-statistic confidence intervals must be considered approximate rather than exact, with the approximation getting appreciably worse as the confidence coefficient increases—as α gets small (Hahn and Meeker, 1991, p. 65).

Without invoking the Central Limit Theorem, Land (1972) developed a procedure to translate confidence intervals in log units back into intervals around the mean in original units. Land's method is discussed in the section on maximum likelihood. However, for small sample sizes and skewed data the estimate of standard deviation in log units is often poor. A poor estimate of this standard deviation causes Land's method to produce a confidence interval that is too wide (Singh et al., 1997). Bootstrapping (Efron and Tibshirani, 1986) is a newer and more efficient method than Land's for constructing confidence intervals on the mean of skewed data. Bootstrapping is a nonparametric method discussed in later sections.

7.2   NONPARAMETRIC INTERVALS

Rather than computing intervals using parameter estimates, traditional nonparametric intervals are based on the values of one or more observations in the data set. Observations are chosen to be interval endpoints by their positions in the data set, called their order statistics. First the data are ordered from low to high. Interval endpoints are chosen at specific order statistics based on the sample size n, the desired confidence coefficient $(1 - \alpha)$ or $(1 - \alpha/2)$, and the type of interval (confidence, prediction, or tolerance) to be computed. The values of observations located at each endpoint define the shape and width of the nonparametric interval.

Nonparametric intervals do not depend on an assumption of a normal distribution for their validity. The shape of the interval will reflect the shape of the data set. The trade-off for this flexibility or robustness is that nonparametric intervals will be wider than parametric intervals when data do follow the assumed distribution. The distributional assumption is another piece of information used to construct a parametric interval. That additional information shortens the interval length when the data follow the assumed distribution. It will be misleading information producing misleading intervals when the data do not follow the distribution assumed by the process. In the latter case a nonparametric interval or an interval following appropriate transformation will provide better results.

A newer approach to computing nonparametric intervals is called bootstrapping. With bootstrapping the targeted statistic (mean, median, percentile, etc.) is repeatedly computed and the estimates stored. A thousand or more replications is typical. The collection of estimates approximates the distribution of the target statistic. The mean or median of estimates becomes the bootstrapped estimate at the center of the interval, and the appropriate low and high percentiles of the estimates forms the interval endpoints. For a 95% confidence interval, the 2.5th and 97.5th percentiles are used, leaving a total of 5% of the computed estimates outside the interval.

7.3   INTERVALS FOR CENSORED DATA BY SUBSTITUTION

As shown in Chapter 1, substitution of an arbitrary constant may widely miss the mark when estimating a standard deviation. Computations of parametric t-intervals based on this substitution standard deviation will vary widely in quality. For example, Table 7.1 shows 95% confidence t-intervals for the mean of the Oahu data set

TABLE 7.1   Confidence Intervals (95%) for the Mean of the Oahu Data Set Using Substitution

Substitution Method	Lower Limit	Upper Limit
Zero	0.19	0.94
One-half RL	0.70	1.30
RL	1.12	1.76

following substitution. The three substitutions result in three quite different intervals, with the lower end of the RL (substitution of the reporting limit) interval being higher than the upper end of the Zero-substitution interval. Yet there is no valid justification for arguing that one of these intervals is any better than another based on the data at hand. There is no knowledge that any of the substitution standard deviations comes close to the underlying standard deviation of the data. Substitution has added external, invasive data into the computations whose values are strongly influenced by the operating characteristics of the laboratory (the reporting limits they settle on), interferences from other analytes, the size of the sample submitted, and by the choice of an arbitrary fraction of that reporting limit to substitute, rather than by the concentration of the target chemical that was in any given sample.

Substituting the possible extremes of zero and the reporting limit will not necessarily produce intervals that bracket the range of possible interval widths. Though the mean varies monotonically as the substitution value changes, the standard deviation does not (see Chapter 1). For the Oahu data, the standard deviation resulting from substitution of one-half DL was smaller than that when substituting zero or the reporting limit. Interval widths of t-intervals using substitution follow the same pattern, and so will not change monotonically. The maximum or minimum interval width may occur at an unknown substituted value somewhere between zero and the reporting limit. So it is not possible to easily "bracket the extremes" of intervals using substitution.

Substitution should be avoided when computing interval estimates. There are better ways.

7.4   INTERVALS FOR CENSORED DATA BY MAXIMUM LIKELIHOOD

Parametric intervals can be computed using maximum likelihood estimates of the mean and standard deviation of a censored data distribution, placing these into the standard formulae for interval endpoints. The assumed distribution for the MLE is specified within the software. Example computations using Minitab® for many types of intervals follow. The data used are lead concentrations in the blood of herons in Virginia (Golden et al., 2003). The data are found in bloodlead.xls.

7.4.1   Confidence Interval for the Mean (Two-Sided) Assuming a Normal Distribution

Assuming the data follow a normal distribution, upper and lower confidence limits for the mean are computed as

$$\bar{x} - t_{(1 - \alpha/2), n - 1} \frac{s}{\sqrt{n}}, \qquad \bar{x} + t_{(1 - \alpha/2), n - 1} \frac{s}{\sqrt{n}} \tag{7.1}$$

where $t_{(1 - \alpha/2), n}$ is the $1 - \alpha/2$th quantile of the t-distribution, n is the sample size, and $s/\sqrt{n}$ is the standard error of the mean. If the interval is computed using data that are not normally distributed, the true probability of including the unknown population mean within this interval will be somewhat lower than $1 - \alpha\%$. The dependence on the normality assumption can be relaxed as sample sizes increase. According to the Central Limit Theorem, the probability of inclusion approaches $1 - \alpha\%$ as n gets "large," where large increases as the skewness of data increases. For the skewness found in environmental data, "large" is often around 50–100 observations.

In Minitab the mean, standard error, and confidence intervals are estimated and printed using maximum likelihood with the menu command:

```
Stat > Reliability/Survival > Distribution analysis (arbitrary
censoring) > Parametric distribution analysis
```

Data are input in interval endpoint format with Blood as the start variable and BloodPb as the end variable. The resulting output is

	Estimate	Standard Error	95.0% Normal CI Lower	Upper
Mean (MTTF)	0.0397452	0.0123321	0.0155748	0.0639156
Standard Deviation	0.0639343	0.0087089	0.0489537	0.0834990
Median	0.0397452	0.0123321	0.0155748	0.0639156
First Quartile (Q1)	−0.0033778	0.0136753	−0.0301810	0.0234253
Third Quartile (Q3)	0.0828682	0.0136438	0.0561268	0.109610
Interquartile Range (IQR)	0.0862460	0.0117481	0.0660376	0.112638

Using NADA for R, the mean function on a censored data object will provide confidence intervals. The confidence coefficient can be changed from the default 95% when first computing the object using the conf.int option. Here is an example of computing the 90% interval:

```
> data(Golden)
> attach(Golden)
> BldPb=cenmle(Blood,BloodCen,conf.int=0.90, dist="gaussian")
> mean(BldPb)
mean        se            0.9LCL      0.9UCL
0.03974518  0.01233205    0.01946075  0.06002960
```

The 95% confidence interval on the mean extends from 0.016 to 0.064 µg/g lead. A normal probability plot (Figure 7.1) shows that the data are not normally distributed; they do not follow a straight line on the probability plot. The data set is of moderate size (27 observations). Therefore, the probability that the population mean is somewhere within the 95% confidence interval is likely to be somewhat lower than 95%.

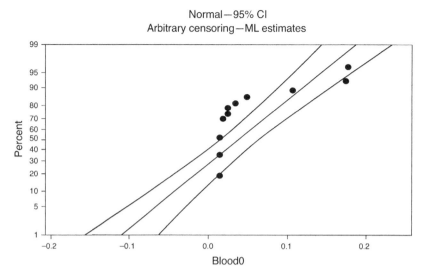

FIGURE 7.1   Probability plot of the censored blood lead data. Note the curvature, indicating non-normality.

7.4.2   Confidence Bound for the Mean (One-Sided) Assuming a Normal Distribution

A one-sided confidence bound on the mean places the entire error probability α on either the upper or lower side. An upper confidence bound is computed using equation (7.2), assuming a normal distribution.

$$\bar{x} + t_{(1 - \alpha/2, n - 1)} \frac{s}{\sqrt{n}} \qquad (7.2)$$

A lower confidence bound is obtained by subtracting rather than adding from the estimate of the mean. A 95% upper confidence bound on the mean, assuming a normal distribution, is computed using censored MLE in Minitab by selecting the "upper bound" option in the "Estimate" dialog box:

	Estimate	Standard Error	95.0% Normal Upper Bound
Mean (MTTF)	0.0397452	0.0123321	0.0600296
Standard Deviation	0.0639343	0.0087089	0.0799908
Median	0.0397452	0.0123321	0.0600296
First Quartile (Q1)	−0.0033778	0.0136753	0.0191161
Third Quartile (Q3)	0.0828682	0.0136438	0.105310
Interquartile Range (IQR)	0.0862460	0.0117481	0.107906

The 95% upper confidence bound on the mean, assuming data follow a normal distribution, is 0.060.

7.4.3 Confidence Interval for the Median Assuming a Normal Distribution

If the data are believed to follow a normal distribution the mean and median are the same. Therefore confidence intervals for the two are identical, as shown in the above Minitab output. A 90% confidence interval on the median is identical to the 90% interval on the mean printed by NADA for R, above. More typically, confidence intervals on the median are computed using the nonparametric process shown in a later section. The nonparametric intervals will have a true 95% probability of enclosing the population median, unlike parametric intervals, when the data are skewed.

7.4.4 Confidence Bound for the Median Assuming a Normal Distribution

A one-sided confidence bound for the median will be identical to that for the mean if data follow a normal distribution. The Minitab results shown above for the one-sided upper bound on the mean have identical values in the row for the median.

7.4.5 Confidence Interval for a Percentile (Quantile) Assuming a Normal Distribution

Confidence intervals around a percentile (also known as a quantile) bracket the range of values within which the true population percentile is expected to be located, with $(1 - \alpha)\%$ confidence. Assuming that data follow a normal distribution, the sample estimate of the pth percentile is

$$\overline{x} + z_p \cdot s \tag{7.3}$$

where z_p is the pth percentile of the standard normal distribution. A two-sided confidence interval around a percentile larger than the median ($p > 0.5$) follows the formula:

$$\overline{x} + g'_{(\alpha/2),p,n} \cdot s, \qquad \overline{x} + g'_{(1-\alpha/2),p,n} \cdot s \tag{7.4}$$

where g' values are the percentiles of a noncentral t-distribution. The noncentral t-distribution is a function of the confidence coefficient $(1 - \alpha)$, the percentage p of the desired percentile, and the sample size n. Tables of the g'-statistic for commonly used values of α and p are found in Tables A.12a–A.12d of Hahn and Meeker (1991). Some statistical software, including Minitab, will also produce values for the noncentral t-distribution.

For percentiles below the median ($p < 0.5$), a two-sided confidence interval is

$$\overline{x} - g'_{(1-\alpha/2),p,n} \cdot s, \qquad \overline{x} - g'_{(\alpha/2),p,n} \cdot s \tag{7.5}$$

Minitab reports confidence intervals around the first quartile (25th percentile) and the third quartile (75th percentile) as part of its MLE distributional analysis. For other percentiles the intervals will need to be computed using equations (7.4) and (7.5).

The sample 90th percentile of the blood lead data is

$$0.0397 + 1.28 \times 0.064, \quad \text{or} \quad 0.122$$

where 1.28 is the 90th percentile of a standard normal distribution ($p = 0.9$). The 95% confidence interval around this estimate for the 90th percentile is

$$(0.0397 + 0.846 \times 0.064, \quad 0.0397 + 1.932 \times 0.064), \quad \text{or}$$
$$(0.094, 0.163)$$

where 0.846 and 1.932 are the 2.5th and 97.5th percentiles of a noncentral t-distribution from Tables A.12a and A.12d of Hahn and Meeker (1991) for $\alpha/2 = 0.025$, $p = 0.9$ and $n = 27$. Therefore, the population 90th percentile of blood lead levels is between 0.094 and 0.163 μg/g with 95% confidence if the data follow a normal distribution.

Using NADA for R, the modeled percentiles or quantiles can be computed by

```
> Pbmle=cenmle(Blood, BloodCen, dist="gaussian")
> quantile(Pbmle)
quantile          value
1       0.05      -0.065417328
2       0.10      -0.042189878
3       0.25      -0.003377829
4       0.50       0.039745176
5       0.75       0.082868181
6       0.90       0.121680231
7       0.95       0.144907680
```

Confidence intervals on the quantiles can also be reported using the "conf.int = TRUE" option. However, a normal instead of the noncentral t-distribution is currently used in computing these intervals.

7.4.6 One-Sided Confidence Bound for a Percentile Assuming a Normal Distribution

One-sided confidence bounds for percentiles are computed in much the same way as the two-sided intervals. In environmental studies the objective for computing this one-sided bound is usually to define a limit that contains below (or above) it a specified proportion of the data. The equation for a one-sided upper confidence bound on a percentile is

$$\overline{x} - g'_{(1-\alpha/2),p,n} \cdot s \qquad (7.6)$$

where g' is the $(1-\alpha) \times 100$th percentile of a noncentral t-distribution from Table A.12d in Hahn and Meeker (1991).

Suppose we want to determine a threshold that is higher than 90% of the blood lead levels in the population of herons. An upper 95% confidence bound on the 90th percentile of lead concentrations will provide a threshold with only a 5% chance that the 90th percentile of the population these data represent is higher than our estimate. This confidence bound for the heron blood lead data, assuming that they follow a normal distribution, is

$$0.0397 + 1.811 \times 0.064, \quad \text{or} \quad 0.156$$

where 1.811 is $g'_{0.95, 0.9, 27}$ from Table A.12d of Hahn and Meeker (1991). We expect that 90% of all lead concentrations in the heron population represented by these birds would be below 0.156 µg/g, if the assumption of a normal distribution were reasonable. However, the data do not appear to follow a normal distribution, so this estimate is likely to be incorrect.

7.4.7 Tolerance Interval to Contain a Central Proportion of the Data

Tolerance intervals bracket values containing a specified proportion of the data. Two-sided tolerance intervals are rarely used in environmental studies, perhaps because there are few applications that attempt to determine the location of a central proportion of data, with allowable exceedances at both high and low ends. Assuming the data follow a normal distribution, a two-sided tolerance interval follows the formula:

$$\overline{x} - g_{(1-\alpha/2),p,n} \cdot s, \qquad \overline{x} + g_{(1-\alpha/2),p,n} \cdot s \qquad (7.7)$$

where tables of the g-statistic (different than the g'-statistic) are found in Table A.10 of Hahn and Meeker (1991). Note that the tables are set up to use α rather than $\alpha/2$ for determining the g-statistic for each end of the $(1-\alpha)\%$ confidence interval. The g-statistic is also a function of the proportion p of the distribution to be included within the interval and the total sample size n. Neither Minitab nor other standard statistical software computes tolerance intervals for a central proportion directly.

A tolerance interval expected to contain the central 90% of the blood lead data with 95% confidence is

$$(0.0397 - 2.184 \times 0.064, \quad 0.0397 + 2.184 \times 0.064), \quad \text{or}$$
$$(-0.100, 0.179)$$

where 2.184 is the g-statistic from Table A.10a of Hahn and Meeker (1991) for $p = 0.9$ and $n = 27$. No more than 5% of the population of blood lead concentrations are expected to be less than the lower limit, and no more than 5% greater than the upper limit, with 95% confidence. The negative lower limit of the interval is a reminder that these data do not follow a normal distribution, and that a transformation should have been applied prior to computing this interval. This will be done in the later section on intervals for lognormal distributions.

7.4.8   Tolerance Bound for a Proportion of the Data

One-sided upper tolerance bounds estimate a value that exceeds $p\%$ of the population with $(1 - \alpha)\%$ confidence. The percent of data (p) designed to be below the bound is often called the coverage. A one-sided tolerance bound is identical to the one-sided confidence limit for the equivalent (pth) percentile. So a 95% upper tolerance bound covering at least 90% of the data is identical to the 95% upper confidence bound for the 90th percentile. Both are higher than $p\%$ of the data with $(1 - \alpha)\%$ confidence. The equation for a one-sided tolerance bound with coverage p is the same as equation (7.6), above.

A 95% upper tolerance bound with 90% coverage of blood lead levels, assuming data follow a normal distribution, is

$$0.0397 + 1.811 \times 0.064, \quad \text{or} \quad 0.156$$

where 1.811 is the noncentral t-statistic $g'_{0.95,\,0.9,\,27}$ from Table A.12d of Hahn and Meeker (1991). This is the same value obtained previously for the 95% upper confidence bound on the 90th percentile. We would expect that at least 90% of blood lead concentrations in the population these data represent lie below 0.156, with 95% confidence, assuming these data follow a normal distribution.

7.4.9   Prediction Interval for One New Observation, Assuming a Normal Distribution

Prediction intervals bracket the range of locations for one or more new observations not currently in the data set. Two-sided intervals are of interest if both extreme high and extreme low values of new observations are of concern. Obtaining a new observation beyond the limits of the prediction interval should happen only $\alpha\%$ of the time if nothing has changed and the new observation(s) come from the same distribution as did the existing data, in this case a normal distribution.

A two-sided prediction interval for normal distributions that covers the likely values for one new observation with $(1 - \alpha)\%$ confidence is

$$\bar{x} - t_{(1-\alpha/2, n-1)} \cdot \sqrt{1 + \frac{1}{n}} \cdot s, \qquad \bar{x} + t_{(1-\alpha/2, n-1)} \cdot \sqrt{1 + \frac{1}{n}} \cdot s \qquad (7.8)$$

where t is from a Student's t-distribution with $n - 1$ degrees of freedom. Note this is similar to the equation for the confidence interval around a mean, except that an additional term (a 1) appears under the square root sign. The uncertainty in prediction for a single new observation includes both the variability of the data (the standard deviation s) and the variability of the estimated mean (the standard error $\sqrt{(1/n)} \cdot s$). While the width of a confidence interval is determined only by the standard error, both terms contribute to the width of a prediction interval. Unlike a confidence interval, as sample sizes increase the width of a prediction interval goes no lower than $t \cdot s$.

A 95% prediction interval for the range of probable values for a new blood lead observation is

$$(0.0397 - 2.056 \times \sqrt{1 + \tfrac{1}{27}} \times 0.064, \quad 0.0397 + 2.056 \times \sqrt{1 + \tfrac{1}{27}}$$
$$\times 0.064), \quad \text{or} \quad (-0.094, \, 0.174)$$

where 2.056 is the 0.975th quantile of a t-distribution with 26 degrees of freedom. The unrealistic negative lower end of the interval is a signal that these data do not fit a normal distribution well, and that a normal-theory prediction interval should not be used without prior transformation of the data.

7.4.10   Prediction Interval for Several New Observations, Assuming a Normal Distribution

An approximate prediction interval that covers the likely range of values for m new observations with $(1 - \alpha)\%$ confidence is

$$\bar{x} - t_{1-\alpha/(2m),n-1} \cdot \sqrt{1 + \frac{1}{n}} \cdot s, \qquad \bar{x} + t_{1-\alpha/(2m),n-1} \cdot \sqrt{1 + \frac{1}{n}} \cdot s \qquad (7.9)$$

Including multiple predicted observations within the interval is accomplished by dividing α for the t-statistic by $2m$ rather than by 2. The t-statistic increases in value as m increases, widening the prediction interval. Tables for more exact and slightly smaller prediction intervals than those using equation (7.9) are found in Hahn and Meeker (1991).

A 95% prediction interval that should include values for three new blood lead observations is

$$(0.0397 - 2.577 \times \sqrt{1 + \tfrac{1}{27}} \times 0.064, \quad 0.0397 + 2.577 \times \sqrt{1 + \tfrac{1}{27}}$$
$$\times 0.064), \quad \text{or} \quad (-0.128, \, 0.208)$$

where 2.577 is the 0.992th quantile $(1 - 0.05/6)$ of a t-distribution with 26 degrees of freedom.

Prediction intervals get very wide very quickly as m increases. Users often decide rather than to accept such wide intervals to use a tolerance interval instead. In this case, the tolerance interval can be interpreted as giving the range of values that will include $p\%$ of all new observations, rather than all m observations, with $(1 - \alpha)\%$ confidence.

7.4.11 Prediction Bound for New Observations, Assuming a Normal Distribution

A one-sided prediction bound states the probable extreme in one direction for one or more new observations not currently in the data set. One example of its application is to determine a limit not likely to be exceeded by a new observation, based on an existing set of observations. For example, concentrations are measured in field blanks representing contamination due to the sampling and analytical processes. We might like to define a limit that, if exceeded, would indicate that the concentration in a new observation was greater than those in a blank. A one-sided 95% prediction interval would have only a 5% chance of being exceeded by a new observation similar to the blanks. That is a sufficiently small probability that an exceedance would be grounds for declaring that the concentration in a new observation was not simply due to contamination.

An upper prediction bound (assuming a normal distribution) that exceeds the likely values for m new observations, with $(1 - \alpha)\%$ confidence, is

$$\bar{x} + t_{(1 - \alpha/m), n - 1} \cdot \sqrt{1 + \frac{1}{n}} \cdot s \qquad (7.10)$$

A one-sided lower bound can be found by changing the plus sign following the mean in equation (7.10) to a minus sign.

For example, an upper prediction bound that should not be exceeded by one new blood lead observation with 95% confidence is

$$(0.0397 - 1.706 \times \sqrt{1 + \tfrac{1}{27}} \times 0.064), \quad \text{or} \quad 0.151$$

where 1.706 is the 0.95th $(1 - 0.05/1)$ quantile of a t-distribution with 26 degrees of freedom.

All of the above intervals, confidence, prediction, and tolerance, were for the situation where the data can be assumed to come from a normal distribution. We turn our attention now to the situation where the data do not appear to do so. First this situation is addressed by transforming the data to better fit a normal distribution, prior to using the equations already presented for a normal distribution. Later, nonparametric intervals are created that require no assumption about the shape of the data distribution.

7.5 INTERVALS FOR THE LOGNORMAL DISTRIBUTION

The lognormal distribution is a skewed distribution of a variable x whose natural logarithms $y = \ln(x)$ follow a normal distribution. The lognormal distribution has

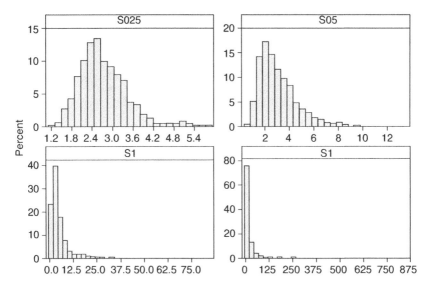

FIGURE 7.2   Histograms of four lognormal distributions with increasing skewness. Lowest skew at top left; highest skew at bottom right.

been used by many investigators in environmental studies, as well as in other disciplines, to describe the shapes seen when a lower limit of zero is combined with the occurrence of infrequent yet annoyingly regular large outliers. The distribution has a flexible shape, appearing similar to a normal distribution when the skew is small (Figure 7.2, upper left), and much like an exponential decay function when skew is large (Figure 7.2, bottom right). To evaluate whether data follow a lognormal distribution the logarithms y are plotted on a normal probability plot, and the resulting pattern tested to see if it follows a straight line using the probability plot correlation coefficient, an analog to the standard Shapiro–Wilks test for normality (Looney and Gulledge, 1985).

Tolerance and prediction intervals follow a simple process for lognormal distributions. The data are log-transformed, intervals are computed in the transformed units, and the interval endpoints are retransformed back into the original units. The resulting intervals can be directly interpreted as prediction and tolerance intervals in the original units. Unfortunately, confidence intervals are not so simply computed.

Each of the intervals for a lognormal distribution uses estimates of the mean and standard deviation of the logarithms. In Minitab these are calculated for censored data using MLE with the menu command

```
Stat > Reliability/Survival > Distribution analysis
(arbitrary censoring) > Parametric distribution analysis
```

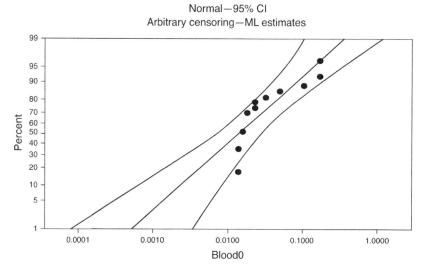

FIGURE 7.3 Probability plot of logarithms of the censored blood lead data.

while selecting lognormal as the fitted distribution. Results for the blood lead data are plotted in Figure 7.3 and listed in the following output:

Parameter	Estimate	Standard Error	95.0% Normal CI Lower	Upper
Location	−4.26595	0.357922	−4.96746	−3.56443
Scale	1.40747	0.313299	0.909849	2.17728

Log-Likelihood = 7.548

Characteristics of Distribution

	Estimate	Standard Error	95.0% Normal CI Lower	Upper
Mean (MTTF)	0.0377996	0.0153122	0.0170874	0.0836181
Standard Deviation	0.0944994	0.0786706	0.0184844	0.483118
Median	0.0140384	0.0050247	0.0069607	0.0283127
First Quartile (Q1)	0.0054329	0.0027094	0.0020443	0.0144385
Third Quartile (Q3)	0.0362750	0.0112877	0.0197124	0.0667537
Interquartile Range (IQR)	0.0308422	0.0100925	0.0162407	0.0585712

The mean and standard deviation of the (natural) logarithms are reported as the location and scale parameter estimates, respectively. So the mean of the logarithms $\bar{y}$ is estimated as −4.26 and the standard deviation of the logarithms s_y as 1.407. Estimates listed under "Characteristics of Distribution" are in the original units. The mean (0.0377) was estimated by (same as equation 6.1)

$$\bar{x} = \exp\left(\bar{y} + \frac{s_y^2}{2}\right) \tag{7.11}$$

and the standard deviation (0.094) as

$$s = \overline{x} \cdot \sqrt{\exp\left(s_y^2\right) - 1} \tag{7.12}$$

The median (0.014) is the mean of the logarithms, retransformed

$$\text{median}_x = \exp(\overline{y}) \tag{7.13}$$

Similarly, a two-sided confidence interval for the mean (by the Cox method, equation 7.14) for an assumed lognormal distribution is computed in NADA for R by typing

```
> Pbmlelog=cenmle(Blood, BloodCen, dist="lognormal")
> mean(Pbmlelog)
```

mean	sd	0.95LCL	0.95UCL
0.03779965	0.0944994	0.0170874	0.0836181

Equations (7.11) and (7.12) are correct for large samples, but do not perform well for small ($n < 50$) sample sizes. In particular, the estimate for the standard deviation of the logarithms s_y is inaccurately estimated by MLE for small samples. The result is a biased estimate of the mean; its value is generally too large when data are right-skewed. Various methods have been devised for correcting this bias (e.g., see Shumway et al., 2002; Cohn, 1988) but these are not implemented in standard statistical software.

7.5.1   Confidence Interval for the Mean Assuming a Lognormal Distribution

If the data follow a lognormal distribution, Cox's method (Olsson, 2005) can be used to calculate upper and lower confidence limits for the mean of lognormal data. The equation for Cox's method is shown as equation (7.14).

$$\exp\left[\overline{y} + \frac{s_y^2}{2} - z\sqrt{\frac{s_y^2}{n} + \frac{s_y^4}{2(n-1)}}\right], \quad \exp\left[\overline{y} + \frac{s_y^2}{2} + z\sqrt{\frac{s_y^2}{n} + \frac{s_y^4}{2(n-1)}}\right] \tag{7.14}$$

This is the most commonly reported method for computing confidence intervals for lognormal data. Olsson (2005) proposed a slight modification by using a t-statistic rather than a normal z-statistic in computing the intervals, resulting in a slightly wider interval. This modification improved the coverage of the interval for smaller sample sizes. However, the modification is not yet commonly found in software, including Minitab and NADA for R.

Gibbons and Coleman (2001) report several approximate or less satisfactory methods for producing confidence intervals of lognormal data. All of them either assume that the distribution of a statistic is normal when it is known not to be, or assume that the standard deviation of the logarithms is known, when in reality it is only estimated by s_y. This mis-specification of the standard deviation introduces large errors when sample sizes are small. One of those methods commonly used in the past is Land's method (Gibbons and Coleman, 2001). Land's upper and lower two-sided confidence limits are computed as

$$\exp\left(\bar{y} + \frac{s_y^2}{2} + \frac{s_y \cdot H_\alpha}{\sqrt{n-1}}\right), \quad \exp\left(\bar{y} + \frac{s_y^2}{2} + \frac{s_y \cdot H_{1-\alpha}}{\sqrt{n-1}}\right) \tag{7.15}$$

where tables of Land's H-statistic are provided in an appendix by Gibbons and Coleman (2001). Land's H-statistic is a function of both the desired confidence level α and of the standard deviation s_y of natural logarithms. When sample sizes are smaller *or* when the coefficient of variation is greater than 1 (even for large sample sizes) these H-limits will not perform well (Singh et al., 1997). This includes most of the cases found in practice in environmental sciences, and therefore Land's method is rarely the best procedure. Singh et al. (1997) conclude that "for samples of size 30 or less, the H-statistic-based UCL results in unacceptably high estimates of the threshold levels such as the background level contamination."

Currently, bootstrapping (Efron, 1981) is the most satisfactory method for computing confidence limits around the mean of lognormal data. Bootstrapping involves repeated computations of the same statistic thousands of times, each time on a temporary set of data chosen with replacement from the original data set. The mean of the computed estimates is the bootstrapped estimate of that statistic. Though there are several ways to compute a bootstrapped confidence interval, the method that assumes nothing about the distribution of the estimates is to take the 2.5th and 97.5th percentiles of the estimates as the 95% confidence interval endpoints. This was called the "percentile method" by its developer (Efron, 1981). Singh et al. (1997) strongly recommended the bootstrap or other nonparametric methods over Lands' H-statistic when computing confidence intervals for lognormal means.

Bootstrapped two-sided 95% confidence intervals for the mean of lognormal data can be computed by

1. From the original set of n observations, sample with replacement to obtain a temporary set of n observations. Because some observations will be chosen more than once, the temporary set is rarely identical to the original data set.
2. Compute an MLE (or other method) estimate of the mean of the temporary set of data.
3. Save the estimate and repeat the process many times with new temporary sets of data. One thousand to ten thousand replicates is a commonly used range.
4. Compute the mean of the replicate estimates for the mean. This is the bootstrapped estimate of the mean. It has no advantage over the mean of the original n observations.

5. Locate the 2.5th and 97.5th percentiles of the estimates of the mean. These are
   the endpoints of the two-sided 95% confidence interval for the mean. Generating confidence intervals when data do not necessarily follow a standard
   distribution is one of the major contributions of the bootstrapping process.

Five percent of the estimates (2.5% on each side) are outside of the bootstrapped
interval endpoints. The confidence interval may be asymmetric around the mean,
reflecting that the distribution of the mean for small skewed data sets may not
approach a normal distribution. More detail on bootstrapping is given in Section 7.8.
MLE bootstrapping for the mean of left-censored data can be implemented with the
Minitab macro BootMLE, available online at www.practicalstats.com in the package
of Minitab macros for NADA (NADA for Mtb). For the blood lead data, the
bootstrapped 95% two-sided confidence interval for the mean, assuming a lognormal
distribution, is computed by

```
> %bootmle 'blood Pb' 'blood LT1'
```

producing both the written output below and Figure 7.4, which shows the two-sided
interval superimposed on a histogram of all of the bootstrapped estimates of the mean.
Note that the distribution of means in Figure 7.4 is skewed and unlike a normal
distribution. Therefore the Central Limit Theorem is unlikely to apply here, and the
bootstrapped interval, with an upper bound further from the mean than the lower
bound reflecting the skewness of the data, will provide a more accurate interval than
by using the standard t-interval formula.

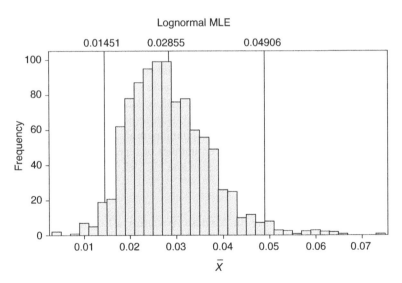

FIGURE 7.4   Bootstrapped lognormal MLE estimates and 95% confidence interval on the
mean of censored blood lead data using the BootMLE macro for Minitab.

```
MLE(lognormal) mean = 0.028549
*****************************
Bootstrap estimate of the 95% confidence interval around the mean
Row LWR95       UPPR95
1   0.0145136  0.0490597
*****************************
Bootstrap estimates of one-sided upper confidence bounds on the mean
UCL95 = Upper 95% conf bound  0.0441785
UCL99 = Upper 99% conf bound  0.0578734
*****************************
```

7.5.2   Confidence Bound for the Mean Assuming a Lognormal Distribution

Either the original or modified Cox's method can be used to compute an upper confidence bound on the lognormal mean. The original Cox's formula for a one-sided bound is given in equation (7.16).

$$\exp\left[\bar{y} + \frac{s_y^2}{2} + z\sqrt{\frac{s_y^2}{n} + \frac{s_y^4}{2(n-1)}}\right] \qquad (7.16)$$

A lower confidence bound is obtained by substituting a minus for the plus sign prior to the z-statistic of equation (7.16). The modified Cox's method (Olsson, 2005) replaces the z-statistic with a t-statistic for the chosen confidence coefficient and degrees of freedom $(n-1)$. If other methods are chosen, the cautions of the previous section including the comments by Singh (1997) on Land's method apply equally well to one-sided bounds as to two-sided confidence intervals for lognormal data.

A more robust method for computing confidence bounds is bootstrapping. The one-sided upper 95% bound is the 0.95 quantile of the estimates of the means produced by the bootstrapping procedure. The lower 95% confidence bound is found at the 0.05 quantile of the bootstrapped means. Figure 7.5 illustrates the bootstrapped 95% upper confidence limit on the mean for the blood lead data. Note that the values for the UCL95 and bootstrapped mean are slightly different than for Figure 7.4 and its associated text. This is inherent in bootstrapping methods, and gives an idea of the precision that can be obtained for a given number of repetitions. To obtain a more precise and consistent estimate of the mean and UCL95, perform a larger number of repetitions.

7.5.3   Confidence Interval for the Median Assuming a Lognormal Distribution

If the data follow a lognormal distribution the same central value is both the mean and median of the logarithms. The mean of the logarithms retransformed back into original units is the geometric mean, an estimate for the median of a lognormal distribution. A confidence interval around this median is calculated by

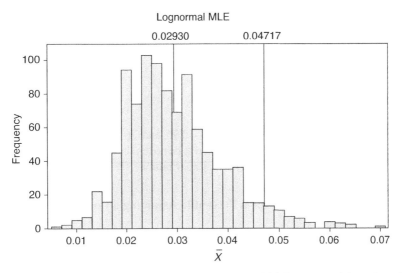

FIGURE 7.5 Bootstrapped lognormal MLE estimates and 95% upper confidence bound on the mean of censored blood lead data using the BootMLE macro for Minitab.

retransforming the confidence interval for the mean of the logarithms back into original units.

$$x_{lo} = \exp\left(\bar{y} - t_{(1-\alpha/2),n-1} \cdot \frac{s_y}{\sqrt{n}}\right), \qquad x_{hi} = \exp\left(\bar{y} + t_{(1-\alpha/2),n-1} \cdot \frac{s_y}{\sqrt{n}}\right) \quad (7.17)$$

This interval has probability $(1 - \alpha)$ of enclosing the population median as long as the data reasonably follow a lognormal distribution.

A confidence interval for the median blood lead concentration, assuming a lognormal distribution, is

$$x_{lo} = \exp\left(-4.266 - 2.056 \times \frac{1.407}{\sqrt{27}}\right), \qquad x_{hi} = \exp\left(-4.266 + 2.056 \times \frac{1.407}{\sqrt{27}}\right), \text{ or}$$

$$(0.008, 0.024)$$

where 2.056 is the t-statistic for $(1 - \alpha/2) = 0.975$ and 26 degrees of freedom.

7.5.4   Confidence Bound for the Median Assuming a Lognormal Distribution

An upper confidence bound around the median is computed by retransforming the upper confidence bound on the mean of the logarithms.

$$\left\{ x_{hi} = \exp\left(\bar{y} + t_{(1-\alpha),n-1} \cdot \frac{s_y}{\sqrt{n}}\right) \right\} \qquad (7.18)$$

The lower confidence bound substitutes a subtraction sign for the plus sign following $\bar{y}$. As before, the y values are logarithms and the x values are data in their original units.

An upper 95% confidence bound on the median blood lead concentrations is

$$x_{hi} = \exp\left(-4.266 + 1.706 \cdot 1.407/\sqrt{27}\right), \text{ or}$$

$$(0.022)$$

where 1.706 is the t-statistic for $(1 - \alpha) = 0.95$ and 26 degrees of freedom. The median blood lead concentration is expected to be no higher than 0.022 µg/g with 95% confidence if the data follow a lognormal distribution.

7.5.5 Confidence Interval for a Percentile Assuming a Lognormal Distribution

Confidence intervals around a percentile of a lognormal distribution are computed in log units using the method for a normal distribution, and then retransformed back into original units. The sample estimate of the pth percentile is computed as

$$x_p = \exp(\bar{y} + z_p \cdot s_y) \tag{7.19}$$

where z_p is the pth percentile of the standard normal distribution.

A two-sided confidence interval around a percentile larger than the median $(p > 0.5)$ follows the formula:

$$\exp(\bar{y} + g'_{(\alpha/2),p,n} \cdot s_y, \ \bar{y} + g'_{(1-\alpha/2),p,n} \cdot s_y) \tag{7.20}$$

where tables of the g'-statistic may be found in Tables A.12a–A.12d of Hahn and Meeker (1991). The g'-statistic is based on a noncentral t-distribution, and is a function of the confidence coefficient $(1 - \alpha)$, the percentage p corresponding to the desired percentile, and the sample size n. For percentiles less than the median $(p < 0.5)$, the interval is

$$\exp(\bar{y} - g'_{(1-\alpha/2),p,n} \cdot s_y, \ \bar{y} - g'_{(\alpha/2),p,n} \cdot s_y) \tag{7.21}$$

For example, the sample 90th percentile of the blood lead data, assuming the data are lognormal, is

$$\exp(-4.266 + 1.28 \times 1.407), \quad \text{or} \quad (0.085)$$

where 1.28 is the 90th percentile of the standard normal distribution $(p = 0.9)$. The 95% confidence interval around this estimate for the 90th percentile is

$$\exp(-4.266 + 0.846 \times 1.407, \ -4.266 + 1.932 \times 1.407), \quad \text{or}$$

$$(0.046, 0.213)$$

where 0.846 and 1.932 are the g'-statistics from Tables A.12a and A.12d of Hahn and Meeker (1991) for $\alpha/2 = 0.025$, $p = 0.9$, and $n = 27$. Therefore, the 90th percentile of the population of blood lead levels lies between 0.046 and 0.213 µg/g, with 95% confidence, if the data follow a lognormal distribution.

7.5.6 Confidence Bound for a Percentile Assuming a Lognormal Distribution

One-sided confidence bounds for percentiles are computed in much the same way as the two-sided intervals. A confidence bound is appropriate if the objective is to define a limit that exceeds (or is lower than) a specified proportion (p) of the data, with a specified confidence α. The equations for one-sided confidence bounds on a percentile are

$$x_{\mathrm{hi}} = \exp(\bar{y} + g'_{(1-\alpha),p,n} \cdot s_y) \quad \text{upper bound for } p > 0.5$$
$$x_{\mathrm{lo}} = \exp(\bar{y} - g'_{(1-\alpha),1-p,n} \cdot s_y) \quad \text{lower bound for } p < 0.5$$

(7.22)

where g' is from Tables A.12a–A.12d in Hahn and Meeker (1991).

Suppose a threshold must be determined that is higher than 90% of the blood lead levels in herons represented by the sample data. An upper 95% confidence bound will reflect the uncertainty in the true 90th percentile of lead concentrations. The confidence bound, assuming that the data follow a lognormal distribution, is

$$\exp(-4.266 + 1.811 \times 1.407), \quad \text{or} \quad (0.179)$$

where 1.811 is $g'_{0.95,\,0.9,\,27}$ from Table A.12d of Hahn and Meeker (1991). We expect that at least 90% of all lead concentrations in blood of herons represented by this study will be below 0.179 µg/g, if the assumption of a lognormal distribution is reasonable for these data.

7.5.7 Tolerance Interval to Contain a Central Proportion of Lognormal Data

Two-sided tolerance intervals to contain a specified central proportion of the data, rarely used in environmental studies, provide allowable exceedances at both high and low ends. Assuming the data follow a lognormal distribution, a two-sided tolerance interval follows the formula:

$$\exp(\bar{y} - g'_{(1-\alpha/2),p,n} \cdot s_y), \quad \exp(\bar{y} + g'_{(1-\alpha),p,n} \cdot s_y)$$

(7.23)

where tables of the g-statistic (different than the g'-statistic) are found in Hahn and Meeker (1991). Note that the tables are set up to use α rather than $\alpha/2$ for determining the g-statistic for each end of the $(1-\alpha)$% confidence interval. The g-statistic is also a function of the proportion p of the distribution to be included within the interval and of the total sample size n. Neither Minitab nor other standard statistical software computes tolerance intervals for a central proportion directly.

A tolerance interval expected to contain the central 90% of the heron blood lead data with 95% confidence is

$$\exp(-4.266 - 2.184 \times 1.407), \quad \exp(-4.266 + 2.184 \times 1.407), \quad \text{or}$$
$$(0.0006, 0.303)$$

where 2.184 is the g-statistic from Table A.10 of Hahn and Meeker (1991) for $p = 0.9$ and $n = 27$. No more than 5% of blood lead concentrations in the population of herons are expected to be less than the lower limit of the interval, and no more than 5% above the upper limit, with 95% confidence. Unlike the two-sided tolerance interval for a normal distribution, the lower end of a lognormal interval is not negative. The lognormal distribution therefore is a more reasonable assumption than the normal distribution for these data. But perhaps a better fitting distribution than the lognormal could be found, improving the resulting intervals.

7.5.8 Tolerance Bound for a Proportion of Lognormal Data

One-sided upper tolerance bounds estimate a value that exceeds $p\%$ of the population values with $(1 - \alpha)\%$ confidence. The percent of data (p) designed to be below the bound is often called the coverage. A one-sided tolerance bound is identical to a one-sided confidence limit for the equivalent (pth) percentile. So a 95% upper tolerance bound covering at least 90% of the data is identical to the 95% upper confidence bound for the 90th percentile. Both contain $p\%$ of the data with $(1 - \alpha)\%$ confidence. The equation for a one-sided tolerance bound with coverage p is simply equation (7.22), above.

A 95% upper tolerance bound below which is at least 90% of blood lead levels, assuming that the data follow a lognormal distribution, is

$$\exp(-4.266 + 1.811 \times 1.407), \quad \text{or} \quad (0.179)$$

where 1.811 is $g'_{0.95,\, 0.9,\, 27}$ from Table A.12d of Hahn and Meeker (1991). This result is the same value obtained previously for the 95% upper confidence bound on the 90th percentile.

7.5.9 Prediction Interval for One New Observation, Assuming a Lognormal Distribution

Prediction intervals bracket the range of locations for one or more new observations not currently in the data set. Two-sided intervals are of interest if both extreme high and extreme low values of new observations are of concern. Obtaining a new observation beyond the limits of the prediction interval should happen only $\alpha\%$ of the time if nothing has changed and the new observation(s) come from the same distribution as did the existing data.

A lognormal prediction interval that covers the likely values for one new observation with $(1 - \alpha)\%$ confidence is

$$\exp\left(\bar{y} - t_{(1-\alpha/2), n-1} \cdot \sqrt{1 + \frac{1}{n}} \cdot s_y\right), \quad \exp\left(\bar{y} + t_{(1-\alpha/2), n-1} \cdot \sqrt{1 + \frac{1}{n}} \cdot s_y\right) \quad (7.24)$$

where t is from a Student's t-distribution with $n-1$ degrees of freedom. See the discussion on prediction intervals in the normal distribution section for more detail.

A 95% prediction interval for the range of probable values for one new blood lead observation is

$$\exp\left(-4.266-2.056\cdot\sqrt{1+\frac{1}{27}\cdot1.407}\right),\quad \exp\left(-4.266+2.056\cdot\sqrt{1+\frac{1}{27}\cdot1.407}\right),$$

or $(0.0007, 0.267)$

where 2.056 is the 0.975th quantile of a t-distribution with 26 degrees of freedom. The lower end of a lognormal prediction interval will not go below zero, avoiding one of the primary problems for prediction intervals when assuming data follow a normal distribution.

7.5.10 Prediction Interval for Several New Observations, Assuming a Lognormal Distribution

An approximate prediction interval that covers the likely range of values for m new observations with $(1-\alpha)\%$ confidence is

$$\overline{x}-t_{1-\alpha/(2m),n-1}\cdot\sqrt{1+\frac{1}{n}}\cdot s,\quad \overline{x}+t_{1-\alpha/(2m),n-1}\cdot\sqrt{1+\frac{1}{n}}\cdot s \qquad (7.25)$$

assuming that data follow a lognormal distribution. Tables for more exact interval coefficients are found in Hahn and Meeker (1991).

A prediction interval with 95% confidence of including values for three new blood lead observations is

$$\exp\left(-4.266-2.577\cdot\sqrt{1+\frac{1}{27}\cdot1.407}\right),\quad \exp\left(-4.266+2.577\cdot\sqrt{1+\frac{1}{27}\cdot1.407}\right),$$

or $(0.0003, 0.563)$

where 2.577 is the 0.992th quantile $(1-0.05/6)$ of a t-distribution with 26 degrees of freedom.

7.5.11 Prediction Bound for *m* New Observations, Assuming a Lognormal Distribution

A one-sided prediction bound is used to determine a limit not to be exceeded by (or lower than) one or more new observations, based on an existing set of observations. Exceedance signifies that the new observation represents a different population than the existing data, with $(1 - \alpha)\%$ confidence. See the section on normal distribution prediction bounds for more detail.

A one-sided upper prediction bound for a lognormal distribution that exceeds the likely values for m new observations, with $(1 - \alpha)\%$ confidence is

$$x_{hi} = \exp\left(\bar{y} + t_{(1 - \alpha/m),n - 1} \cdot \sqrt{1 + \frac{1}{n}} \cdot s_y\right) \qquad (7.26)$$

A one-sided lower bound x_{lo} can be found by changing the plus sign to a minus sign following $\bar{y}$, the mean of the logarithms in equation (7.26).

An upper prediction bound that will likely not be exceeded by one new blood lead observation from the same population as the original data, with 95% confidence, is

$$\exp\left(-4.266 + 1.706 \cdot \sqrt{1 + \frac{1}{27}} \cdot 1.407\right), \quad \text{or} \quad (0.162)$$

where 1.706 is the 0.95th $(1 - 0.05/1)$ quantile of a t-distribution with 26 degrees of freedom.

7.5.12   Using Other Transformations

Other transformations may be used to construct intervals similar in purpose to those listed for the lognormal distribution, above. The general procedure is to

1. Find a transformation that produces data close to a normal distribution.
2. Compute an interval on the transformed data.
3. For confidence intervals on percentiles, prediction intervals, and tolerance intervals, retransform the interval endpoints directly back into original units. Confidence intervals on the mean and standard deviation are best performed using bootstrapping (see later section on bootstrapped intervals). Alternatively, complicated procedures based on the mathematics of the transformation itself may be possible, though the limited success with Land's method for small to moderate sized lognormal data sets should caution against using those type of methods.

Several methods can be used to determine which transformation best produces data that follow a normal distribution. Modern statistical software will easily produce a probability plot of the transformed data, comparing percentiles of the transformed data set to percentiles of a normal distribution. If the transformed data follow a normal distribution their points will plot on a straight line. In addition to choosing the units that visually produce the straightest data, there are several numerical measures for judging the adequacy of alternate transformations.

A. The probability plot correlation coefficient (PPCC—see Looney and Gulledge, 1985) increases to a value of 1.0 as data on the probability plot approach a straight line. Choosing the transformation that produces a PPCC closest to 1.0 is the measure most closely associated with the probability plot itself. The coefficient is also used in a test for normality.

B. Two other popular tests for normality are the Anderson–Darling (Stephens, 1974) and Shapiro–Wilk (Shapiro and Wilk, 1965) tests. In each case the null hypothesis is that the data follow a normal distribution. The transformation with the least-significant test statistic (largest p-value) would be produced from the transformation closest to a normal distribution by this measure.

C. After transforming data to approximate normality, compute the MLE for estimating mean and standard deviation assuming a normal distribution. The highest log-likelihood statistic will result from the transformation producing data closest to a normal distribution (Shumway et al., 2002).

Any of these criteria should lead to a reasonable transformation for the data. Several authors recommend limiting Box–Cox power transformations to values on the Ladder of Powers—transformations easy to interpret. Shumway et al. (2002) recommend considering only the log and square root transformations when computing summary statistics and interval estimates of environmental data, because these transformations mitigate the effects of the severity of right-skewness commonly seen in these data.

7.6 INTERVALS USING "ROBUST" PARAMETRIC METHODS

Methods other than maximum likelihood may also be used to estimate the mean and standard deviation of data or of its transformed values. Equations from the previous sections are then used with these estimates to compute the intervals. Huybrechts et al. (2002) found that robust methods performed better than MLE for the small, skewed data sets common to environmental studies. Using either the robust ROS method described in Chapter 6, or the "robust MLE" method described by Kroll and Stedinger (1996) should produce estimates for interval calculations that are as good as or better than MLE for small (<50), skewed data sets.

Using the %BootROS macro for Minitab, a bootstrapped 95% confidence interval on the mean as estimated by robust ROS can be computed.

```
MTB > %bootros 'blood Pb' 'blood LT1'
```

Results are given below, along with the plot of all the bootstrapped estimates in Figure 7.6. Again notice that the distribution of estimated means is not a normal distribution, but skewed, signifying that the Central Limit Theorem cannot be invoked, and that the t-interval formula will not provide a good estimate of the confidence bounds. Bootstrapping is the only method to compute ROS confidence

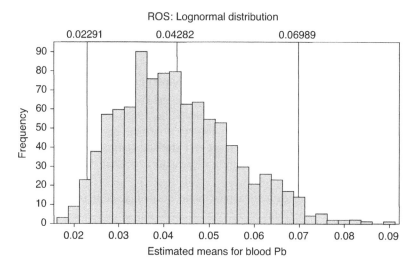

FIGURE 7.6   Bootstrapped ROS estimates and 95% confidence interval on the mean of censored blood lead data.

intervals other than using the *t*-interval formula and assuming a normal distribution of data. For more on bootstrapping, see Section 7.8.

```
ROS (lognormal) mean for blood Pb   0.042818
*****************************
Bootstrap estimate of the 95% confidence interval around the mean
Row LWR95      UPR95
1   0.0229111  0.0698897
*****************************
Bootstrap estimates of one-sided upper confidence bounds on the mean
UCL95 = Upper 95% conf bound   0.0659566
UCL99 = Upper 99% conf bound   0.0740089
*****************************
```

7.7   NONPARAMETRIC INTERVALS FOR CENSORED DATA

Nonparametric intervals assume no shape for the underlying data when computing locations of interval endpoints. The interval shape generally reflects the shape of the observed data, whatever that may be. The primary benefit of a nonparametric interval is that the probabilities of the targeted measure being within and outside the interval are correct regardless of the shape of the underlying data. A 95% nonparametric confidence interval for the median will have a 5% probability of not including the true value, whether the underlying data were normal, lognormal, or some other distributional shape.

The primary drawback of a nonparametric interval is that it makes no use of the information present in a distributional assumption. Therefore if the data do follow a known distribution, the width of a nonparametric interval will be larger than necessary—larger than the width of a parametric interval based on the correct data distribution at the same confidence level. The choice of whether to use a nonparametric interval should be made based on how uncertain the analyst is that the data follow a specific distribution. If the data do not fit the assumed distribution well, the parametric interval may be wider, and will be less accurate, than a nonparametric interval.

The primary method for computing nonparametric intervals is to order the data from smallest to largest, counting in from the ends a specific number of observations. The number of observations is determined by binomial probabilities. The ordered observations are called the "order statistics" of the data set, and for one reporting limit are known as well as for any data that contain ties. For example, counting in approximately 20 observations from each end of a data set of 50 observations, selecting the 20th and 30th smallest observations, produces an approximate 90% confidence interval on the median. If the low end of the interval drops below the one reporting limit, the low end may always be specified as "<RL" without making any unfounded assumptions or statements. For multiple reporting limits a simple nonparametric interval may always be obtained after recensoring all observations below the highest reporting limit (HRL). Any interval endpoints below that threshold are called "<HRL." However, nonparametric intervals for multiply censored data can be obtained with more precision by using methods based on Kaplan–Meier (KM) statistics rather than by censoring at the highest reporting limit.

As the width of nonparametric intervals jumps from point to point, the associated confidence coefficients jump as well. The result is an interval with only approximately the same confidence level as what is desired. If a 95% confidence interval around the median is desired, the closest interval may either be a 96% interval when counting in five points from each end, or a 93% interval when counting in six points. A typical decision is to use the interval that has no more than an α% error. So the 96% interval with a 4% error rate would be chosen. Alternatively some software will interpolate between the two sets of endpoints in order to provide a pseudo-95% interval. Jumping from one confidence coefficient to the next is most severe when sample sizes are small; for larger sample sizes a set of observations can usually be found that are quite close to the desired level of confidence.

7.7.1 Nonparametric Binomial Confidence Interval for the Median

Nonparametric interval estimates for the median and other percentiles can be computed using binomial probabilities. Interval endpoints are chosen using binomial tables where the proportion p in the table is the percentage corresponding to the targeted percentile at the center of the interval. For an interval surrounding the median, the binomial table is entered at $p = 0.5$. Binomial tables are also programmed into most statistical software.

A two-sided confidence interval on the median is computed using

$$[x_L + 1, \ n - x_L] \tag{7.27}$$

where x_L is the entry in a binomial table with $p = 0.5$ whose probability is closest to but not exceeding $\alpha/2$. The confidence interval endpoints are data values $x_L + 1$ in from both ends of the list of ordered observations.

For the heron blood lead data there are 15 censored observations at <0.02 as well as 4 uncensored observations below 0.02, so the smallest 19 observations are all <0.02. The sample median of the 27 observations is the $(n + 1)/2 = 14$th smallest observation, and so <0.02. A 95% confidence interval on this median is found using a binomial table with $p = 0.5$ (representing the median), $\alpha/2 = 0.025$ and $n = 27$. In Minitab the binomial table is accessed by the command

```
Calc > Probability distributions > Binomial
```

Tabled entries are obtained by specifying input values as shown in Figure 7.7.
   The Minitab output is

```
Binomial with n = 27 and p = 0.5
x    P ( X < = x )    x    P ( X < = x )
7   0.0095786    8    0.0261195
```

Minitab reports two possible values for x_L, either 7 or 8, because neither produce an exactly 95% confidence interval. If 7 is used the resulting significance level will be $2 \cdot (0.0096) = 0.019$, or a confidence level of 98.1%, larger than the desired level. If 8 is used the resulting significance level will be $2 \cdot (0.026) = 0.052$, or a 94.8%

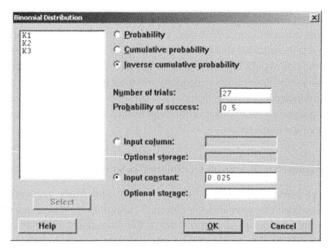

FIGURE 7.7   Minitab dialog box to obtain the order statistics value for a 95% nonparametric confidence interval on the median ($p = 0.5$) of 27 blood lead observations.

confidence interval, slightly less than the desired 95% level. Most practitioners would choose the latter as being quite close to 95%. To obtain interval endpoints, count in $8 + 1 = 9$ observations from each end of the ordered data set. These would be the 9th and $(27 - 9) = 18$th smallest observations, with values of [<0.02, <0.02]. Using binomial probabilities the nonparametric interval estimate states only that the median is somewhere below 0.02 with 95% confidence.

Minitab also provides a nonlinear interpolation between these values to obtain an approximate 95% confidence interval, using the

```
Stat > Nonparametrics > 1-sample sign
```

procedure. This interval around the median is directly related to the sign test. Observations outside the $(1 - \alpha)$% interval are sufficiently far from the center that the sign test would declare their value to be significantly different than the observed median at the α% level. Values within the interval are not significantly different. The interpolated 95% confidence interval is given as [<0.02, 0.020], as shown in the output below.

```
Sign confidence interval for median
                                           Confidence
                                            Interval
             N    Median    Achieved    Lower      Upper      Position
                            Confidence
blood Pb    27   0.01999    0.9478      0.01999    0.02000    9
                            0.9500      0.01999    0.02009    NLI
                            0.9808      0.01999    0.02352    8
```

"NLI" stands for nonlinear interpolation. Both the values of 0.01999 and 0.02000 should be read as actually <0.02—the sign test does not read less-than signs. The upper interval endpoint of 0.02009 appears to be above the reporting limit of 0.02, but is actually an interpolation between values of <0.02 and 0.02352, one of the uncensored observations. Whenever the interpolation is between a censored and uncensored observation, the reported value is not exactly known. A value of 0.02 was used as the low end of the interpolation, which is too high. Unless both interval endpoints are above the reporting limit, using an interpolated interval remains inexact. Use the closest exact interval, here the 94.8% interval, rather than the interpolation.

Nonparametric intervals based on binomial probabilities work reasonably well for one reporting limit, but cannot be used for more than one limit except by censoring all values below the highest reporting limit as <HRL. If an upper bound is all that is desired, and binomial probabilities result in an upper bound that is higher than the HRL, all is fine. However, nonparametric intervals that account for multiple reporting limits can also be produced using Kaplan–Meier statistics. KM methods have the advantage of being the standard methods in medical statistics, and do not require recensoring to the highest limit. KM intervals are illustrated in the next section.

7.7.2   Nonparametric Confidence Intervals Based on Kaplan–Meier Methods

Kaplan–Meier methods estimate the survival function (or edf or percentiles) of the data without assuming that data follow any particular distribution. KM methods are coded in statistics software assuming that data will be right-censored. Left-censored environmental data must first be flipped into right-censored format (see Chapter 2) prior to computing the estimates. The blood lead data were flipped to a right-censored format by subtracting the lead concentrations from 1.0, a value larger than the maximum concentration. The flipped data are processed using the Kaplan–Meier procedure, invoked in Minitab using

```
Stat > Reliability/Survival > Distribution analysis (right-
censoring) > Nonparametric Distribution Analysis
```

A plot of the survival function with confidence bands around the function is shown in Figure 7.8.

The KM table of survival probabilities and their 95% confidence intervals are shown below. Added to the table is a column of retransformed blood lead values (Lead), computed as (1-Time) where Time represents the flipped data. The Kaplan–Meier estimate for the pth percentile is the observation with the largest survival probability $\leq p$. For this example, the KM estimate of the median is at the observation with a survival probability of 0.35, the largest probability ≤ 0.5. The median is therefore at a Time of 0.984, or a lead concentration of $(1 - 0.984) = 0.016$.

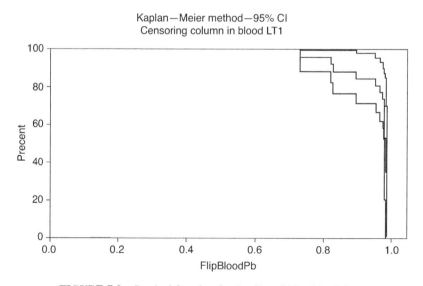

FIGURE 7.8   Survival function for the flipped blood lead data.

```
                  Standard         95.0% Normal CI
     Mean(MTTF)    Error        Lower       Upper
     0.957310      0.0125752    0.932663    0.981957

     Median = 0.984483
     IQR = 0.0102426   Q1 = 0.975472   Q3 = 0.985714
```

```
                          Kaplan-MeierEstimates

               Number  Number  Survival     Standard    95.0% Normal CI
   Time    Lead  at Risk Failed Probability  Error       Lower     Upper
   0.731034 0.261 27      1       0.962963    0.036345   0.891729  0.99999
   0.822951 0.177 26      1       0.925926    0.050401   0.827142  0.99999
   0.825926 0.174 25      1       0.888889    0.060481   0.770348  0.99999
   0.893939 0.106 24      1       0.851852    0.068367   0.717854  0.98584
   0.950847 0.049 23      1       0.814815    0.074757   0.668294  0.96133
   0.966038 0.034 22      1       0.777778    0.080009   0.620963  0.93458
   0.975472 0.025 21      1       0.740741    0.084337   0.575443  0.90603
   0.976471 0.024 20      1       0.703704    0.087877   0.531468  0.87593
   0.981356 0.019 4       1       0.527778    0.166001   0.202422  0.85312
   0.984483 0.016 3       1       0.351852    0.181330   0.000000  0.70724
   0.985714 0.014 2       1       0.175926    0.153932   0.000000  0.47762
   0.986275 0.014 1       1       0.000000    0.000000   0.000000  0.00000
```

The "95.0% Normal CI" KM intervals are confidence intervals on survival probabilities rather than on data values. They represent vertical distances on a survival plot such as Figure 7.8, rather than the horizontal interval that would provide a confidence interval around the median or other percentiles. Klein and Moeschberger (2003) note that these probability confidence intervals ("linear confidence intervals") are quite inaccurate for "small" data sets of less than 200 observations! See Klein and Moeschberger (2003) for more information if intervals around survival probabilities are of interest. Confidence intervals around the median or other percentiles are provided in the next sections.

7.7.3 Nonparametric Confidence Intervals for the Median Based on Kaplan–Meier

Three methods of computing intervals for the median of a multiply censored variable are presented below. All avoid assumptions about the shape of the data distribution, though the first method assumes that the variation in estimates of a single survival probability is asymptotically normal. The second method is based on an adaptation of the sign test for multiply censored data. It is fully nonparametric, but uses a large-sample normal approximation for the test statistic. The third method is bootstrapping, which makes no assumptions about the distribution of data or any test statistic. Bootstrapping is discussed in its own section. The bootstrap and sign-test methods provide reliable nonparametric estimates of confidence intervals for percentiles of data with censored observations.

The first type of nonparametric interval based on Kaplan–Meier statistics uses Greenwood's formula for the standard error of the survival function, s.e.$[\hat{S}]$ (equation 6.6), as the measure of the vertical variability around the survival curve. Greenwood's standard error is usually printed in output from KM software as a column of standard errors, one value for each survival probability. The standard error for a percentile, the variability in a horizontal direction on a plot of the survival curve, is related to Greenwood's standard error by estimating the probability density function or pdf at that percentile (Collett, 2003).

The standard error of the median is related to Greenwood's standard error s.e.$[\hat{S}]$ by equation (7.28) (Collett, 2003).

$$s.e.[\text{median}] = \frac{s.e.[\hat{S}]}{\text{pdf}[\text{median}]} \tag{7.28}$$

where pdf[median] is the probability density function evaluated at the median. The pdf is approximated using equation (7.29) (Collett, 2003).

$$\text{pdf}[\text{median}] = \frac{\hat{S}(T^+(0.55)) - \hat{S}(T^-(0.45))}{T^-(0.45) - T^+(0.55)} \tag{7.29}$$

where $T^+(0.55)$ is the largest survival time whose estimated survival probability exceeds 0.55, $T^-(0.45)$ is the smallest survival time whose estimated survival probability is less than or equal to 0.45, and $\hat{S}$ is the estimated survival probability. The $(1 - \alpha)\%$ confidence interval around the sample median $\hat{T}(0.5)$ is then

$$\hat{T}(0.5) - z_{1-\alpha/2} \cdot s.e.[\text{median}], \quad \hat{T}(0.5) + z_{1-\alpha/2} \cdot s.e.[\text{median}] \tag{7.30}$$

As an example, the 95% confidence interval on the median blood lead concentration computed with equations 7.28–7.30 is

$$\text{pdf}[\text{median}] = \frac{\hat{S}(T^+(0.55)) - \hat{S}(T^-(0.45))}{T^-(0.45) - T^+(0.55)} = \frac{0.704 - 0.352}{0.984 - 0.976} = 44$$

$$s.e.[\text{median}] = \frac{s.e.[\hat{S}]}{\text{pdf}[\text{median}]} = \frac{0.1813}{44} = 0.0041$$

where 0.1813 is the Greenwood estimate of the standard error at the observation selected to be the median.

Once the standard error is estimated, a z-interval is computed as the confidence interval for the median survival time (the median of flipped data). The 95% confidence interval using a t-statistic of 1.96 is computed as

$$[0.016 - 1.96 \times 0.0041, \quad 0.016 + 1.96 \times 0.0041] = [0.008, \quad 0.024]$$

Klein and Moeschberger (2003) discourage use of the Greenwood pdf-Z interval, calling equation (7.29) a "crude" estimate of the pdf, sufficiently unreliable for small sample sizes that they avoid its use and compute intervals by the inverted sign test method. This second type of interval inverts a sign test for multiply censored data to avoid estimating the pdf, was originally developed by Brookmeyer and Crowley (1982), and so is called the B–C sign method.

The B–C sign method computes an estimate of the sign test statistic for multiply censored data as a ratio whose variation is approximately normal (equation (7.31)). This ratio at the center of equation (7.31) is computed for each detected observation in the KM table. All observations whose ratios lie between the critical Z-statistics at each end are considered inside the sign-test confidence interval. The extreme observations still within the limits of the Z-statistics become the endpoints of the $(1 - \alpha)\%$ interval. The B–C sign equation (Klein and Moeschberger, 2003) is

$$-z_{1-\alpha/2} \leq \frac{\hat{S} - (p)}{s.e.[\hat{S}]} \leq +z_{1-\alpha/2} \tag{7.31}$$

where $\hat{S}$ is the estimated survival probability for each detected observation, p is the percentage of the target percentile at the center of the interval, and s.e.$[\hat{S}]$ is Greenwood's standard error given in equation 6.7. $\hat{S}$ and s.e.$[\hat{S}]$ are printed for each observation by KM software. It should be noted that the percentage p is in the same direction as the survival probabilities, and as are the original concentration data prior to flipping. So the 25th percentile of concentration is the observation with a 25% survival probability, and is the $(1 - p) = 75$th percentile of the flipped Time variable. Table 7.2 lists the B–C sign test statistic for the median in the column "B–C sign," calculated at each uncensored observation (here the flipped blood lead data) using equation (7.31).

TABLE 7.2 B–C Inverted Sign Test Statistics for Determining Confidence Intervals for the Median of the Blood Lead Concentrations

Time	Lead	Survival Probability	Standard Error	B-C Sign
0.731034	0.261	0.962963	0.036345	12.7380
0.822951	0.177	0.925926	0.050401	8.4507
0.825926	0.174	0.888889	0.060481	6.4299
0.893939	0.106	0.851852	0.068367	5.1465
0.950847	0.049	0.814815	0.074757	4.2112
0.966038	0.034	0.777778	0.080009	3.4718
0.975472	0.025	0.740741	0.084337	2.8545
0.976471	0.024	0.703704	0.087877	2.3180
0.981356	0.019	0.527778	0.166001	0.1673
0.984483	0.016	0.351852	0.181330	-0.8170
0.985714	0.014	0.175926	0.153932	-2.1053
0.986275	0.014	0.000000	0.000000	*

For a two-sided 95% confidence interval, 2.5% of the error is placed on each side of the interval. The $(1 - \alpha/2)$ Z (standard normal) statistic is Z(0.975) or 1.96. Any observation with a value in the B–C sign column between and including -1.96 and $+1.96$ is within the 95% inverted sign test confidence interval for the median. Because the B–C sign statistic jumps in value from one detected observation to the next, the confidence interval should also include part of the region extending to the first observation with a B–C sign statistic greater in absolute value than 1.96 (unless the statistic for the endpoints were exactly equal to 1.96). To account for this, Brookmeyer and Crowley (1982) use the convention that the interval shall include the first observation on the high Time side with absolute value of its statistic >1.96. This is the low side for concentration. From Table 7.2, the set of observations with B–C sign statistics less than 1.96 in absolute value are the lead concentrations between 0.016 and 0.019 µg/L. Including the next observation on the high Time (low concentration) side, the lead concentration of 0.014 with B–C sign statistic of -2.1, the 95% confidence interval on the median lead concentration is [0.014, 0.019] using the inverted sign test.

Klein and Moeschberger (2003) suggest transforming the sign test ratio when determining which observations are within the $(1 - \alpha)$% interval. Two alternative transformations are log–log and arcsine transformations (Borgan and Liestøl, 1990). The choice of these transformations arose out of the shape of hazard functions in survival analysis for producing confidence intervals on probabilities, rather than intervals for survival times. Borgan and Liestøl (1990) claim that more accurate coverage probabilities $(1 - \alpha)$ are obtained using one of these two transformed test statistics for small data sets that follow a Weibull distribution. Their applicability has not yet been demonstrated for use in intervals for environmental data sets, whose shape is generally close to a lognormal distribution. So the original B–C sign test method should be used for environmental data until these variations have been tested further.

Table 7.3 summarizes the Greenwood pdf-Z and B–C sign test results for a 95% two-sided confidence interval around the median for the multiply censored blood lead data. Also shown is the "binomial" confidence interval for singly censored data, applied to the blood lead concentrations by treating all data below the highest reporting limit of 0.02 as simply <0.02. The bootstrap confidence interval is discussed in a later section. The B–C sign interval is shorter than the Greenwood pdf-Z interval, is fully nonparametric, and does not require a highly variable estimate of the pdf of the

TABLE 7.3   Nonparametric Two-Sided 95% Confidence Intervals on the Median Blood Lead Concentrations

Method	Lower limit	Median	Upper Limit
Binomial (94.6%)	<0.02	<0.02	<0.02
Greenwood pdf-Z	0.008	0.016	0.024
B–C sign	0.014	0.016	0.019
Bootstrap KM	0.014	0.016	0.019

The binomial interval only handles one reporting limit.

distribution as does Greenwood. The usefulness and precision of the binomial interval is greatly diminished by its requirement of increased censoring—the only information is that the entire interval is less than 0.02. Of the three, the B-C sign interval provides the greatest precision and information content for multiply censored data.

7.7.4 Nonparametric Confidence Intervals for Percentiles Other than the Median

Nonparametric interval estimates for percentiles other than the median can be computed by the same methods used for the median—intervals based on binomial probabilities, Greenwood pdf-Z, the B–C inverted sign interval, and bootstrapping. With binomial intervals all data below the highest reporting limit are again treated as <HDL. The binomial table provides interval endpoints, where p is the percentage related to the percentile of interest at the center of the interval.

For other percentiles than the median the Greenwood standard error (equation 7.28) becomes equation (7.32)

$$s.e.[p] = \frac{s.e.[\hat{S}_p]}{\mathrm{pdf}[p]} \tag{7.32}$$

where p is the survival probability or percentage for the percentile at the center of the interval. The pdf is estimated using probabilities slightly larger and smaller than p, analogous to equation (7.29). The variability in $\hat{S}_p$ is assumed to follow a normal distribution as in equation (7.30). This will be a poorer approximation at extreme percentiles near 0 and 1.

The B–C inverted sign test interval of equation (7.31) can be used without modification for confidence intervals on other percentiles. $\hat{S}$ is evaluated at the pth percentile location, not at the median. The percentage p is in the same scale as are survival probabilities and original concentrations, and will correspond to the $(1 - p)$th percentile of the flipped Time variable.

These intervals are briefly illustrated by computing a one-sided upper 95% confidence bound on the 90th percentile of the lead data. This is a value in which there is 95% confidence of being exceeded in no more than 10% of the population.

Endpoint positions for the binomial interval are found by entering a binomial table with probability $p = 0.9$, along with the sample size and confidence coefficient. For a 95% upper bound the input constant is the confidence coefficient, 0.95. The output from Minitab's table is

```
Binomial with n = 27 and p = 0.9
x      P ( X < = x )    x    P ( X < = x )
26     0.941850        27 1
```

There is a 94.18% probability of being less than or equal to the 26th observation, and so a $(1 - 0.9418) = 6\%$ chance of exceeding this value. If this is close enough to 5%, the 26th observation from the low end, or 0.177, is the endpoint for a 94%

nonparametric upper confidence bound on the 90th percentile. As this lead concentration is well above the highest reporting limit, no confusion is caused by the censoring, and more complex intervals are not necessary. But to illustrate the other intervals we go on.

The KM estimate of the 90th percentile is the observation with the highest survival probability ≤ 0.9. This is the observation with $\hat{S} = 0.888$, and lead $= 0.174$. A 95% upper bound on this value using Greenwood's pdf-Z interval is

$$s.e.[p = 0.9] = \frac{s.e.[\hat{S}]}{\text{pdf}[0.9]} = \frac{0.0605}{0.6820} = 0.0887$$

where

$$\text{pdf}[0.9] = \frac{\hat{S}(T^+(0.95)) - \hat{S}(T^-(0.85))}{T^-(0.85) - T^+(0.95)} = \frac{0.9630 - 0.8519}{0.8939 - 0.7310} = \frac{0.1111}{0.1629} = 0.6820$$

and so the 95% upper bound on the 90th percentile is

$$\hat{T}(0.9) + z_{1-\alpha} \cdot s.e.[\text{median}] = 0.174 + 1.64 \times 0.0887 = 0.319$$

This bound is much higher than the binomial endpoint, lending support to caution when using this method for smaller sample sizes and percentiles close to the endpoints of the distribution.

The B–C sign endpoint is computed using a one-sided version of equation (7.31) and $p = 0.9$ rather than 0.5 for the median.

$$\frac{\hat{S} - (0.9)}{s.e.[\hat{S}]} \leq + z_{1-\alpha}$$

From Table 7.4, comparing the B–C sign statistic to the $(1 - \alpha)$ normal quantile $Z_{0.95} = 1.64$, the one-sided 95% upper confidence bound on lead is at the observation whose B–C sign statistic is the largest value < 1.64, or the second highest observation of 0.177. This agrees with the binomial interval.

7.8   BOOTSTRAPPED INTERVALS

An alternative method of obtaining unbiased nonparametric estimates of parameters and their confidence intervals is called bootstrapping (Efron, 1981). Bootstrap estimates are produced by computing statistics on repeated random samples taken with replacement from the observed data. The repeated samples have the same number of observations as in the observed data set. Censored and uncensored observations are equally available for sampling. The summary statistic is computed for each random sample either by Kaplan–Meier or another censored data procedure. The distribution of the statistic is then estimated by the percentiles of the collection of

TABLE 7.4   Calculations for a One-Sided Upper Bound on the 90th Percentile Using the B–C method

Time	Lead	Survival Probability	Standard Error	B-C Sign
0.731034	0.261	0.962963	0.036345	1.73237
0.822951	0.177	0.925926	0.050401	0.51438
0.825926	0.174	0.888889	0.060481	-0.18370
0.893939	0.106	0.851852	0.068367	-0.70425
0.950847	0.049	0.814815	0.074757	-1.13949
0.966038	0.034	0.777778	0.080009	-1.52759
0.975472	0.025	0.740741	0.084337	-1.88836
0.976471	0.024	0.703704	0.087877	-2.23375
0.981356	0.019	0.527778	0.166001	-2.24228
0.984483	0.016	0.351852	0.181330	-3.02292
0.985714	0.014	0.175926	0.153932	-4.70384
0.986275	0.014	0.000000	0.000000	*

computed values. For example, if 1000 means were computed by performing robust ROS 1000 times, a two-sided 95% confidence interval for the mean could be determined from the 2.5th and 97.5th percentiles of these means, the 25th and 975th of the 1000 mean values when ordered from low to high. Akritas (1986) and Efron (1981) demonstrated the utility of bootstrap estimates for censored data.

To continue our example of estimating the median blood lead level in herons (Golden et al., 2003), the median along with a 95% confidence interval is estimated without assuming a distributional shape for the data by bootstrapping. To begin with, the median of the original 27 observations can be calculated using the KM procedure in Minitab. The output was listed in Table 7.2. The median is the observation whose survival probability is the largest value less than or equal to 0.5. For these 27 observations, that corresponds to a lead concentration of 0.016 after rounding.

The same method can be repeatedly used on random samples from these sample data in order to provide a bootstrap estimate of the median, and a 95% nonparametric confidence interval around this estimate. Bootstrapping methods are beginning to be provided in statistical software, and if not present, are easy to add with a macro or script. The Minitab macro for bootstrapping Kaplan–Meier estimates of median is called KMBoot, and is invoked by the command

```
%KMBoot c1 c2
```

where c1 is the column of left-censored observations and c2 is the column of censoring indicators. One thousand random samples from the data in c1 are created by sampling with replacement, and temporarily stored within the macro. That is, a sample of 27 observations is selected from the original 27 data values. Each of the 27 values has an equal chance of being selected at every step, so some of them are chosen more than once for the new sample. For each of the 1000 new samples, an estimate of

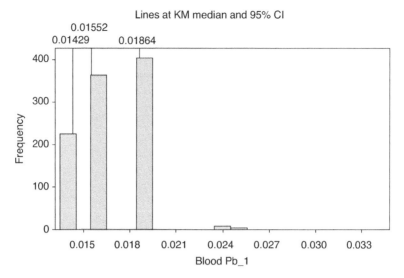

FIGURE 7.9 Histogram of 1000 bootstrap estimates of the median blood lead concentrations using Kaplan–Meier.

the median is computed using Kaplan–Meier. The median of these 1000 estimates is the bootstrap estimate of the KM median, and as shown by the output below is approximately equal to 0.016, the original sample median. The 95% confidence interval for the bootstrap KM estimate is formed from the 25th and 975th smallest estimates out of the 1000 estimates computed. The results of the KMBoot macro are printed below, and shown in Figure 7.9

```
Endpoints of 90%, 95%, 99% confidence intervals
  based on bootstrap samples
    of the KM median

Bootstrap Kaplan-Meier median = 0.015517

*****************************
Bootstrap estimates of the 90% confidence interval of the median
  UPPER90 also = UCL95, the upper 95% CI on the median.

Row   LOWER90     UPPER90
1     0.0142857   0.0186441
*****************************
Bootstrap estimates of the 95% confidence interval of the median

Row   LOWER95     UPPER95
1     0.0142857   0.0186441
*****************************
```

```
Bootstrap estimates of the 99% confidence interval of the median

Row   LOWER99      UPPER99
1     0.0137255  0.0235294
*****************************
```

Note that because Kaplan–Meier selects an observed value as its estimate of the median rather than interpolating, the possible values for the median are few in number. This is characteristic of procedures that do not assume a distribution for the sample statistic, but instead use the observed data repeatedly. It is especially true for data sets with a relatively modest number of observations, as with the 27 observations available here. The 95% confidence interval on the median spans values between 0.014 and 0.019 µg/L. This agrees well with the B–C inverted sign test interval presented previously (Table 7.3). Bootstrapping is certainly expected to be a better procedure than either the Greenwood pdf-Z method, or the binomial procedure that requires censoring to the highest reporting limit. Either the bootstrap or B–C inverted sign test methods can provide efficient nonparametric estimates for the median and other percentiles, along with their confidence intervals, for multiply censored data.

Kaplan–Meier may also be used to provide bootstrapped estimates of confidence intervals around the mean. The procedure is similar to that for the median above, and to bootstrapping ROS and MLE confidence bounds presented earlier in the chapter. Bootstrapping Kaplan–Meier estimates was one of the best of numerous methods for estimating the UCL95 evaluated in the large simulation study by Singh et al. (2006). They found that this method has less error in the coverage, the width of the confidence bound for a given confidence level, than did estimates using ROS, MLE (for small data

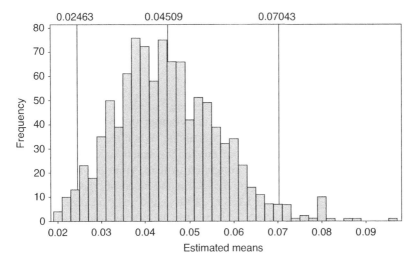

FIGURE 7.10 Histogram of 1000 bootstrap estimates of the mean and 95% confidence intervals for the Golden blood lead concentrations using Kaplan–Meier.

sets), and far less error than substitution procedures. To compute bootstrapped KM intervals for the mean in Minitab, use the BootKM macro:

```
%bootKM 'blood Pb' 'blood LT1'
```

and for the Golden blood lead data, the results are below and in Figure :

```
ENDPOINTS OF 90%, 95%, 99% CONFIDENCE INTERVALS
  BASED ON 1000 BOOTSTRAP SAMPLES OF THE
    K-M MEAN (Efron bias correction)

Kaplan-Meier mean = 0.045087
*****************************
Bootstrap estimate of the 90% confidence interval around the mean
Row   LWR90        UPR90
1      0.0269158   0.0649108
*****************************
Bootstrap estimate of the 95% confidence interval around the mean
Row   LWR95        UPR95
1      0.0246337   0.0704344
*****************************
Bootstrap estimate of the 99% confidence interval around the mean
Row   LWR99        UPR99
1      0.0210309   0.0805647
*****************************
Bootstrap estimates of one-sided upper confidence bounds on
the mean
  UCL95 = Upper 95% conf bound, UCL99 = Upper 99% conf bound
UCL95   0.0649108
UCL99   0.0794907
*****************************
```

7.9 FOR FURTHER STUDY

Confidence intervals for Kaplan–Meier estimates are an active area of research. The probability intervals printed by Minitab's output are "pointwise" intervals to be used only around a single probability estimate. When the entire pattern of the survival function is of interest, joint confidence bands analogous to multiple comparisons for ANOVA must be computed instead. More information on joint confidence bands is found in Chapter 4 of Klein and Moeschberger (2003), as well as in Nair (1984), Weston and Meeker (1991), and Jeng and Meeker (2001). Methods for computing intervals by inverting likelihood-ratio tests and other procedures have been developed by Emerson (1982), Simon and Lee (1982), Murphy (1995), and Slud et al. (1984), among others. Akritas (1986) discusses bootstrapping to compute KM confidence intervals in greater detail than given here.

EXERCISES

7-1 Using the zinc data from the Alluvial Fan zone of Millard and Deverel (1988) found in the data set CuZn, compute a 95% confidence interval on the mean zinc concentration, assuming the data follow a normal distribution. Based on a probability plot, is this assumption reasonable? If not, estimate a 95% confidence interval using bootstrapping and by using Land's method. How do these two intervals compare?

7-2 Estimate a 90% nonparametric confidence interval around the median of the zinc data using the binomial method (first recensoring at the highest reporting limit). Then compute the interval using the B–C sign method. How do these intervals compare? Which would you choose to use?

7-3 Construct a flow chart of the methods of this chapter for computing confidence intervals. Ignore methods known to be inadequate, such as the Greenwood pdf-Z method. Make sure that determining whether the data follow a normal or lognormal distribution figures prominently in your chart. Which method could appear throughout the chart, and work well regardless of the shape of the data distribution?

8 What Can Be Done When All Data Are Below the Reporting Limit?

Measured concentrations may be consistently low in comparison to analytical reporting limits, and all observations recorded as below the reporting limit. Some scientists have felt in these circumstances that they have no data to work with, because they have no single numbers reported to them. Is this correct? Absolutely not! For certain questions there is a lot of information in the fact that data are consistently below one or more thresholds. This information is contained in the proportion of data below versus above the specified thresholds. Methods based on binomial probabilities look at the probabilities of being above and below one or more thresholds. They can be used to answer several relevant questions about censored data.

For example, suppose 2 years of monthly samples of drinking water delivered by a supply system were tested for arsenic concentrations, and all 24 samples had concentrations reported as $<3\,\mu g/L$. Three types of questions for which these data provide answers are as follows:

1. *Point Estimates.* What is an estimate of the typical arsenic concentrations being delivered in these waters?
2. *Exceedance Probability of the Reporting Limit.* How often might a detected concentration above $3\,\mu g/L$ be measured in next year's samples?
3. *Exceedance Probability for a Standard Higher than the Reporting Limit.* How confident can we be that if current conditions are maintained, the drinking water standard of $10\,\mu g/L$ will not be exceeded in more than 5% of samples?

Each of these three types of questions can be addressed when the data set consists of all censored observations.

With the exception of the example in the next section, there is no distinction made in this chapter between detection and reporting limits. The terms "censored observations" and "uncensored observations" refer to data below and above the reporting threshold. This is consistent with their (overly relaxed) use by most environmental scientists.

Statistics for Censored Environmental Data Using Minitab® and R, Second Edition.
Dennis R. Helsel.
© 2012 John Wiley & Sons, Inc. Published 2012 by John Wiley & Sons, Inc.

8.1   POINT ESTIMATES

There are at least two ways to compute an estimate of the mean or median for data sets that are entirely censored observations. A point estimate for the mean or median *can* (but should not) be computed using MLE when all data are below reporting thresholds. The MLE estimate will be extremely unreliable. As a second approach, a nonparametric estimate of the median can be calculated by stating that the median is below the reporting limit.

The (artificial) data set AsExample.xls consists of 13 measurements below the reporting limit of 1, and so reported as within the interval of 0–1, and 11 measurements between the detection and quantitation limits and reported as within the interval of 1–3. Using maximum likelihood and assuming a lognormal distribution, summary statistics estimated for these data include

	Estimate	Standard Error	95.0% Normal CI Lower	Upper
Mean (MTTF)	0.994679	30255.3	0	*
Standard Deviation	0.121099	1839229	0	*
Median	0.987388	188559	0	*
First Quartile (Q1)	0.909821	1293760	0	*
Third Quartile (Q3)	1.07156	1114491	0	*
Interquartile Range (IQR)	0.161747	2408251	0	*

Summary statistics for the same data, assuming a normal distribution, are

	Estimate	Standard Error	95.0% Normal CI Lower	Upper
Mean (MTTF)	0.988121	25147.3	−49286.8	49288.8
Standard Deviation	0.113527	240337	0	*
Median	0.988121	25147.3	−49286.8	49288.8
First Quartile (Q1)	0.911549	187252	−367006	367008
Third Quartile (Q3)	1.06468	136957	−268430	268433
Interquartile Range (IQR)	0.153145	324209	0	*

The most important numbers to notice in these tables are the confidence intervals. When all observations are below the reporting limit the reliability of MLE parameter estimates is very poor. Confidence intervals are either undefined or extremely wide, showing that in reality there is insufficient information to compute reliable estimates of parameters such as the mean and standard deviation. Estimates will strongly depend on which distribution has been selected, even though observed data give no information on which distribution might be most appropriate. If an estimate known only as somewhere between −49,000 and +49,000 is considered too imprecise, individual parameter estimates by MLE for such data should be avoided.

The median of data that are all censored observations can always be stated as being <RL, where RL is the median reporting limit when ranked in order from low to high

limits. A similar procedure is possible for other percentiles. For the AsExample data, the median of 24 observations would fall at the value midway between the 12th and 13th ranked observations. Since there are 13 values known to be <1, the median of these 24 observations is <1. No assumptions are made with this approach, and what is known about a single value is accurately presented.

In general, it is not very helpful to calculate point estimates for data that are entirely censored observations. Instead, inferences can be made concerning probabilities of exceeding the reporting limit(s), based on binomial probabilities. And if one group of data is entirely censored observations while another has some uncensored observations, tests of differences in the probabilities of exceedance can be performed.

8.2   PROBABILITY OF EXCEEDING THE REPORTING LIMIT

When all data are censored observations the observed proportion of data exceeding the reporting limit is 0%. However, this is a sample percentage, only an estimate of the percentage above the reporting limit for the underlying population. If p is the proportion of the population below the reporting limit, $1 - p$ is the proportion above. The certainty with which p is known is a function of the number of observations n. As n increases, the confidence interval around p decreases. The methods of this section involve estimating a confidence interval around the population proportion p, based on binomial probabilities.

Answers to the following four questions illustrate what can be accomplished when all data are censored observations. These questions come in pairs, with questions 1 and 3 estimating a confidence interval for p, and questions 2 and 4 testing to see if a specific value for p is within a confidence interval. The first two questions deal with the proportion p, while the second two questions deal with numbers of occurrences ($p \cdot n$).

1. What is the range of possible proportions of censored observations actually in the population from which these data came?

The sample estimate for the proportion of data below the reporting limit is $p = c/n$, where c is the number of censored observations and n is the sample size. The estimated proportion above the reporting limit is $1 - (c/n)$. When all measured values are below the reporting limit, $c = n$ and the observed proportion below is $p = 1$. A two-sided confidence interval around p is constructed using quantiles of the F distribution (Hahn and Meeker, 1991).

$$[\text{LL, UL}] = [(1 + (n - c + 1) F_1/c)^{-1}, \ (1 + (n - c)/(c + 1)F_2)^{-1}] \quad (8.1)$$

where LL is the lower limit of the confidence interval, $= 0$ when $c = 0$; UL is the upper limit of the confidence interval, $= 1$ when $c = n$; $F_1 = F_{(1 - \alpha/2, \ 2n - 2c + 2, \ 2c)}$; and $F_2 = F_{(1 - \alpha/2, \ 2c + 2, \ 2n - 2c)}$.

When all measurements are censored observations, so that $c = n$ and the observed probability equals the maximum of 1, all error must be placed on one side of the interval, and $F_{1-\alpha}$, rather than $F_{1-\alpha/2}$ is used in the above equations.

For the example of $c = 24$ censored observations in 2 years of monthly measurements of arsenic in drinking water ($n = 24$), a two-sided 95% confidence interval for the proportion of arsenic concentrations below the reporting limit of $3\,\mu g/L$ is

$$\begin{aligned}[\text{LL, UL}] &= [(1 + (24 - 24 + 1)F_1/24)^{-1}, (1 + (24 - 24)/(24 + 1)F_2)^{-1}]\\ &= [(1 + F_{(0.95,\,2,\,48)}/24)^{-1}, (1)^{-1}]\\ &= [(1 + 3.19/24)^{-1}, (1)]\\ &= [0.88, 1]\end{aligned}$$

At a 95% confidence level, between 88 and 100% of arsenic concentrations in the population of delivered drinking waters can be expected to be below the reporting limit of $3\,\mu g/L$, based on the data collected. No more than 12% of measurements are expected to exceed $3\,\mu g/L$.

In Minitab®, this confidence interval can be obtained with the command

```
Stat > Basic Statistics > 1 proportion
```

entering $n = 24$ as the number of "trials," and $c = 24$ as X, the number of "events":

```
Sample   X    N    Sample p    95% CI                      Exact
                                                            P-Value
1        24   24   1.000000    (0.882654, 1.000000)        0.000
```

The 95% confidence interval states that based on the evidence in these data, the true proportion of censored observations lies (with 95% confidence) between 88 and 100%. The p-value reported by Minitab is for a test of whether the proportion of censored observations is significantly different than 0.5. The small p-value shows that it is different, but this is not a question of particular interest here.

2. Are fewer than 10% (or some other proportion) of values in the population above the reporting limit?

The test of whether $(1 - p)$, the proportion of uncensored observations, equals 10% or less is identical to a test for whether p, the proportion of censored observations, is 90% or greater. The motivation for doing so is often the presence of some legal standard or corporate guideline—10% or more uncensored observations is considered too high for some reason, and if true a change would need to be implemented to lower the proportion. The null hypothesis is assumed to be true until proven otherwise, and could be set to assume there is a problem unless the data prove otherwise (it also could be set as the reverse). Assuming the proportion is high until proven otherwise, the null hypothesis would state that there are 10% or more uncensored observations in the underlying population, and therefore fewer than 90% of observations below $3\,\mu g/L$. The alternative hypothesis is the statement to be demonstrated—there are fewer than 10% uncensored observations, or greater than 90% of observations in the population are below $3\,\mu g/L$. This one-sided test is conducted by Minitab with the same command as above, after setting a directional option for the alternative hypothesis in the Options dialog box. The percentile of censored observations p is set to 0.9, and the alternative is set to greater than 0.9. The data measured ($c = 24$) were the number of censored observations, so the null and alternative hypotheses are stated in terms of the proportion of censored observations expected. The result is

```
Test of p = 0.9 vs p > 0.9

                                        95%
                                        Lower        Exact
Sample      X     N     Sample p        Bound        P-Value
1           24    24    1.000000        0.882654     0.080
```

This tells us again that the lower bound of the 95% confidence interval for p is 88%, and a p-value for the test is 0.08. Because the proportion 0.9 was entered, this test determines whether the proportion of censored observations equals 0.9, just as stated in the first line of the output. Interpreting the p-value, 24 out of 24 nondetections would occur about 8% of the time when the true proportion of censored observations is 0.9. Eight percent is larger than the traditional 5% significance level, so the null hypothesis that there are 10% or more detections cannot be rejected. If we need to prove that there are fewer than 10% detections for legal or other purposes, more data are required. For example, after 6 more months and with all 30 observations below the reporting limit, the result for the same test would be

```
Test of p = 0.9 vs p > 0.9

                                        95%
                                        Lower        Exact
Sample      X     N     Sample p        Bound        P-Value
1           30    30    1.000000        0.904966     0.042
```

and since $p < 0.05$ the null hypothesis is rejected. As stated by the alternative hypothesis, fewer than 10% uncensored observations are expected from this population. Also note the direct correspondence between the p-value and the confidence bound. When the 95% (0.95) lower confidence bound is below (and so includes) the tested proportion of 0.9 as a possible proportion, the p-value will be above $(1 - 0.95)$ $= 0.05$. When as with the 30 observations the confidence bound is higher than (does not include) the tested proportion of 0.9, the p-value is below 0.05.

3. How many values above $3\,\mu g/L$ can be expected in a new set of 12 observations from this same population?

This one-sided prediction interval for the number of possible exceedances is computed using the hypergeometric distribution, a variation on the binomial distribution for sampling without replacement (Hahn and Meeker, 1991, page 113). The question is rephrased as "out of 36 total observations (N), how many exceedances could be expected to occur and yet none occur in the first $n = 24$ measurements?" Find the smallest number of exceedances that has no more than a 5% chance of producing 0 exceedances in the first 24 measurements. Using Minitab, if there were 3 exceedances in 36, the chance of observing 0 exceedances in the first 24 is 0.0308

```
MTB > InvCDF 0;
SUBC> Hypergeometric 36 3 24.

Hypergeometric with N = 36, M = 3, and n = 24
x  P( X <= x )       x  P( X <= x )
0          0         0    0.0308123
```

If there were 2 exceedances out of 36, the chance of observing 0 exceedances in the first 24 is 0.1047.

```
MTB > InvCDF 0;
SUBC> Hypergeometric 36 2 24.

Hypergeometric with N = 36, M = 2, and n = 24
x  P( X <= x )       x  P( X <= x )
0          0         0    0.104762
```

Therefore we could expect a 95% probability (actually a $1 - 0.03 = 97\%$ with this discrete method) that no more than 3 of the 12 new samples would have a concentration above $3\,\mu g/L$.

4. What is the probability that more than 1 (or some other number of) detected value(s) occur in the next 12 (or some other number of) samples?

This question specifies the number of uncensored observations as a "standard" or "quality control measure." One detect out of the next 12 will be allowed; more than 1 will not be. Of interest is whether the probability of failing the standard is less than a specified risk level α. If based on the 24 existing observations the probability of failing the standard in the next 12 observations exceeds α, some treatment might be implemented right away. Assume α is the traditional risk level of 5%. The probability of measuring 2 or more uncensored observations out of 12 new samples can be calculated with binomial tables.

To compute the probability of getting 2 or more uncensored observations, first compute the 95% confidence interval on the proportion of uncensored observations using the 24 existing values. From the answer to question 1, between 0 and 12% of new data can be expected to be uncensored observations, with 95% probability. The two percentages 0 and 12, the endpoints of the 95% confidence interval on $(1 - p)$, will be used to compute the risk of failing the standard.

For the percentage $(1 - p) = 12\%$ uncensored observations, the probability of measuring 2 or more observations out of the next 12 measurements is

$$\text{Prob}\,(x \leq 1) \; = \text{Binomial}\,(1,\, n_{\text{NEW}},\, [1-p]) = \text{Binomial}\,(1,\, 12,\, 0.12)$$

where the function Binomial is the binomial cumulative distribution function (equation 8.2):

$$\text{Binomial}\,(x',\, n,\, p) = \text{Prob}\,(x \leq x') = \sum_{i=0}^{x'} \left(\frac{n!}{i!(n-i)!} \right) p^i (1-p)^{n-i} \qquad (8.2)$$

where $n! = n(n-1) \times \cdots \times 1$.

Both of the 95% interval endpoints, 0 and 0.12, are used as p in equation 8.2 to determine the range of probabilities. Therefore, the probability that there will be 1 or fewer uncensored values in the next 12 samples is between

$$\sum_{i=0}^{1} \left(\frac{12!}{i!(12-i)!} \right) 0.12^i (0.88)^{n-i} = (1)1(0.88)^{12} + (12)0.12(0.88)^{11} = 0.569 \quad \text{for } p = 0.12$$

$$\sum_{i=0}^{1} \left(\frac{12!}{i!(12-i)!} \right) 0^i (1)^{n-i} = (1)1(1)^{12} + (12)0(1)^{11} = 1 \quad \text{for } p = 0$$

and so the probability of compliance, of getting 1 or fewer uncensored observations out of the 12 new observations, is between 57 and 100%. There is somewhere between a 0 and 43% probability of noncompliance. This is understandable given that if the true proportion of uncensored observations is as high as 12% (the upper end of the confidence interval), the expected number of uncensored observations is 1.4 out of 12, and so it should not be very unusual to see 2 uncensored observations. But the 43% probability is higher than the traditional risk level of 5%.

More information on computing binomial confidence, tolerance, and prediction intervals is found in Hahn and Meeker (1991).

8.3 EXCEEDANCE PROBABILITY FOR A STANDARD HIGHER THAN THE REPORTING LIMIT

Observing that all data fall below the reporting limit provides information about the likelihood of exceeding a legal limit at or higher than the reporting limit. If the value is below the reporting limit, it is also below the legal standard. Two approaches may be taken. In the first and simplest, the data are recorded as either exceeding or not exceeding the legal limit and the binomial methods of the previous section applied. The advantage of doing so is that no distribution need be assumed for the data. The disadvantage is that the proportion of data exceeding the reporting limit is only an upper bound on the proportion of data exceeding the standard. If there are no more than 12% uncensored observations, as in the previous section, then the probability of exceeding a higher legal limit is also no more than 12%. But it may be considerably smaller than 12%. How much smaller than 12% is unknown unless it is reasonable to assume a distribution for the data, allowing the difference in probabilities between exceeding the reporting limit and exceeding the legal limit to be modeled. This is the second approach.

8.3.1   Binomial (Nonparametric) Tests

The binomial test determines whether a percentile of the data distribution exceeds the legal limit at a stated confidence level. For the special case of testing the median or 50th percentile the test is also called the sign test, and is discussed in detail in Chapter 9.

To test whether the median of 24 arsenic concentrations, all below $3\,\mu g/L$, exceeds the legal limit of $10\,\mu g/L$, the null hypothesis is stated as the proportion of data above the legal limit of 10 is 0.50. In other words, the median is at the legal limit. The binomial test uses the number of times ("events") the data exceed the limit out of the total number of "trials." If there were many more exceedances than 50%, the null hypothesis would be rejected. In the arsenic example, all of the 24 "trials" are below both the reporting limit and the legal limit—there are 0 exceedances. Using the 1 Proportion routine within Minitab and testing the proportion of uncensored observations $= 0.5$ versus the alternative that it is greater than 0.5, the test results are

				95% Lower	Exact
Sample	X	N	Sample p	Bound	P-Value
1	0	24	0.000000	*	1.000

With a p-value of essentially 1, the binomial test states that there is no reason to reject the null hypothesis. There is no evidence that the median concentration is above the legal limit of 10. With all the observations below the reporting limit, this is not a surprising result.

If instead of the median a regulation states that another percentile of the distribution shall not exceed the standard, the quantile test (Conover, 1999) may be used. The quantile test is the binomial test applied to a proportion other than 50%, where the quantile $=$ the percentile/100. So the 50th percentile is the 0.5 quantile. Suppose that a regulation asserts that 90% or more of observations must be below the limit—the proportion of exceedances must be no more than 10%. The null hypothesis (of compliance) states that there are 10% or fewer exceedances, $p = 0.1$, with the alternative hypothesis that there are greater than 10% exceedances. The binomial test is run again using the proportion tested equal to 0.1 (10% exceedances) and the "event" being an exceedance, of which there are none. The result is

				95% Lower	Exact
Sample	X	N	Sample p	Bound	P-Value
1	0	24	0.000000	*	1.000

and so there is also insufficient evidence to reject the null hypothesis and declare that the legal standard of no more than 10% uncensored observations has been violated. If the true proportion were 10% exceedances, 2.4 exceedances would be expected. From the width of a binomial confidence interval, 5 exceedances would need to be observed to have enough evidence to prove that p was at least 10%, with 24 observations.

Of course the test against a standard could also be run assuming noncompliance, by reversing the direction of the alternate hypothesis. The burden of proof would be to determine if 0 out of 24 exceedances were enough evidence to declare that the probability of exceedance is less than 10%. The Minitab output for this perspective is

```
Test and CI for One Proportion
Test of p = 0.1 vs p < 0.1
                                      95%
                                      Upper        Exact
Sample      X     N     Sample p      Bound        P-Value
1           0     24    0.000000      0.117346     0.080
```

The p-value for the test is 0.08. This indicates that the probability of observing 0 out of 24 exceedances just due to chance is 8% when the true proportion of exceedances is 10%. If the acceptable error rate α is 0.05, the null hypothesis of noncompliance is not rejected. The small p-value would indicate a preference toward compliance, but the strength of the evidence is insufficient. If there were six more samples all of which were censored observations, the test becomes

```
Test and CI for One Proportion
Test of p = 0.1 vs p < 0.1
                                      95%
                                      Upper        Exact
Sample      X     N     Sample p      Bound        P-Value
1           0     30    0.000000      0.095034     0.042
```

and noncompliance is rejected at the acceptable error rate α of 0.05.

8.3.2   Parametric Tests of Exceedance

The parametric approach to comparing all censored observations to a standard is to estimate the exceedance probability of a standard as a function of the difference between the reporting limit and the legal standard. This requires an assumption about the shape of the data distribution. Smith and Burns (1998) present a method to estimate exceedance probabilities of a legal standard assuming a normal distribution when all data are censored observations. The assumption of a normal distribution is not commonly adhered to by environmental data. However, they state that it is more applicable for composite samples, where observations measured in the laboratory are the mean values of several individual samples composited together prior to analysis. Typically, 10–20 individual samples are combined into one composite sample in environmental studies. For the large skewness found in environmental data, composites of a small number of individual samples can still exhibit considerable skewness, so the assumption of a normal distribution for composite samples should be demonstrated.

A more reasonable assumption might be that data follow a lognormal distribution. In a brief side comment, Smith and Burns (1998) suggest a method for estimating exceedance probabilities when assuming data follow a lognormal distribution. The estimated lower bound for the percentile associated with the legal standard is a function of the difference between the standard (abbreviated Std) and the reporting limit (DL), divided by the standard deviation σ_L, all in log units. This standardized distance is subtracted from the normal quantile of the binomial exceedance probability, the probability calculated in question 1 for exceeding the reporting limit. Their estimator is

$$\Phi\left[\Phi^{-1}(\alpha^{1/n}) = \frac{(\ln[\text{Std}] - \ln[\text{DL}])}{\sigma_L}\right] \tag{8.3}$$

where Φ is the cumulative distribution function for the standard normal distribution and $(1 - \alpha)$ is the desired confidence level for the lower bound. This equation assumes that data follow a lognormal distribution, but more importantly that a reasonable estimate of σ_L can be obtained. When all data are below the reporting limit, obtaining a reasonable estimate of σ_L is unlikely. The Smith and Burns estimator in equation 8.3 comes from an unrefereed proceedings document rather than from a refereed journal article, and so should be approached with caution until validated by further work. Equation 8.3 agrees with their proceedings document. A minus sign rather than a plus sign appears between the two main components in Smith and Burns (1998); the minus sign is an error (D.E. Smith, personal communication, 2004).

Note that if a reasonable estimate of the standard deviation is doubtful, the first part of equation 8.3 is the binomial estimate of the lower confidence bound for the proportion of the population less than the reporting limit, previously produced by the 1 proportion software:

$$\Phi[\Phi^{-1}(\alpha^{1/n})] = \alpha^{1/n} = 0.05^{1/24} = 0.8826$$

This estimate requires no distributional assumption and so can be used instead of equation 8.3 as a lower bound for the proportion of the population less than the legal standard.

8.4   HYPOTHESIS TESTS BETWEEN GROUPS

The subject of hypothesis tests for differences in mean or median between two groups is covered at length in Chapter 9, and for three or more groups in Chapter 10. When one of the groups contains only censored observations, it can be a strong indication that there are differences among the groups. If so, the methods in those chapters can make that determination. If all groups consist entirely of censored observations, it is evidence that the distributions of data must be considered similar, at least within the analytical precision available to the scientist. With many censored observations it is difficult to determine whether data follow any specific distribution, so nonparametric tests are useful. Contingency tables (see Chapter 10) test differences between the proportions of uncensored observations among groups. These tests work well for one reporting limit. The score tests of Chapters 9 and 10 determine differences in the distribution functions of data with multiple reporting limits, even if one group contains all censored observations. For testing among groups there is no need to avoid using a nonparametric test because of "too many" censored observations. If the proportion of censored observations differs significantly between groups, nonparametric score and contingency table tests will respond to that difference.

8.5 SUMMARY

When all observations are recorded below the reporting limit, methods based on binomial probabilities can produce confidence intervals and hypothesis tests concerning the proportion and number of uncensored or censored observations in the population being measured. Though estimates of mean or median are not really available, statements about the probability of exceeding the reporting limit or other threshold can be made. The quantile test can determine whether a specified percentile is proven to be above or below a legal standard, even when all observations are censored observations. When testing data of all censored observations against other data containing some uncensored observations, binary and ordinal nonparametric methods such as contingency tables, the sign test, or the Kruskal–Wallis test (see Chapters 9 and 10) provide considerable power for determining whether one group generally produces higher values than another.

EXERCISES

8-1 Thurman et al. (2002) measured concentrations of antibiotics in discharges from fish hatcheries across the United States. A summary of the data are found in hatchery.xls. Twenty-five samples contained no concentrations of tetracycline above the reporting limit of $0.05 \, \mu g/L$. Two samples did contain detectable concentrations, but these were believed to be analytical artifacts from another compound, and the observations were discounted. Based on 25 censored observations, and assuming that these hatcheries represent the conditions found at others to be sampled in the future, what is the likelihood of getting at least one detection in the next 15 samples analyzed?

8-2 Assuming these 25 locations reasonably represent fish hatcheries across the United States, estimate a 90% confidence interval on the proportion of concentrations of tetracycline below $0.05 \, \mu g/L$ in waters draining fish hatcheries in the United States.

8-3 Use a contingency table analysis (see Chapter 10) to determine if the proportion of detections for oxytetracycline is significantly different than that for tetracycline.

8-4 MTBE in groundwater is a concern for drinking water supplies in states where the compound has been used as a gasoline additive. If in a survey of a county's drinking water supply wells, all 36 measurements have been recorded as below the reporting limit of 3 ppb, the data are assumed to be lognormal, and the standard deviation of the logarithms (based on other data) is estimated to be 1.0, what is an estimated probability of exceeding the "level of concern" of 13 ppb in groundwater?

9 Comparing Two Groups

In their classic paper on methods for censored data, Millard and Deverel (1988) studied copper and zinc concentrations in shallow groundwaters from two geologic zones underneath the San Joaquin Valley of California. Throughout this chapter these data will be used to illustrate whether methods can determine that zinc concentrations differ between the two zones. Zinc concentrations were subject to two reporting limits at 3 and 10 µg/L, with a number of values detected between these limits during the time when the lower reporting limit was in effect. Using nonparametric methods discussed later in this chapter, Millard and Deverel found that zinc concentrations were different in the two zones. Methods that work reasonably well should also find this difference. Censored boxplots of the two data sets are shown in Figure 9.1.

Comparing two groups is a basic design in many environmental studies. In some cases, a "treatment" group is compared to a "control." The control group represents background, or historical conditions. The treatment group represents conditions where, for example, contaminant concentrations are suspected to be higher, or numbers of healthy organisms lower. Differences are tested in one direction—treatment conditions are suspected to be worse in comparison to the control. Because a difference is expected in only one direction, these are called one-sided tests, tests where the direction of difference is specified as part of the study design. One-sided tests are also appropriate where the expected direction is an improvement over existing conditions—a new lab method with more accuracy, air concentrations following implementation of new scrubber technology, and so on. The key to a one-sided test is not in which direction the change is expected, but that there is only one direction expected.

In other cases, the two groups are inspected for differences where either may be better or worse than the other. Neither group can be labeled as a control group. The interest is truly in whether measurement levels in the two groups are the same, or different. These are two-sided tests—differences are investigated in two directions. Two-sided tests are appropriate for comparing the concentrations in different locations, for example, where if either location has significantly higher values the outcome is of interest. The zinc data are like this—the question stated was whether the two groups had *different* concentrations of zinc—no direction was specified. For either one to be higher is of interest. This is a two-sided test.

Statistics for Censored Environmental Data Using Minitab® and R, Second Edition.
Dennis R. Helsel.
© 2012 John Wiley & Sons, Inc. Published 2012 by John Wiley & Sons, Inc.

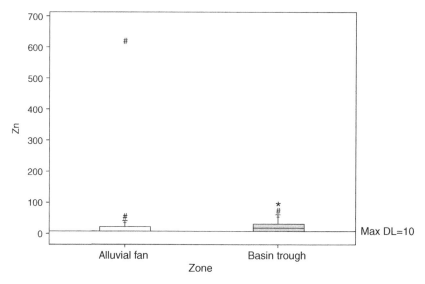

FIGURE 9.1   Boxplots of zinc (Zn) concentrations. Note that a greater percentage of the Basin Trough data are above the higher 10 μg/L limit.

For both one and two-sided tests, the procedures used when there is no censoring are familiar to data analysts—the parametric two-sample t-test, and the nonparametric Mann–Whitney (or its alternate name, the Wilcoxon rank-sum) test. When censored values are present, options for analysis include binary methods, ordinal nonparametric methods, and both parametric and nonparametric methods of survival analysis.

9.1   WHY NOT USE SUBSTITUTION?

The only way that a standard two-sample t-test can be run on data with censored observations is to fabricate (substitute) values prior to computing the test. One approach used in some environmental reports (please do not claim that it is supported here!) is to run the test twice, first substituting zero, and then the reporting limit. The argument goes that if the results of the two tests agree, then perhaps these are the correct results. Perhaps, but this is far from a sure thing. Though the mean will change monotonically as the substituted value goes from the reporting limit down to zero, the standard deviation will not (see Chapter 1). The t-test statistic is a function of both the mean and the standard deviation, and will vary in an indeterminate pattern when different values are substituted. Regardless of which value is substituted, the process implies that more "information" (the invasive pattern caused by substituted values) is known about the data than is truly known by the analyst. Because this "information" is not correct, the test result will likely not be correct either.

Setting all censored observations equal to zero, the t-test does not indicate a difference between zinc concentrations within the two zones ($p = 0.995$):

```
Two-sample T for Zn0

Zone                        N       Mean         StDev        SE Mean
Alluvial Fan                67      21.2         75.0         9.2
Basin Trough                50      21.3         19.3         2.7

Difference = mu (Alluvial Fan) - mu (Basin Trough)
Estimate for difference: −0.06
95% CI for difference: (−19.08, 18.97)
T-Test of difference = 0 (vs not =): T-Value = −0.01 P-Value = 0.995
DF = 77
```

Setting all censored observations equal to their reporting limits (some 3, some 10 µg/L), the *t*-test again indicates no difference between the groups ($p = 0.869$):

```
Two-sample T for Zn_dl

Zone                        N       Mean         StDev        SE Mean
Alluvial Fan                67      23.5         74.4         9.1
Basin Trough                50      21.9         18.7         2.6

Difference = mu (Alluvial Fan) - mu (Basin Trough)
Estimate for difference: 1.57
95% CI for difference: (−17.29, 20.43)
T-Test of difference = 0 (vs not =): T-Value = 0.17 P-Value = 0.869
DF = 76
```

Does this mean that there is no true difference between the two groups of zinc concentrations? No, it certainly does not. It was previously mentioned that the correct result is that the distribution of concentrations does differ between the two groups. Neither extreme of substituted value is likely to be close to the true concentrations actually present in the samples. As a third attempt at substitution, the common method of setting all censored observations equal to one-half of their reporting limits is tried. This time, logarithms are taken following substitution to address the skewness of the data. Data with values close to zero are usually skewed, lowering the power of parametric *t*-tests to detect differences. A *t*-test on logarithms following one-half DL substitution again produces a nonsignificant *p*-value. The *t*-test fails to find differences in the mean of the logarithms following substitution ($p = 0.109$):

```
Two-sample T for lnZn1/2

Zone                        N       Mean         StDev        SE Mean
Alluvial Fan                67      2.444        0.816        0.10
Basin Trough                50      2.707        0.911        0.13

Difference = mu (Alluvial Fan) - mu (Basin Trough)
Estimate for difference: −0.263101
95% CI for difference: (−0.586284, 0.060081)
T-Test of difference = 0 (vs not =): T-Value = −1.62 P-Value = 0.109
DF = 98
```

Clearly, if the true result is that there are differences between the zinc concentrations in these two geologic zones, substitution followed by a *t*-test is an inadequate procedure for detecting it, regardless of the value being substituted.

9.2   SIMPLE NONPARAMETRIC METHODS AFTER CENSORING AT THE HIGHEST REPORTING LIMIT

In Chapter 6, a small example data set of 11 observations was presented, along with the binary coding of being less than (LT) or greater than or equal to (GE) the highest reporting limit of 5:

<1	<1	3	<5	7	8	8	8	12	15	22
LT	LT	LT	LT	GE	GE	GE	GE	GE	GE	GE

Now consider a second data set of 12 observations and its binary classification at the reporting limit of 5. Are these data sets significantly different?

<1	<1	2	3	3	<5	<5	<5	<5	7	8	10
LT	LT	LT	LT	LT	LT	LT	LT	LT	GE	GE	GE

9.2.1   Binary Methods

The percent of "uncensored observations"—detects at a specified single threshold—in two or more groups can be tested to determine whether they are the same (null hypothesis), or not the same (alternate hypothesis). This test of binomial proportions is given either the name "test of proportions" or "contingency table test." The test determines whether the proportion of data above versus below the threshold is contingent upon (changes with) the group classification.

There are 3 out of 12 or 25% uncensored observations above 5 in the second data set. Is this 25% significantly different than the 64% of the first group (Figure 9.2)? The test of proportions can be computed using the %detects macro, part of the NADA for Minitab® collection.

```
Pearson Chi-Square = 3.486, DF = 1, P-Value = 0.062
Likelihood Ratio Chi-Square = 3.576, DF = 1, P-Value = 0.059
```

The *p*-values just above 0.05 indicate that there is not enough of a difference between the two proportions in this small collection of data. The proportions are almost, but not quite, discerned as different. Comparisons of proportions between three and more groups can also be performed with this test and macro, without substitution.

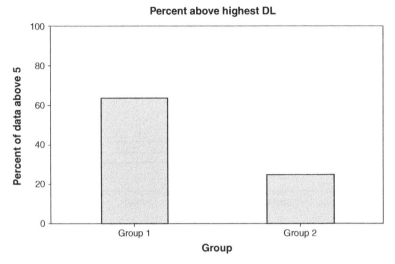

FIGURE 9.2   The percent of observations above 5 in two groups.

9.2.2 Ordinal Methods

The standard two-group nonparametric test is the Mann–Whitney test, also called the rank-sum test. It is found in most introductory statistics texts. Like other nonparametric methods it can be directly applied to censored data once all values below the highest reporting limit are considered tied. The first step in performing the Mann–Whitney test is to jointly rank the data from 1 to $n + m$, where n is the number of data in first group and m in the second group. For the two small data sets previously listed, their 23 observations produce the following joint ranks.

Group 1 Data:	<1	<1	3	<5	7	8	8	8	12	15	22
Ranks:	7	7	7	7	14.5	17.5	17.5	17.5	21	22	23

Group 2 Data:	<1	<1	2	3	3	<5	<5	<5	<5	7	8	10
Ranks:	7	7	7	7	7	7	7	7	7	14.5	17.5	20

All 13 observations below 5 are ranked as tied with each other, and so are assigned ranks of 7, the median (and mean) of numbers 1–13. The detected values above 5 then receive differing ranks, unlike the ordinal test where they are considered identical. The two detected 7s are also tied. They would receive the ranks 14 and 15 if they could be distinguished from each other, but since not they both are assigned a rank of 14.5. The four detected 8s are similarly assigned the median of ranks 16—19, or 17.5. Above this the remaining values are untied, and receive individual ranks 20 through 23. The Mann–Whitney test sums the ranks in one of the groups, and if the sum is unexpectedly low or high, will reject the null hypothesis. If the sum is a moderate number, where moderate is a function only of the sample size, the null hypothesis will not be rejected. With ties, commercial software should include a tie correction for computation of the p-value. Minitab's output for these data

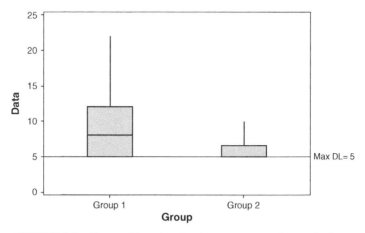

FIGURE 9.3   Censored boxplots for the two groups of example data.

```
W = 161.0
Test of ETA1 = ETA2 vs ETA1 not = ETA2 is significant at 0.0794
The test is significant at 0.0520 (adjusted for ties)
```

shows a p-value of 0.052 after correcting for ties. The null hypothesis of similarity between the data in these two groups cannot be rejected at the usual alpha of 0.05, though the p-value is extremely close to the rejection criteria. No data were fabricated to perform this test, unlike substitution followed by something like a t-test. The data are illustrated with censored boxplots in Figure 9.3.

To account for the variation in data below the highest reporting limit of 5, survival analysis methods will be discussed in sections 9.3 and 9.4. If simplicity is paramount, however, this ordinal method is simple and yet, unlike the t-test, does not require you to fabricate invasive values.

9.2.3   Ordinal Methods Versus Substitution by ROS Prior to Hypothesis Testing

When there is only one reporting limit, standard (ordinal) nonparametric tests such as the rank-sum test can be computed directly from the data. When the test converts data to their ranks, censored observations are represented as a tied rank lower than the rank for the lowest detected observation. These ranks will efficiently capture the information in the data, including the proportion of censored observations, accurately representing what is actually known about the data. Test results are reliable, not based on "information" that is not known, and not dependant on the substitution of arbitrary values.

Evaluations of methods for testing differences between groups of censored data are scarce. However, in a noteworthy paper, Clarke (1998) evaluated 10 methods for comparing two groups of singly censored data. Sample sizes were small, less than 10 observations in each group. Methods included substitution, and several "robust" methods estimating single values for censored observations by MLE or ROS (presumably using normal scores, but details were not included). Following the creation of artificial values for each censored observation, the two groups of data were evaluated by the least significant difference (LSD) multiple comparison test, essentially a two-group t-test. In addition to the standard normal-theory test, the LSD test was computed on the logarithms of the data, and on the "rankits," a rank transformation. The best test in the greatest number of cases was one using rankits after substituting tied values (1/2RL) for nondetects.

Given the small sample sizes of the study, MLE methods may not have been expected to work well. But the good performance of substitution of a single number over "robust" methods seems to contradict the findings of Chapter 6. Are they contradictory? No, but the choice of method names in Clarke's study was unfortunate. The "rankit substitution" method was in essence the rank-sum test.

First, simple substitution of a single tied low value for censored observations followed by converting to rankits and then testing with LSD is in essence the Mann–Whitney rank-sum test, the ordinal nonparametric method recommended here. Parametric tests on ranked data approximate their equivalent nonparametric procedures (Conover and Iman, 1981). Substitution of a single value followed by a nonparametric procedure *is the ordinal method* when there is one reporting limit. The problems with substitution occur either when different values are assigned to observations that are all "<1," or parametric tests are used. Parametric tests require an estimate of how far less a <1 is below a 5. Nonparametric methods do not.

Second, all the other methods that performed poorly somehow assigned unequal values to censored data that were known only to be equal. In essence, Clarke found that any method of assigning unequal values to individual samples observed as "<RL" could produce a signal that was not in the original data. These fabricated values could also obscure a signal that was present. They are invasive data. This includes the "placeholder" estimates used internally to estimate summary statistics by ROS. Millard and Deverel (1988) stated a similar finding when methods generated uniform placeholder values between 0 and the reporting limit, prior to hypothesis testing. Clarke's paper is strong evidence for using nonparametric methods on small censored data sets, and for NOT using methods such as ROS or uniform distributions to assign unequal numbers to identical censored samples followed by parametric hypothesis tests.

9.2.4   Example: Rank-Sum Test for Data with One Reporting Limit

Dissolved organic carbon (DOC) concentrations were measured by Junk et al. (1980) in background wells and in other wells affected by cropland irrigation, in September 1978 (Figure 9.4). Of interest is whether concentrations are higher in the wells affected by irrigation. Three values are censored at the single reporting limit of 0.2 µg/L. The data are found in doc.xls and shown in Table 9.1.

The three censored observations would have had the ranks of 1 through 3 if analytical precision had allowed their concentrations to be quantified. However, with the available precision the three are tied with each other. Therefore each is assigned the rank of 2, the median of ranks 1 through 3. The Mann–Whitney (or rank-sum) test can be easily applied to these data, without any changes. This should be set up as a one-sided test—the question was "Are concentrations *higher* in wells affected by irrigation?"—so a difference in only one direction is of interest. The results below

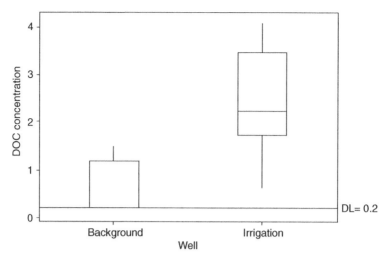

FIGURE 9.4   DOC concentrations in two types of wells. (Data from Junk et al., 1980.)

show that DOC concentrations are higher in irrigation-influenced wells, with a
p-value of 0.0064. No substitutions or assumptions of a distributional shape for
these data were required to run the test.

```
Mann-Whitney Test and CI: Irrigation, Background
Irrigation                          N = 10            Median =    2.250
Background                          N = 4             Median =   < 0.200
Point estimate for ETA1-ETA2 is     2.000
96.0 Percent CI for ETA1-ETA2 is    (0.601,3.700)
W = 93.0

Test of ETA1 = ETA2 vs ETA1 > ETA2 is significant at 0.0067
The test is significant at 0.0064 (adjusted for ties)
```

TABLE 9.1   DOC Concentrations in Two Sets of Wells

Background	Irrigation	Background Rank	Irrigation Rank
< 0.2	3.4	2	12
1.5	1.9	6	7.5
< 0.2	3.7	2	13
< 0.2	2.1	2	9
	3.2		11
	2.4		10
	1.2		5
	4.1		14
	1.9		7.5
	0.6		4

From Junk et al. (1980).

9.3   MAXIMUM LIKELIHOOD ESTIMATION

Standard parametric tests for differences between groups of uncensored data, such as analysis of variance (ANOVA) and t-tests, can be computed using simple linear regression where the explanatory variables are coded to indicate group membership. For the simple case of testing the difference between the mean of two groups, only one explanatory variable (X) is needed. X is a binary variable, coded as a 0 if the data come from the first group, and a 1 if the data are from the second group. Solving the regression produces an estimate for the slope of the X variable. This slope equals the difference between the two group means. The regression t-test of significance for this slope (assuming equal group variances) is the test to determine whether this difference equals zero. In this way, a two-sample t-test can be computed using regression software. This type of procedure will now be used to test for differences in groups of censored data.

Computation of parametric hypothesis tests between group means of censored data is accomplished using software for censored regression. Censored regression methods use maximum likelihood estimation to compute estimates of slope and intercept, and to conduct hypothesis tests on the significance of the slope coefficient. The benefits in using MLE methods include the fact that they work for data with multiple reporting limits, and do not require substitution of fabricated data in order to perform the tests. The caution with MLE methods is that the validity of their results depends on selecting the correct distribution. For a small amount of censoring, the fit of data to the distribution can be evaluated with probability plots or hypothesis tests. For larger amounts of censoring, it is difficult to judge which type of distribution the data might have come from. In this case, use of parametric MLE methods requires that the choice of distribution be based on other knowledge, such as distributions for similar data in previous studies. For environmental data, the lognormal distribution is most often assumed when evidence is not available in the data themselves.

As noted in Chapter 6, the usefulness of MLE methods is limited if the sample size is small. In addition to the difficulty in determining whether small data sets follow a particular distribution, the optimization procedures of MLE do not have enough information with small data sets to accurately estimate parameters. As shown by many simulation studies (including Helsel and Cohn, 1988; Gleit, 1985) errors associated with MLE methods increase dramatically as sample sizes decrease below 50 or so observations. For smaller data sets, methods other than MLE are recommended.

To estimate a slope and intercept using maximum likelihood, possible values for the two parameters are adjusted until the values most likely to produce the observed measurements are determined. Matching of possible parameters to observations is done through the likelihood function. Parameters chosen by MLE maximize the likelihood function for censored regression, evaluated by setting the derivatives of the likelihood function L with respect to each parameter equal to zero, simultaneously solving the two equations:

$$\frac{d\,L(\beta)}{d\beta_0} = 0 \qquad \frac{d\,L(\beta)}{d\beta_1} = 0 \qquad\qquad (9.1)$$

where β_0 is the intercept and β_1 is the slope.

MLE also estimates the standard errors of both parameters, so that confidence intervals around each can be constructed. Standard errors of coefficients are the square root of the entries on the main diagonal of the covariance matrix (Allison, 1995, p. 84). The distributional assumption for the procedure has a strong influence on the resulting confidence intervals, much as the normality assumption does for intervals around the parameters for simple least-squares regression. If a normal distribution is assumed, symmetry of the input data is critical for the results to make sense. If the input data are skewed, variance estimates will be too large, tests for parameters will be too often insignificant, and confidence intervals may have negative lower bounds even when this is physically impossible. When these difficulties occur, a log or other transformation should be considered before using MLE.

Due to the skewness of many environmental variables, the assumption of normality for data is often not a good one. In many cases, the logarithms of data more closely follow a symmetric distribution than do the original data. In hypothesis testing, taking logarithms prior to conducting the test is a common practice, for uncensored as well as censored data. Yet doing so changes the hypothesis being tested. For two-group tests in original units, the null hypothesis is that means of two distributions are identical. The alternative hypothesis is that they differ by an additive constant. With logarithms, the null hypothesis becomes a statement concerning the means of the logarithms of the data. The alternative hypothesis that the mean logarithms are offset by an additive constant translates to multiplication in the original units. When logarithms are used in a two-group test, the null hypothesis is therefore that the ratio of the geometric means (the mean logarithms exponentiated back to original units) equals one. The alternative is that this ratio is not equal to one, and so the groups differ by a multiplicative constant. Finally, if the distribution of the logarithms is symmetric, as would be hoped in order to assume normality, the geometric mean estimates the median of the original units, not the mean. A two-group parametric test on the logarithms becomes a test for equality of medians. This is true whether the test is a standard t-test for uncensored data, or a censored regression with parameters computed by maximum likelihood.

9.3.1 Example: The Zn Data

Censored regression is found in the survival analysis section of statistical software. Correct results for left-censored data require their entry as arbitrarily or interval-censored values using the interval endpoints format. The endpoints span the range of possible values for the censored data. In Minitab this is accomplished by entering two columns as response variables. The explanatory variable is a column named GeoZone with 0 for data from the Basin Trough and 1 for data from the Alluvial Fan. The censored regression procedure, found in Minitab under the Reliability/Survival menu, tests whether the mean zinc concentration in groundwater of the two zones differs, using maximum likelihood. The output from the procedure is

```
Estimation Method: Maximum Likelihood
Distribution: Normal
Relationship with accelerating variable(s): Linear

Regression Table
                      Standard                    95.0% Normal CI
Predictor   Coef      Error       Z      P        Lower      Upper
Intercept   21.6125   8.11997     2.66   0.008    5.69774    37.5274
GeoZone     0.762958  10.7311     0.07   0.943    -20.2697   21.7956
Scale       57.4123   3.75329                     50.5078    65.2607

Log-Likelihood = -596.254
```

The slope coefficient value of 0.76 estimates the difference between the two group means, and the Z statistic (0.07) tests whether 0.76 is significantly different from zero. The test is equivalent in function to the t-test for explanatory variables in simple least-squares regression. The p-value of 0.943 indicates that the null hypothesis of no difference cannot be rejected—there is insufficient evidence of a true difference in mean Zn concentrations between the two zones using this test. "Using this test" is an important qualifier.

This test assumed that both groups of Zn concentrations follow a normal distribution. That assumption can be checked using a probability plot, after the group mean is subtracted from both group's data. Differences between the observed data and their group mean are called residuals, and it is these residuals that are assumed to follow a specific distribution by parametric tests. The probability plot of residuals in Figure 9.5 appears nonlinear, indicating that the data do not follow a normal distribution. Therefore, the test might have failed to indicate a true difference because it had low power resulting from non-normal data.

Note that for software that does not include an interval-endpoint format for use with left-censored data, the data could be flipped and run as right-censored values. However, this will not produce the appropriate test. Right-censored survival analysis methods assume data have no upper bound. Values up to positive infinity convert back to concentrations down to negative infinity on the original scale; the lower bound of zero is not recognized. Without this lower bound, test results are incorrect. When using MLE procedures assuming a normal distribution, left-censored data must be entered as interval-censored values. However, as shown later, this is not the case for tests assuming a lognormal distribution.

Given that these data are right skewed, a log transformation will be employed prior to running another two-group test. The test will determine whether the mean of logarithms (the geometric mean) differs between the two groups. Minitab and other statistical software can do this step automatically by creating a likelihood function for the lognormal distribution. The assumed distribution in the dialog box is changed from normal to lognormal. For software that is unable to perform maximum

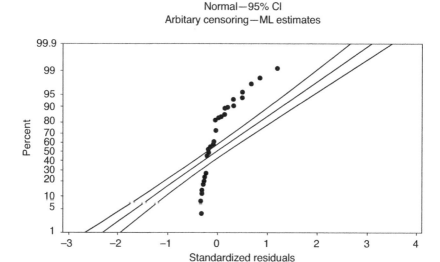

Normal—95% CI
Arbitary censoring—ML estimates

FIGURE 9.5 Probability plot of residuals from the censored regression of Zn concentrations against group membership. The data do not appear to follow a normal distribution.

likelihood for a lognormal distribution, data can be log-transformed and the normal distribution assumed for the logarithms.

For the zinc example the lognormal distribution is selected as the assumed distribution, and the results for the censored regression procedure are

```
Estimation Method: Maximum Likelihood
Distribution: Lognormal
Relationship with accelerating variable(s): Linear

Regression Table
                    Standard                  95.0% Normal CI
Predictor  Coef     Error      Z      P       Lower      Upper
Intercept  2.72375  0.120325   22.64  0.000   2.48792    2.95958
GeoZone    -0.257408 0.161211  -1.60  0.110   -0.573377  0.0585604
Scale      0.842529 0.0618053                 0.729698   0.972806

Log-Likelihood = -407.296
```

The Z statistic for GeoZone has a much lower p-value (0.11) than when the normal distribution was assumed, but this value still leads to a conclusion of no difference between the two groups' geometric means. A probability plot (Figure 9.6) indicates that most data follow the straight line representing a lognormal distribution, though one or two points are outside the 95% confidence interval boundaries for the location of the distribution. Concern that there may be an effect due to violating the distributional assumption could lead to either use of yet another distribution, or to using a nonparametric approach.

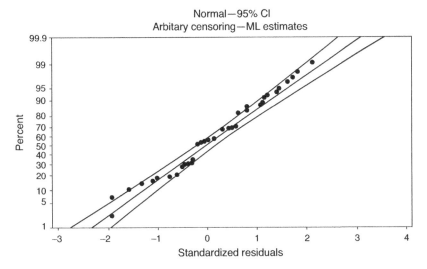

FIGURE 9.6 Probability plot for residuals from a regression of the Zn data versus group membership, compared to a lognormal distribution (center straight line).

When software does not allow input of interval-censored data, log-transformed data can be flipped and the procedure run. The result will be identical to a test on log-transformed data (or when assuming a lognormal distribution) had the interval endpoints format been available. The procedure is to take logarithms, flip the data by subtracting from a value larger than the largest logarithm, and run the right-censored regression. The lack of an upper boundary for right-censored data will map into negative infinity when the data are reflipped back to the original log scale. Negative infinity for logarithms becomes a lower limit of zero when data are exponentiated back into units of concentration. For data with a lower bound of zero, censored regression can be correctly performed by software that does not allow interval-censored input by first taking logarithms, and then flipping to produce the required right censoring.

To illustrate the process when regression on right-censored data is the only option, the natural logs of Zn are computed, and then subtracted from 7, a number larger than the maximum logarithm, to produce right-censored data. The values are stored in a column named "FliplnZn." The indicator column format is used by the software to designate which values of FliplnZn are censored. The MLE regression output for the flipped natural logs of Zn includes

```
Distribution: Normal
Response Variable: FliplnZn
Regression Table
                         Standard                        95.0% Normal CI
Predictor    Coef        Error      Z        P          Lower      Upper
Intercept    4.2763      0.1204     35.52    0.000       4.0403     4.5122
GeoZone      0.2575      0.1613     1.60     0.110      -0.0587     0.5736
Scale        0.84292     0.06188                         0.72996    0.97335
```

The slope coefficient for GeoZone and test results agree with those above when untransformed data were input as interval endpoints and a lognormal distribution was assumed.

In summary, for data with one or multiple censoring thresholds a two-group parametric test can be computed using censored regression, where a binary 0/1 variable is the only explanatory variable. Care must be taken, however, to judge the validity of the test's assumptions of a normal distribution and equal variance. Censored data are frequently non-normal. Violating the normality assumption will result in a loss of power to detect differences between the two groups. Environmental data are often fit better by lognormal distributions, and taking logarithms has the added benefit of allowing software with input only of right-censored data to correctly test data with a zero lower bound. However, the user must be aware that transforming data to logarithms (or assuming a lognormal distribution) changes the basic form of the test, determining whether the two groups' geometric means (medians) differ from a ratio of one.

The same test is available using NADA for R, where CuZn is one of the data sets that comes with the package. The cenmle function can be used to perform the parametric test of whether there is a significant difference in the mean of the zinc concentrations in the two zones. Internally, it performs a censored regression on a binary group as the explanatory variable, estimating the slope using maximum likelihood. Note that now the grouping variable (Zone) must be specified as the explanatory variable in the third argument.

```
> cenmle(Zn, ZnCen, Zone, dist="gaussian")
                Value       Std. Error       z           p
(Intercept)     22.376      7.0160           3.1892      0.00143
ZoneBasinTrough -0.763      10.7312          -0.0711     0.94332
Log(scale)      4.050       0.0654           61.9549     0.00000

Scale = 57.4

Gaussian distribution
Loglik(model) = -596.3  Loglik(intercept only) = -596.3


Loglik-r: 0.006572786
Chisq= 0.01 on 1 degrees of freedom, p= 0.94
Number of Newton-Raphson Iterations: 1
n =117 (1 observation deleted due to missingness)
```

The likelihood-ratio p-value of 0.94 (a two-sided p-value) indicates that the means for the two groups are not different. Using logarithms, the same procedure finds no significant difference in the mean of the logarithms, with a p-value of 0.11

```
> cenmle(Zn, ZnCen, Zone, dist="lognormal")
                Value       Std. Error       z           p
(Intercept)     2.466       0.1078           22.88       6.65e-116
ZoneBasinTrough 0.257       0.1613           1.60        1.10e-01
Log(scale)      -0.171      0.0734           -2.33       1.99e-02

Scale = 0.843
```

```
Log Normal distribution
Loglik(model) = -407.3  Loglik(intercept only) = -408.6
Loglik-r: 0.1467494
Chisq= 2.55 on 1 degrees of freedom, p= 0.11
Number of Newton-Raphson Iterations: 3
n =117 (1 observation deleted due to missingness)
```

The slope coefficient for Basin Trough zone of 0.257 measures the difference in the mean logarithm between the two groups. So in natural log units, the mean of the Basin Trough group is 0.257 log units higher than the mean of the Alluvial Fan group. The group not written out in the output is the "base" group to which the named group is being compared. This translates into a ratio of their geometric means, so that the Basin Trough group has a geometric mean (median) averaging $e^{0.257} = 1.29$ times that for the Alluvial Fan group. However, the nonsignificant hypothesis test states that this increase between geometric means of 29% is not significantly different from no change, which is the null hypothesis.

9.4   NONPARAMETRIC METHODS

Nonparametric methods do not require an assumption that data follow a specific distribution. They use no information on the shape of the distribution in conducting tests. Instead, data are ranked, providing information on the relative positions of each observation. Tests determine whether one group generally has more frequent high or low values, a test for whether percentiles of the data differ between the groups. For censored data, positions are represented by scores, which are ranks of the data adjusted for the information missing because some values are censored. Nonparametric tests using these methods are called score tests.

9.4.1   Ordinal Nonparametric Test After Censoring at the Highest Limit

An ordinal nonparametric test can always be computed on censored environmental data by censoring all values below the highest reporting limit to a common value. The advantage of this procedure is that it uses standard software. The disadvantage is that it loses information in comparison to score tests, described below, and so has less power than those more complicated procedures. However, when there are sufficient differences between groups, this simple method may provide all the power needed to reject the hypothesis of no difference. As with other nonparametric methods, the assumptions of a normal distribution and equal variance are not necessary. These assumptions are difficult to check with censored data, as the entire distribution of data cannot be determined.

For the zinc data from the San Joaquin Valley, all <3s, all <10s, and all detected values between 3 and 10 are considered to be <10, resulting in tied ranks for all, and the rank-sum test performed. Censoring data in this way uses the same information portrayed in the censored boxplot of Figure 9.1, where only the frequencies of observations below 10 in each group, and not the values themselves, are used to compute the positions of the box. Each of the 20 observations (30%) in the Alluvial Fan zone below 10 µg/L, and the 12 observations (24%) in the Basin Trough zone below 10, is assigned a rank of 16.5, the median of numbers 1 through 32. Then the Mann–Whitney test is computed. It is essential that software be able to compute tie corrections to test statistics for the many tied values resulting from censored observations.

The Mann–Whitney test on these recensored zinc data finds a definite and reliable difference in the distributions of zinc concentration between the two zones ($p = 0.0185$). Median concentrations are higher in the Basin Trough zone. Unlike MLE methods, no distributional assumption was required, and therefore no loss of power results if the data are non-normal (which they are).

```
Mann-Whitney test for ZnMaxDL

Alluv_Zn          N =        67      Median =       10.00
Basin_Zn          N =        50      Median =       18.50
Point estimate for ETA1-ETA2 is -5.00
95.0 Percent CI for ETA1-ETA2 is (-10.00,-0.00)
W = 3533.5
Test of ETA1 = ETA2 vs ETA1 not = ETA2 is significant at 0.0210
The test is significant at 0.0185 (adjusted for ties)
```

9.4.2 Survival Analysis: Score Tests

Score tests are nonparametric tests that determine whether distribution functions differ among groups of censored data. The distribution function of each group is pictured by its edf or survival function. Some score tests are direct extensions of the rank-sum test, including the "generalized Wilcoxon test," the "Peto–Prentice test," and the "Gehan test." If uncensored data are input to these Wilcoxon-type tests, the results are similar to the Wilcoxon rank-sum (Mann–Whitney) results. Score tests were designed to handle data censored at multiple reporting limits, using the information contained in uncensored values between reporting limits in addition to the information in the proportion of values below each reporting limit. The survival function for left-censored environmental data estimates the cumulative distribution function i/n, where values for i are the ranks of data from smallest to largest. For values below a detection threshold, the survival probability i/n cannot be calculated because the ranks of these values are not known. But the scores or ranks of detected observations are computed by taking into consideration the presence of censored observations below them.

9.4.3 The Gehan Test

To illustrate exactly how a score test works, the Gehan test statistic (Gehan, 1965) will be manually computed for cadmium concentrations in fish livers (Cd.xls) in the Southern Rocky Mountain and Colorado Plateau physiographic regions. This data set is small and therefore easy to use for demonstrating how the Gehan test works. The rocks of the Southern Rocky Mountains are more mineralized, and contain considerably more trace metals including cadmium, than do rocks of the Colorado Plateau. Of interest is whether fish livers show this same pattern, indicating that concentrations in the host rock show up in stream waters, and subsequently in the biota of the two regions. The null hypothesis is that there is no difference in cadmium concentrations in livers of fish from the two regions. The alternative hypothesis is that concentrations in fish from the Southern Rocky Mountains are higher than those from the Colorado Plateau, a one-sided test.

There are 9 observations from the Colorado Plateau region, and 10 observations in the Southern Rocky Mountain region. The data are listed in the top row and left column of Table 9.2.

For the $n = 9$ values x_1 to x_n in the Colorado Plateau, and the $m = 10$ values y_1 to y_m in the Southern Rocky Mountains, there are $n \cdot m = 90$ possible comparisons. These comparisons, shown in the center cells of the table, are called U_{ij}, where

$$U_{ij} = \begin{array}{ll} -1 & \text{for} \quad x_i > y_j \ (y_j \text{ may be a nondetect}) \\ +1 & \text{for} \quad x_i < y_j \ (x_i \text{ may be a nondetect}), \text{ and} \\ 0 & \text{for} \quad x_i = y_j \text{ or for indeterminate comparisons } (< 10 \text{ to a } 5) \end{array} \quad (9.2)$$

The Gehan test statistic W is the sum of the values for U_{ij}, or

$$W = \sum_{i=}^{n} \sum_{j=}^{m} U_{ij} \quad (9.3)$$

For Table 9.2, there are 75 + 1s and 12 − 1s, so $W = 63$. If the null hypothesis were true, approximately half of these comparisons would be positive, and half negative, and W would be something close to 0. After computing W, it is standardized by dividing by a measure of its standard error (the square root of the variance),

$$Z = \frac{W}{\sqrt{\text{Var}[W]}} \quad (9.4)$$

TABLE 9.2   Comparison of Cadmium Concentrations in Fish Livers of the Rocky Mountain Region

	Southern Rocky Mountains									
CO Plateau	81.3	4.9	4.6	3.5	3.4	3	2.9	0.6	0.6	<0.2
1.4	1	1	1	1	1	1	1	−1	−1	−1
0.8	1	1	1	1	1	1	1	−1	−1	−1
0.7	1	1	1	1	1	1	1	−1	−1	−1
<0.6	1	1	1	1	1	1	1	1	1	0
0.4	1	1	1	1	1	1	1	1	1	−1
0.4	1	1	1	1	1	1	1	1	1	−1
0.4	1	1	1	1	1	1	1	1	1	−1
<0.4	1	1	1	1	1	1	1	1	1	0
<0.3	1	1	1	1	1	1	1	1	1	0

and the ratio Z is compared to a standard normal distribution. When the two distributions are not equal, U, W, and Z are large in absolute value, and the null hypothesis is rejected. Note that if there were no censoring, all $n \cdot m$ comparisons could be quantified, and U would equal the rank-sum test statistic.

Two alternative formulae for the variance are common, the permutation and the hypergeometric estimates. Most software uses the hypergeometric variance. Permutation estimates are simpler, but are invalid when the censoring rate differs between the groups (one group has generally higher reporting limits than the other).

The permutation variance of W is:

$$\text{Var}[W] = \frac{mn\Sigma h^2}{(m+n)(m+n-1)} \tag{9.5}$$

In this equation, the sample sizes m and n are the numbers of observations in the two groups. h^2 is the sum of squared values for h, the u-score related to the ranks of the data. To compute h (Table 9.3), count the number of observations known to be greater than each observation (G), and the number known to be less than each observation (L), when the data are sorted in ascending order (see Table 9.3). The difference between these two numbers is the u-score, h,

$$h = G - L \tag{9.6}$$

TABLE 9.3   Calculations for the Variance of the Gehan Test Statistic

Cd	G	L	h	h^2
<0.2	15	0	15	225
<0.3	15	0	15	225
<0.4	15	0	15	225
0.4	12	3	8	64
0.4	12	3	8	64
0.4	12	3	8	64
<0.6	12	0	12	144
0.6	10	7	3	9
0.6	10	7	3	9
0.7	9	9	0	0
0.8	8	10	−2	4
1.4	7	11	−4	16
2.9	6	12	−6	36
3	5	13	−8	64
3.4	4	14	−10	100
3.5	3	15	−12	144
4.6	2	16	−14	196
4.9	1	17	−16	256
81.3	0	18	−18	324
			$\Sigma =$	2169

which when squared and summed over all data, equals $\sum h^2$. For the cadmium data, $\sum h^2$ equals 2169. The variance of W is $(9 \times 10 \times 2169)/((19)(18)) = 570.79$. The standard error of W is the square root of the permutation variance, or 23.89. Gehan (1965) provides a more complicated formula when ties occur, as they do here. Due to the three 0.4s and two 0.6s, the tie-corrected standard error becomes 23.53.

The Gehan test statistic Z is therefore

$$Z = \frac{63}{23.53} = 2.68$$

producing a p-value from the standard normal distribution $= 0.0037$. The conclusion is therefore to reject the null hypothesis of equality, finding that cadmium concentrations in fish livers are higher in the Southern Rocky Mountains than in the Colorado Plateau.

9.4.4 The Generalized Wilcoxon Test

Peto and Peto (1972) proposed a modification to the Gehan test called the "generalized Wilcoxon test." Prentice (1978) and Prentice and Marek (1979) elaborated on its properties, so the test is also called the Peto–Prentice or Peto-Peto test. Scores for the generalized Wilcoxon test are a weighted version of the Gehan test, adjusting the U-scores of $+1$ or -1 by the survival function (edf) at that observation to create a new score. The U-score for the generalized Wilcoxon test is

$$U_{ij} = S(t_i) + S(t_{i-1}) - 1 \quad \text{for all uncensored observations } t$$
$$S(t_{i-1}) - 1 \quad \text{for all censored observations } t* \tag{9.7}$$

where $S(t_{i-1})$ is the value of the survival function for the previous uncensored observation. For the first observation in the data set $i = 1$, and the value of $S(t_0)$ equals 1. There is a 100% probability of exceeding a value smaller than the smallest observation in the data set. Klein and Moeschberger (2003) note that S is often multiplied by $n/(n+1)$.

The scores for one group are summed to obtain the test statistic W:

$$W = \sum_{i=1}^{n} U_i \tag{9.8}$$

Dividing W by the square root of the variance for this statistic produces a Z statistic that can be compared to a table of the standard normal distribution. The permutation variance of W (assuming no ties) is

$$\text{Var}[W] = \frac{mn\Sigma U^2}{(m+n)(m+n-1)} \tag{9.9}$$

where the sample sizes m and n are as before.

To continue the cadmium in fish livers example, computations for the generalized Wilcoxon test are listed in Table 9.4. Scores for the Colorado Plateau region are

TABLE 9.4   Computation of the Generalized Wilcoxon Test for the Cadmium Data

Cd	Censoring	Region	FlipCd	Number At Risk	$S(t)*n(n+1)$	U
81.3	0	S RKY MT	18.7	19	0.903	0.903
4.9	0	S RKY MT	95.1	18	0.855	0.758
4.6	0	S RKY MT	95.4	17	0.808	0.663
3.5	0	S RKY MT	96.5	16	0.760	0.568
3.4	0	S RKY MT	96.6	15	0.713	0.473
3	0	S RKY MT	97	14	0.665	0.378
2.9	0	S RKY MT	97.1	13	0.618	0.283
1.4	0	COLO PLT	98.6	12	0.570	0.188
0.8	0	COLO PLT	99.2	11	0.523	0.093
0.7	0	COLO PLT	99.3	10	0.475	−0.003
0.6	0	S RKY MT	99.4	9	0.404	−0.144
0.6	0	S RKY MT	99.4	8	0.404	−0.144
0.6	1	COLO PLT	99.4	*	*	−0.620
0.4	0	COLO PLT	99.6	6	0.271	−0.403
0.4	0	COLO PLT	99.6	5	0.271	−0.403
0.4	0	COLO PLT	99.6	4	0.271	−0.403
0.4	1	COLO PLT	99.6	*	*	−0.783
0.3	1	COLO PLT	99.7	*	*	−0.783
0.2	1	S RKY MT	99.8	*	*	−0.783
					W(CO Plateau) =	−3.12
					Std Error =	1.165
					Z =	−2.677

summed to produce the test statistic. If the score for the S. Rocky Mt. had been selected instead, U would be the same magnitude, with the opposite sign.

Flipping the Cd data into a right-censored variable (the "FlipCd" column) produces values that look like t or "time to censoring" of traditional survival analysis. The "Number At Risk" column lists the number of observations greater than or equal to the value of t. This equals the ranks of the cadmium observations, assuming no ties. The survival function $S(t)$ is the probability of surviving beyond each observation of FlipCd. This survival function is the empirical distribution function of the original data, and equals $i/(n+1)$, where i is the rank of the original observations from low to high. Here tied observations were assigned tied ranks, as is standard in hypothesis testing. For example, the three uncensored observations at a cadmium concentration of 0.4 would have had the ranks of 3, 4, and 5 had there been enough precision in the measurements to tell the observations apart. Without that precision, any of the three observations could be the highest or lowest. All three are therefore given a rank of 4, the median of the three possible ranks. In the survival analysis literature, tied values often follow another convention, assigning the minimum value for S, rather than the median value used here. Using the median assures that the sum of ranks for data with ties is the same as it would have been without ties, an important property for hypothesis tests.

If the null hypothesis is true, observations for each group will be randomly scattered through the list in Table 9.4, with about half of the scores positive and half negative. So W, the sum of the scores, will be near zero. If the null hypothesis is not true, the data from one group will be predominately near the top, or the bottom, of the list in Table 9.4. Consequently the absolute value of W will be larger than zero. From Table 9.4, the test statistic Z equals -2.677, and from a table of the standard normal distribution the associated one-sided p-value is 0.0037. The null hypothesis is soundly rejected, and it is concluded that cadmium concentrations in fish livers in the Southern Rocky Mountains are higher than those in fish from streams in the Colorado Plateau.

Using the right-censored flipped data, Minitab's survival analysis software computes the generalized Wilcoxon (Gehan) test for the cadmium data. Notice that the test statistic for the test is 7.17 (see below). Minitab uses an alternate form of the test statistic that follows a chi-square distribution rather than the normal distribution. The value of the chi-square test statistic will equal the square of the test statistic computed using the normal approximation. For the cadmium data, 2.67^2 equals the 7.17 produced by Minitab. The p-values for the two forms of the test will either be the same or very similar. Note that the generalized Wilcoxon test can also be used to compare three or more distributions, analogous to the Kruskal–Wallis test. Therefore the p-values for the chi-square test statistic in Minitab are always two-sided, as they would be for a Kruskal–Wallis test. To obtain a one-sided p-value when comparing two groups, first check that the observed differences are in the same direction as expected by the alternate hypothesis. If so, divide the reported p-value (0.0074) by two to obtain the one-sided p-value (0.0037). The formula for the chi-square version of the test is easier to follow, and found in Klein and Moeschberger (2003) and other texts.

```
Comparison of Survival Curves

Test Statistics
Method          Chi-Square      DF          P-Value
Log-Rank        5.5260          1           0.0187
Wilcoxon        7.1707          1           0.0074
```

9.4.5 Score Test for the Zn Data from Two Geologic Zones

To test the multiply censored zinc data of Millard and Deverel (1988) for differences between geologic zones, each Zn observation is subtracted from a large number in order to produce right-censored data. This large number could be any value larger than the maximum Zn observation. The maximum observation is $620\,\mu g/L$, so 623 is arbitrarily chosen. The flipped data are stored in a new column labeled "FlipZn," where FlipZn $= 623 - $ Zn.

The generalized Wilcoxon score test determines whether the survival distribution of FlipZn is the same in the two zones. The test is therefore determining whether there are differences in the cumulative distributions of the original data. If the p-value for the test is small, the distribution of zinc concentrations differs between the groups. The test is computed in Minitab using the command:

```
Stat  >  Reliability/Survival  >  Distribution Analysis (Right
censoring) > Nonparametric Distribution Analysis
```

where the grouping variable "Zone" is entered in the "By variable" dialog box. The procedure results in the following output:

```
Distribution Analysis: FlipZn by Zone
Comparison of Survival Curves
Test Statistics
Method          Chi-Square        DF        P-Value
Log-Rank        2.84260           1         0.092
Wilcoxon        5.54396           1         0.019
```

The Wilcoxon test statistic is 5.54, and the two-sided p-value is 0.019. The null hypothesis that the two groups have zinc concentrations with the same distribution is rejected. Note that the p-value is essentially the same as for the rank-sum test on data censored at the highest reporting limit, reported earlier in this chapter. The additional information in these data attributable to the arrangement of uncensored values below $10\,\mu g/L$ is small, compared to the information in the proportion of observations in each group above and below 10, and in the values for observations above 10 in each group. In other data sets the information contained in the multiply censored pattern might be crucial to detecting differences, producing a p-value for the score test considerably smaller than for the one-threshold rank-sum test.

In Figure 9.7, the empirical distribution functions (survival functions) for the two groups are plotted. The vertical Percent scale tracks the estimated percentiles for each group (median $= 0.5$, etc.). The curves are similar at low concentrations (large flipped

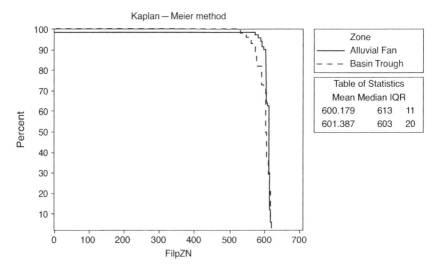

FIGURE 9.7   Survival plot of flipped zinc concentrations. Higher values of "FlipZn" (flipped data) correspond to lower concentrations.

TABLE 9.5 Summary Statistics (Kaplan–Meier Estimates) for Zn Concentrations in the Two Geologic Zones

	Mean	Median	IQR
Alluvial Fan	22.82 (600.18)	10 (613)	11 (11)
Basin Trough	21.61 (601.39)	20 (603)	20 (20)

Statistics for the flipped data are in parentheses.

values), but are different for values around the 70th to 95th percentiles. For these percentiles the Alluvial Fan zone has generally larger flipped values, and thus lower Zn concentrations, than does the Basin Trough zone. This is consistent with what was seen in the boxplots of Figure 9.1.

Estimates of mean, median, and interquartile range are given by Minitab for the flipped data in each group. Location estimates (mean, median) must be rescaled to concentration units by subtracting them from the large constant used to flip the data. This has been done and is presented in Table 9.5. Estimates of variability (standard deviation, variance, interquartile range) do not need to be rescaled; they are the same on both the flipped and original scales.

A Minitab macro named gw performs the generalized Wilcoxon test after internally flipping the data, and plots an edf of the left-censored values in the familiar left to right scale. For the zinc data, edited output from the gw macro is below, along with the generated edf plot in Figure 9.8.

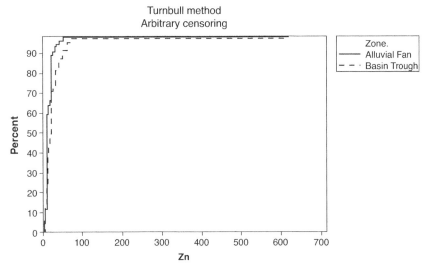

FIGURE 9.8 Empirical distribution function (edf) of censored zinc concentrations. The name "Cumulative Failure Plot" is a term in survival analysis for (1 − survival function), reversing the right-to-left orientation seen in Figure 9.7.

```
MTB > %gw 'Zn' 'ZnLT=1' 'Zone'
Comparison of Survival Curves
Test Statistics

Method          Chi-Square       DF          P-Value
Log-Rank        2.84261          1           0.092
Wilcoxon        5.54397          1           0.019
```

A version of the generalized Wilcoxon test can also be performed using NADA for R. The command is cendiff. Arguments are the same three columns as for the cenmle command: a column of detections plus reporting limits (Zn), a column of logical TRUE/FALSE values where TRUE is a censored observation (ZnCen), and the group identifier (Zone).

```
> Zntest=cendiff(Zn,ZnCen,Zone)
> Zntest
                        N Observed       Expected     (O-E)^2/E     (O-E)^2/V
Zone=Alluvial Fan 67       31.9            38.7          1.20          5.18
Zone=Basin Trough 50       30.1            23.3          1.98          5.18
Chisq= 5.2 on 1 degrees of freedom, p= 0.0228
```

The test results in a *p*-value of 0.0228, so the edfs of the zinc concentrations are found to differ between the two groups. The cenfit command

```
> Znfit=cenfit(Zn,ZnCen,Zone)
> summary(Znfit)
```

will list the Kaplan–Meier percentiles for each group separately, showing that the median of the Alluvial Fan group equals 10, while the median of the Basin Trough group equals 20. The edf for the Basin Trough group is shifted to the right on the edf plot (Figure 9.9) produced by the commands below, showing that concentrations in the Basin Trough group are generally higher than the equivalent percentiles in the Alluvial Fan group, at least above about the 35th percentile. Below that, data in both the groups are censored, and so not able to be distinguished. There is more detail in Figure 9.9 than Figure 9.8 because concentrations are plotted on a log scale in Figure 9.9, minimizing the long tail on the right side due to one outlier.

```
> plot(Znfit, xlab="Zinc concentration")
> legend(50,0.4,legend=c("Alluvial Fan","Basin Trough"),
lty=c(1:2))
```

9.4.6 Comparisons Among Score Tests

The cendiff command computes HF-1, of the Harrington–Fleming class of score tests (Harrington and Fleming, 1982). The user chooses a parameter rho that controls a weighting factor for each detected observation's contribution to the test statistic.

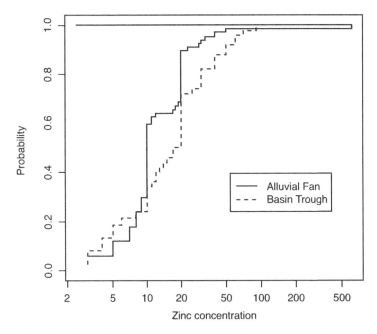

FIGURE 9.9   Empirical distribution functions (edfs) of the censored Zn concentrations, illustrating the results of the two-sample score test.

When rho equals 0 the log-rank test is produced, while rho equaling 1 results in a version of the generalized Wilcoxon test. The cendiff command uses a rho of 1, making it "essentially equivalent to Peto and Peto's (1972) generalization of the Wilcoxon test" (Harrington and Fleming, 1982). A rho of 1 produces the most powerful test for data where differences between groups are of a logistic shape, such as when multiplicative differences between groups occur. Collett (2003) notes that the generalized Wilcoxon test is more appropriate than the log-rank test when proportional hazards do not hold—when hazard functions for the two groups cross. Without going into detail on hazard functions here, this means that the Wilcoxon test is the more generally applicable of the two types of tests for lognormal and near-lognormal shaped data, such as found in environmental applications. Minitab's version of the generalized Wilcoxon test is the Gehan test, and should produce similar though not identical results as the rho = 1 test of the cendiff command. For the zinc data, Minitab's test statistic = 5.54 versus 5.2 from the cendiff command in R, as one example.

Latta (1981) evaluated the power of a number of two-sample score tests under conditions that included unequal sample sizes in the two groups, and unequal censoring (mix of reporting limits) between the groups. The latter is a particularly difficult trait to overcome, because the pattern produced by unequal levels of censoring appears to be a signal instead of a design flaw. As seen in Chapter 1 this is a particularly severe problem for substitution methods. Gehan, Peto–Prentice, log-rank and other score tests were compared in Latta's large Monte Carlo experiment. Several versions of the test statistic variance (denominator of the standardized test statistic) were used, including asymptotic variance and permutation variance estimates. Of these tests, the Gehan and Peto–Prentice tests exhibited the most power when the underlying data were lognormal, the distribution most often used to model environmental data. The test with the overall best performance, including being able to accommodate unequal sample sizes and some measure of unequal censoring mechanisms, was the Peto–Prentice test using the asymptotic variance estimate. Environmental scientists would do well to look for software performing this version of a score test.

Recognizing that water-quality data often exhibit shapes similar to a lognormal distribution, Millard and Deverel (1988) employed a Peto–Prentice score test (Prentice and Marek, 1979) with a permutation variance estimate, and found it to be the "best behaved" score test for censored lognormal data. This test produces a p-value of 0.02 (rounded) for the zinc data used in this chapter, similar to the Peto–Peto test results from of cendiff's HF-1 test reported above. When using other statistical software, look for the Peto–Prentice or Peto–Peto tests to achieve high power for multiply censored environmental data that are shaped something close to a lognormal distribution.

9.4.7 Transformations with Score Tests

Like other nonparametric tests, score test results are invariant when applied to data transformed using power functions such as the square root, logarithm, or inverse. Thus, there is little reason to use transformations when performing score tests. However, one possible reason to transform data prior to a score test is to produce survival and edf plots where the differences between groups are more easily seen. Data with high outliers produce survival functions and edfs having a long "tail," as seen in the Zn concentrations of Figure 9.8. Computing the graphs using logarithms, or using a log scale such as in Figure 9.9, can clarify the tested differences between the grouped data.

9.5 VALUE OF THE INFORMATION IN CENSORED OBSERVATIONS

Some scientists have felt that censored observations carry no information, and so the fewer the uncensored values in a data set, the less information is present. This is not correct. To illustrate, Zn concentrations in the Alluvial Fan zone, the zone with lower median, were altered by changing some of the uncensored values above 10 to a <10, and some of the <10s to <3s. The result is that the Alluvial Fan zone contains 54% censored observations, while the Basin Trough zone remains at 24% censored observations. The overall effect is to lower the zinc distribution in the Alluvial Fan zone, while increasing the proportion of censored observations. The signal in the data should be stronger, and should result in a lower p-value than for the original Zn concentrations, even though there are fewer uncensored values. Using the generalized Wilcoxon test in Minitab on the altered data, the resulting p-value is <0.001, more than an order of magnitude lower than the p-value of 0.019 for the unaltered data.

```
Distribution Analysis: FlipZn2 by Zone
Comparison of Survival Curves

Test Statistics
Method          Chi-Square        DF        P-Value
Log-Rank        17.6936           1         0.000
Wilcoxon        14.4449           1         0.000
```

The increased separation in the altered data is seen in their survival functions (Figure 9.10), where flipped data for the Alluvial Fan zone are now higher than in the original data, and more separate from the Basin Trough zone data. Higher flipped concentrations result from lower Zn concentrations in the original units.

Several guidance documents have recommended that statistical tests not be run with data having a high proportion of censored observations. For example, USEPA (2002b, p. A-9) recommends that if there are more than 40% censored observations in any group, the rank-sum test should not be used. There appears to be little justification for setting these types of rules. If the proportion of censored observations is similar in the two groups, the weight of evidence favors the null hypothesis. If the proportion differs significantly, as with 54% versus 24% censored observations for the altered zinc data, the null hypothesis will likely be rejected. In Chapter 10, TCE concentrations averaging 80% censored observations can be differentiated among three groups using the Kruskal–Wallis test, the multigroup equivalent of the rank-sum test. Tests that efficiently extract information from censored data, such as Wilcoxon score tests, will respond to the information contained in the data. There is no need to limit their use.

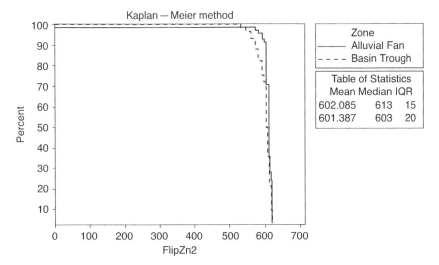

FIGURE 9.10 Survival functions of the altered Zn data with increased proportion of censored observations in the Alluvial Fan zone. Compare with Figure 9.7.

9.6   INTERVAL-CENSORED SCORE TESTS: TESTING DATA THAT INCLUDE (DL TO RL) VALUES

As seen earlier in this chapter, there is a parametric method (MLE) to test data whose lower endpoint is nonzero, primarily data recorded as between the detection and quantitation/reporting limits. All values are expressed as an interval (low, high). Values below a detection limit of 1 are expressed as (0, 1). Values between the detection limit and quantitation limit of 3 are expressed as (1, 3). Detected values possess the same number for both low and high values, so that a detected 5 is (5, 5).

Fay and Shaw (2010) have provided software (the contributed package "interval" for R) for performing interval-censored nonparametric procedures, both log-rank and Wilcoxon type linear rank tests. They apply these procedures to the medical statistics discipline. Peto and Peto (1972) described these tests, but software to accomplish them was not easily available until this contributed R package was released. These procedures extend the rank-sum type score tests to data expressed in the interval endpoints format, allowing environmental scientists to test data sets with "detected but not quantified" data directly. Interestingly, in their paper Fay and Shaw refer to an evaluation by Law and Brookmeyer (1992) on substituting one-half the interval width for interval-censored data—the equivalent of substituting one-half the reporting limit when the lower endpoint is zero. Not surprisingly, it did not work very well. Hence the need for this software to avoid substitution when a disease occurs somewhere between time A and time B. Environmental scientists can use the same procedures to test concentrations that fall between A and B µg/L. In fact, one of the first papers discussing interval-censored tests had applied them to left-censored chemical data, concentrations of PCBs built up in the human body (Self and Grossman, 1986).

Linear rank tests estimate the survival curve (percentiles) and determine if this differs between the two groups of data. Estimates for each group of the probability of exceeding each cut point (detection limits and detected observations) is compared to the overall probability of exceeding those cut points if the null hypothesis is true and the data homogenous among the groups. A score is computed summarizing the differences between the within-group exceedance and the overall exceedance probabilities. One option in the interval package is to use the "wmw" score statistic, producing an interval-censored analog to the generalized Wilcoxon test.

For the zinc data of Millard and Deverel (1988), the interval-censored Wilcoxon test is computed after first computing the low end of the interval. For these data, the low end was zero in each case.

```
CuZn$ZnLow=Zn*(1-ZnCen)
```

The default format in the interval package is that the low end of intervals is just below the range of possible values, while the upper is included in the range of possible values. This is the opposite of the situation for environmental data, where the low end is a possible value, but the high end is just below the censoring limit for censored values. If the high end is a <5 the range of possible values does not include 5. This default format can be changed using the Lin and Rin options if a vector is supplied for each. The Lin vector states for each observation whether the value at the left (low end) of the interval should be included (TRUE) or not (FALSE) in the range of possible values. The Rin vector states the same for the right (high end) of the interval. For censored environmental data the Lin vector should all be TRUE, while the Rin vector should be FALSE for the censored observations. To create the right vector named Rt,

```
CuZn$Rt=Zlow!=0
```

states that for all uncensored observations where Zlow is not equal to 0 (! = 0), Rt is TRUE. Where (Zlow not equal to 0) is false, and so censored, Rt is FALSE. To create the Lt vector of all TRUE values,

```
CuZn$Lt=Zn<1000
```

as all Zn concentrations are lower than 1000. The test is then run with the ictest command:

```
> zntest=ictest(Zlow,Zn,Zone,scores="wmw",Lin=Lt,Rin=Rt)
> zntest
Asymptotic Wilcoxon two-sample test (permutation form)
data: {Zlo,Zn} by Zone
Z=2.2269, p-value=0.02596
alternative hypothesis: survival distributions not equal
                             n Score Statistic*
Alluvial Fan 67                6.799756
Basin Trough 50               -6.799756
* like Obs-Exp, positive implies earlier failures than expected
```

The Z statistic and p-value are very similar to those reported in the original article by Millard and Deveral for the Wilcoxon or Peto–Prentice type procedures, where they found test statistics of 2.37–2.42 and p-values of 0.02. The two will not be identical because the original paper did not employ an interval-censored version of the test. The default type of test with ictest is a permutation procedure. An exact version can be computed using the exact=TRUE option. With this option either all possible permutations of the test statistic are computed, resulting in an exact p-value, or a large number of possible test statistics randomly selected are chosen.

The cadmium data can also be tested with the ictest command. Note that the rho = 1 option is specified below rather than the scores = "wmw" option. These are equivalent and produce the same test. Both are using the rho $= 1$ option of the Harrington–Fleming class of tests.

```
> Cadmium$Rt=!Cadmium$CdCen
> Cadmium$Lt=Cd<1000
> Cadmium$Cdlo=Cd*Cadmium$Rt
> attach(Cadmium)
> cdtest=ictest(Cdlo,Cd,Region,rho=1,Lin=Lt,Rin=Rt,exact=TRUE)
> cdtest
        Exact Wilcoxon two-sample test (permutation form)
data:   {Cdlo,Cd} by Region
p-value=0.007989
alternative hypothesis: survival distributions not equal
                             n                    Score Statistic*
SRKYMT                       10                    -3.289474
COLOPLT                      9                     3.289474
```

* like Obs-Exp, positive implies earlier failures than expected

To demonstrate the procedure on nondetects recorded as (0, 0.4) and values between the detection and reporting limits as (0.4, 0.6), several observations in the data set are changed. Any censored value gets an upper end indicator Rt $=$ FALSE. The data set now looks like:

```
> Cd2
```

	CD	REGION	LT.1	Cdlo	Rt	Lt
1	81.3	S RKY MT	0	81.3	TRUE	TRUE
2	3.5	S RKY MT	0	3.5	TRUE	TRUE
3	4.6	S RKY MT	0	4.6	TRUE	TRUE
4	0.6	S RKY MT	0	0.4	FALSE	TRUE
5	2.9	S RKY MT	0	2.9	TRUE	TRUE
6	3.0	S RKY MT	0	3.0	TRUE	TRUE
7	4.9	S RKY MT	0	4.9	TRUE	TRUE
8	0.6	S RKY MT	0	0.4	FALSE	TRUE
9	3.4	S RKY MT	0	3.4	TRUE	TRUE
10	0.4	COLO PLT	0	0.0	FALSE	TRUE
11	0.8	COLO PLT	0	0.8	TRUE	TRUE
12	0.4	COLO PLT	1	0.0	FALSE	TRUE
13	0.4	COLO PLT	0	0.0	FALSE	TRUE
14	0.4	COLO PLT	0	0.0	FALSE	TRUE
15	0.4	COLO PLT	1	0.0	FALSE	TRUE
16	1.4	COLO PLT	0	1.4	TRUE	TRUE
17	0.6	COLO PLT	1	0.4	FALSE	TRUE
18	0.7	COLO PLT	0	0.7	TRUE	TRUE
19	0.4	S RKY MT	1	0.0	FALSE	TRUE

and the ictest is again run:

```
> cd2test=ictest(Cdlo,CD,REGION,rho=1,Lin=Lt,Rin=Rt,exact=TRUE)
> cd2test
        Exact Wilcoxon two-sample test (permutation form)
```

```
data: {Cdlo,CD} by REGION
p-value = 0.00747
alternative hypothesis: survival distributions not equal
                                n                    Score Statistic*
S RKY MT                       10                    -3.315789
COLO PLT                        9                     3.315789

* like Obs-Exp, positive implies earlier failures than expected
```

The last line obliquely tells the observed difference between the groups. A positive score statistic implies "earlier failures" or lower detected values than a negative score. The Colorado Plateau therefore exhibits generally lower concentrations than the Southern Rocky Mountains. This test successfully used data expressed as intervals, testing differences between the groups in a fully nonparametric mode. No substitution was required; no numbers were estimated between the detection (at 0.4) and quantitation (at 0.6) limits. None was required.

9.7   PAIRED OBSERVATIONS

A variation on the two-group design occurs when observations in each group are purposely paired with one another to block out sources of background noise and focus on the effect being studied (Helsel and Hirsch, 2002, Chapter 6). With this structure, both groups have the same number of observations, and the first observation in the first group is linked to the first observation in the second group. Similarly, the second observation in the first group is linked to the second observation in the second group, the third with the third, and so on. Observations in the first group may be thought of as the "starting point," and in the second group as the "ending point." The test determines if there are differences between the starting and ending points, even though the starting points may differ from pair to pair.

For uncensored data the standard tests for this design are the one-sample (or paired) t-test, and the nonparametric signed-rank test. Differences between each pair of observations are tested to see if their mean (t-test) or median (signed-rank test) difference is significantly different from zero. A binary nonparametric test sometimes used with paired observations is the sign test. The sign test does not compute the magnitude of differences between pairs of observations, but records only whether there is an increase or decrease between the two values. This test determines if the proportion of increases, or decreases, is significantly different than the expected frequency of 50%. Not requiring an estimate of the magnitude of difference makes the sign test very useful for censored data.

With censored data the differences between pairs having one or more censored observations cannot be determined exactly. The same options are open to the scientist for testing differences between paired groups—for binary methods, the sign test (appropriately modified for many tied pairs) may be computed. The ordinal signed-rank test is not very applicable here, as even with one reporting limit the resulting differences between pairs will have magnitudes such as > 3 and > 5 when one value is censored and other not. This is not easily incorporated into the signed-rank test procedure. Happily, there are survival analysis methods, maximum likelihood, and nonparametric score tests that can fully incorporate multiply censored data. As has been presented in other chapters of this book, substitution methods are fraught with problems and are best avoided.

9.7.1 Binary Methods for Paired Data—The Modified Sign Test

The sign test determines whether paired values from one group generally are higher than the values from the other (Helsel and Hirsch, 2002, p. 138). Comparisons between paired observations are recorded only as an increase, a decrease, or a tie. These are shown in the "Sign of Difference" column in Table 9.6. Because the magnitude of the difference is not used, the sign test is directly applicable to paired censored observations. Due to its paired structure, the sign test can be performed whenever one reporting limit is used per x–y pair. Though the test cannot evaluate a

TABLE 9.6 Atrazine Concentration Pairs

June	September	September − June	Sign of Difference
0.38	2.66	2.28	+
0.04	0.63	0.59	+
<0.01	0.59	0.58 to 0.59	+
0.03	0.05	0.02	+
0.03	0.84	0.81	+
0.05	0.58	0.53	+
0.02	0.02	0.00	0
<0.01	<0.01	0.00 to 0.01	+
<0.01	<0.01	−0.01 to 0.01	0
<0.01	<0.01	−0.01 to 0.01	0
0.11	0.09	−0.02	−
0.09	0.31	0.22	+
<0.01	0.02	0.01 to 0.02	+
<0.01	<0.01	−0.01 to 0.01	0
<0.01	0.5	0.49 to 0.50	+
<0.01	0.03	0.02 to 0.03	+
0.02	0.09	0.07	+
0.03	0.06	0.03	+
0.02	0.03	0.01	+
0.02	<0.01	−0.01 to −0.02	−
0.05	0.10	0.05	+
0.03	0.25	0.22	+
0.05	0.03	−0.02	−
<0.01	88.36	88.35 to 88.36	+

Do concentrations increase from June to September?

pair of observations (x, y) at $(<1, <3)$, it can evaluate data where one pair is $(<1, 10)$ and a second pair $(<3, 5)$. Both increase from x to y. Therefore multiple reporting limits can in limited fashion be incorporated into the sign test. Though more powerful and more complicated score tests have been developed for multiple censoring thresholds, the sign test remains the easiest test to employ for the situation of one reporting limit per x–y pair.

As an example of paired data, atrazine concentrations in groundwater were measured at 24 wells in June, and at the same wells again in September, to determine whether concentrations had increased due to application of atrazine at the surface (Junk et al., 1980). Well to well differences are not of concern, and are "blocked out" by the pairing process. All that is of interest is determining whether concentrations have increased for any given well during the time period. Several values are censored at the single reporting limit of 0.01 μg/L. The data are found in atra.xls and shown in Table 9.6.

The sign test determines whether the pattern of pluses and minuses differs from the expected frequency of 50% for each when the null hypothesis is true. This example is a one-sided test—the question was "is there an *increase* from June to September?"—so change in only one direction is of interest. The results below are the standard Minitab output for the sign test, and indicates that there is indeed a difference, at a p-value of 0.0022. No substitutions or assumptions of a distributional shape for these data were required.

```
Sign Test for Median: S-J
Sign test of median = 0.00000 versus > 0.00000
          N      Below     Equal      Above       P         Median
S-J       24     3         5          16          0.0022    0.02500
```

The standard sign test procedure deals with ties by deleting the tied pairs from all calculations. The sample size n is just decreased by the number of tied pairs when computing p-values for the test. In essence, it is testing "for paired values that are not tied, is there a consistency in the pattern of increases and decreases?" For the atrazine example, the test computed the p-value as the likelihood of seeing 16 increases and 3 decreases out of a total of 19 pairs, ignoring the other 5 tied pairs. This may be acceptable for the small proportion of ties typically encountered in uncensored data, but for a larger proportion of ties it is not. Ignoring tied pairs will inflate the Type I error rate, artificially lowering the reported p-values (Fong et al., 2003). Data sets with a large proportion of ties should reflect their greater evidence for similarity than for data without tied values.

Fong et al. (2003) provide two methods for adjusting the sign test in the presence of tied pairs. Their "modified sign test" is implemented in the Minitab macro Csign, producing p-values using the formula

$$p = \frac{\text{Prob}[N \geq \max(n^+, n^-)]}{\text{Prob}[N \geq \langle (n - n^0 + 1)/(2) \rangle]} \tag{9.10}$$

where n is the number of data pairs, n^+ is the number of increases, n^- is the number of decreases, and n^0 is the number of ties. Capital N represents the binomial distribution with n observations evaluated at the central 0.5 proportion, the probability of exceedance (p-values) for the sign test. The angle brackets in the denominator of equation (9.10) represent the floor function, so that $\langle X \rangle$ is the largest integer smaller than or equal to X. The modified p-value adjusted for ties is printed for the atrazine data by the Csign macro as

```
p-value (adjusted for 'Equal' ties) = 0.0448
```

The 5 tied pairs out of a total of 24 pairs of data are evidence favoring the null hypothesis of equality, which when incorporated rather than deleted increase the p-value from 0.002 to 0.04. When computing the sign test for censored data, the modified sign test should be used rather than the default test computed by commercial statistical software so that ties, such as <DL versus <DL, are correctly incorporated into the test results.

While the signed-rank test is not as applicable as is the sign test to censored data due to its need to compute and rank the magnitude of the paired differences, Pratt (1959) provided a modification to the signed-rank test for the case of many tied values.

9.7.2 Survival Analysis—MLE for Paired Data

With maximum likelihood estimation, the mean difference between the starting and ending points is tested to determine if it is significantly different from zero. To do this, a confidence interval is constructed around the mean difference using MLE. If the confidence interval does not include zero, the differences are declared nonzero, and the two columns of data are declared to be different.

For censored data, differences must be calculated as an interval; the differences are interval-censored data. MLE can be used to compute the mean and standard deviation, and therefore a t-interval around the mean difference, for interval-censored data. For paired data, MLE requires an assumption that the paired differences follow a specified distribution. This assumption should first be checked before accepting the results of the MLE procedure.

Consider again the atrazine data of Table 9.6. Twenty-four pairs of concentrations were measured with a reporting limit of 0.01 µg/L. To test whether zero is included in the confidence interval around the mean difference, the differences are calculated (column 3 of Table 9.6). For censored observations there is a range of differences; the lowest and highest possible differences for the pair are stored separately in two columns. For pairs of uncensored observations, the same value for the difference is entered in each column. The two columns become the interval endpoints columns for MLE, with the column of smallest differences designated the Start column and the largest differences the End column. The actual difference is somewhere within that interval. Using maximum likelihood, the mean difference and its 95% confidence interval are estimated; the output appears below. If data follow a normal distribution, this procedure is the equivalent for censored data to the paired t-test. The Minitab macro PMLE.mac computes the confidence interval around the mean, and reports the p-value for the null hypothesis that the mean difference equals zero:

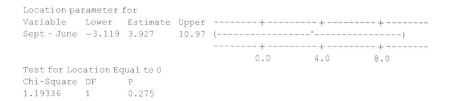

```
Location parameter for
Variable    Lower    Estimate  Upper  --------+---------+---------+--------
Sept - June  -3.119  3.927     10.97  (-----------------*-----------------)
                                      --------+---------+---------+--------
                                          0.0       4.0       8.0
Test for Location Equal to 0
Chi-Square  DF      P
1.19336     1       0.275
```

The 95% confidence interval extends from -3.1 to a 10.97, which includes zero. Therefore, the mean difference September $-$ June is not significantly different from zero at $\alpha = 0.05$. For the test of whether the mean equals zero, the p-value of 0.275 also indicates that no significant difference from zero was found.

These data are severely non-normal, as shown by a probability plot of the differences (Figure 9.11). An assumption of a lognormal distribution may be better than using the normal distribution. The logarithm of each month's data is calculated, and the differences in the logarithms computed and tested. The test for whether the mean difference in log units equals zero is identical to a test for whether the ratio of the two geometric means equals one. This test is appropriate as long as

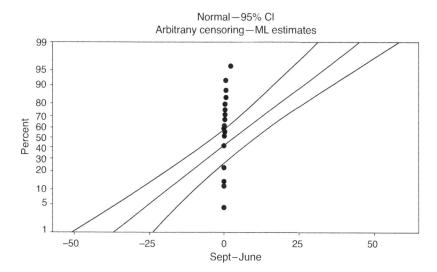

FIGURE 9.11   Probability plot of the paired atrazine differences September $-$ June.

the geometric mean (median) rather than the arithmetic mean is an appropriate measure of the center. If the data must remain in original units, some accommodation for the one large outlier in the last row of Table 9.6, a point that inflates the standard error of the mean and therefore lengthens the confidence interval, must be made.

9.7.3   Nonparametric Method for Multiply Censored Paired Data—The PPW Test

For multiple reporting limits, score tests have been developed for the matched-pair design. These tests use a score statistic as a measure of the position of the observation in a data set. Scores are generally computed to be negative and positive, centered at zero on the midpoint of the data set. These scores are related to the ranks of the data by the equation

$$\text{Rank} = n[0.5 + 0.5 \times \text{Score}] \qquad (9.11)$$

O'Brien and Fleming (1987) introduced the paired Prentice–Wilcoxon, or PPW, test. It uses the same form of scores used in the generalized Wilcoxon test for unpaired data, and is the standard test for the case of censored matched pairs. Akritas (1992) proposed an alternate test, performing a paired t-test on the ranks as defined in equation (9.11). Both tests have similar power (Akritas, 1992) for the situation of skewed data common in environmental studies. Neither the PPW nor the Akritas tests are found in commercial statistical software. However, a Minitab macro (PPW.mac) to perform the paired Prentice–Wilcoxon test is included on the web page that accompanies this book.

To compute the PPW test, the data are stacked into one column, and a Kaplan–Meier estimate of the survival function for the combined data is computed. Scores, the estimated percentiles of the survival function minus 0.5, are computed for each observation, both censored and uncensored. The scores are then split back up into their respective groups. If the null hypothesis is true and the two distributions are the same, differences between pairs of scores should be small, hovering around zero. In other words, the two paired observations should be located at similar places in the combined distribution; therefore their score values should be similar. If the distributions of the two groups differ, the paired observations will be located at different points of the combined survival distribution, with the scores from one data set consistently higher than the paired score from the other. The PPW test computes the differences between the paired scores, and determines whether the sum of these differences is significantly different from zero, using a normal approximation for the test statistic.

Four observations from the atrazine data of Table 9.6 were altered to produce paired data with two reporting limits, at 0.01 and 0.05 μg/L. The altered data are shown in Table 9.7, with the altered observations at <0.05. Alterations were in the direction of stronger differences between the 2 months.

TABLE 9.7   Atrazine Concentration Pairs from Table 9.6 Altered to Add a Second Reporting Limit

June	September	June Score S1	September Score S2	June − September Score S1 − S2
0.38	2.66	−0.67	−0.92	0.24
<0.05	0.63	0.41	−0.84	1.24
<0.01	0.59	0.66	−0.80	1.45
0.03	0.05	0.04	−0.18	0.23
0.03	0.84	0.04	−0.88	0.92
0.05	0.58	−0.18	−0.76	0.57
0.02	0.02	0.32	0.32	0
<0.01	<0.01	0.66	0.66	0
<0.01	<0.01	0.66	0.66	0
<0.01	<0.01	0.66	0.66	0
0.11	0.09	0.55	−0.39	−0.16
0.09	0.31	0.39	−0.63	0.24
<0.01	0.02	0.66	0.32	0.34
<0.01	<0.01	0.66	0.66	0
<0.01	0.50	0.66	−0.71	1.37
<0.01	0.03	0.66	0.04	0.61
0.02	0.09	0.32	−0.39	0.70
<0.05	0.06	0.41	−0.35	0.76
0.02	0.03	0.32	0.04	0.27
0.02	<0.01	0.32	0.66	−0.34
0.05	0.10	−0.18	−0.51	0.33
0.03	0.25	0.04	−0.59	0.64
0.05	<0.05	−0.18	0.41	−0.59
<0.01	88.36	0.66	−0.96	1.62

PPW scores for each pair are also shown in Table 9.7. The scores are computed on flipped data, so their signs are the reverse of what is expected from the original observations.

For uncensored observations:

$$\text{Score} = 1 - 2S$$

where S is the Kaplan–Meier estimate for the percentile of the survival function. For censored observations:

$$\text{Score} = 1 - S_j$$

where S_j is the Kaplan–Meier estimate for the percentile of the survival function for the next smallest (in flipped units) uncensored observation.

Scores can be considered as scaled ranks of observations, although in the reverse order from the original units due to flipping the data. For example, the largest observation of 88.36 has the largest negative score, of −0.96. The second largest observation of 2.66 has the second largest negative score, of -0.92. All concentrations tied at a reporting limit have scores identical to one another. Notice that the difference in scores equals 0 for pairs with tied data. As the number of ties increases, Z will get smaller, providing less evidence against the null hypothesis. Differences in scores are largest for pairs with large differences in ranks of the original observations. For the PPW test, the difference in the paired scores ($d = S_1 - S_2$) is tested using the test statistic

$$Z_{PPW} = \frac{\sum d}{\sqrt{\sum d^2}} \tag{9.12}$$

by comparing Z to a table of the normal distribution. The output below shows that the September atrazine concentrations are significantly higher than their paired June concentrations ($p = 0.001$). This is stronger evidence than that for the sign test, where the p-value was an order of magnitude larger.

```
Paired Prentice-Wilcoxon test
(NonPar test for equality of paired left-censored data)

     Ho: distribution of Sept = June

vs   Ha: greater than

Test Statistic: 2.999
        p value: 0.001
```

The larger differences between groups created when the data were altered is picked up by the PPW test's ability to incorporate multiple reporting thresholds, at least in the case where both groups have data below both reporting limits. The PPW test is not built to handle the case where each group has a predominantly different reporting limit, so that <0.01s dominate one group while <0.05s dominate the other. For this situation, the best procedure is to recensor all data in both groups to the higher reporting limit before computing the test (O'Brien and Fleming, 1987). If this is not done, the values assigned to the higher reporting limit (e.g., <0.05) will be considered a higher score than those assigned to the lower limit (e.g., <0.01). This is more than what is actually known and can contribute falsely to a decision that the paired values are different. Recensoring at the higher reporting limit is easily available in the PPW macro by invoking the HighDL subcommand.

9.7.4   A Second Score Test for Multiply Censored Data—The Akritas Test

For uncensored data, the nonparametric signed-rank test can be approximated by computing a paired t-test on the signed-ranks (Conover and Iman, 1981). The Akritas test (Akritas, 1992) extends this idea to data with censored values. Ranks of uncensored values are computed much as in the PPW test, using Kaplan–Meier statistics for the entire data set. Ranks of censored observations are computed by calculating Kaplan–Meier survival functions separately for each group and averaging the survival probabilities for the pair of observations that includes the censored value. The rank of the censored observation is computed from this averaged probability (Akritas, 1992). A paired t-test is then computed on the calculated ranks to evaluate the similarity of the set of paired observations.

As found by Conover and Iman (1981), rank-transform tests generally produce p-values slightly lower than those of the exact tests for the same situation. Null hypotheses are rejected a little more frequently than they should in simulation studies evaluating these tests. As a rank-transform test, this characteristic may also be true for the Akritas test. More detail on its computation is found in Akritas (1992).

9.7.5   Comparing Data to a Standard Using Paired Tests

Methods for paired observations can be used to determine whether the mean or median of one column of data equals or exceeds a fixed standard. The paired t-test, signed-rank test, and sign test can be used with uncensored data to test compliance to a standard. Similarly for censored data, the MLE confidence interval and nonparametric PPW test can serve the same purpose, even for data with multiple reporting limits.

Instead of two columns of data, comparisons to a standard are made by placing the value of the standard in every entry of the second data column. Differences are computed between the data and the standard, and the mean or median difference tested to determine whether or not it equals zero. Tests for compliance are often set up assuming compliance; the alternate hypothesis is that the mean or median difference exceeds zero. If so, the mean or median of the column of data exceeds the standard.

As an example, consider whether the mean or median of the altered atrazine data for June of Table 9.7 exceeds a standard of 0.05 µg/L. There are two reporting limits, at 0.01 and 0.05. Differences between the data and the standard are interval-censored values. The endpoints of the interval are identical when the data are above the reporting limit. Endpoints differ for censored observations, representing the range of possible differences. Using maximum likelihood (the %PMLE macro) and assuming a normal distribution, the mean concentration does not exceed the standard of 0.05. The chi-square test statistic is a two-sided test: for a one-sided test the p-value is 0.583 divided by 2, or 0.29. Also note that the lower 95% confidence bound extends below zero, so that zero is included as a possible estimate for the difference—this is the confidence interval window into the test procedure.

```
Test for Location Equal to 0

Chi-Square   DF          P
0.300634     1           0.583
Bonferroni 95.0% (indiv 95.00%) Simultaneous Lower Bound
Location parameter for
Variable     Lower       Estimate    --------+---------+---------+--------
June - Std   -0.03388    -0.008472   (-------------------*---------------
                                     --------+---------+---------+--------
                                         -0.024     -0.012     -0.000
```

Using the nonparametric PPW test, the median June atrazine concentration also does not significantly exceed the standard of 0.05; the null hypothesis of equality of medians is not rejected ($p = 1.0$). Both the mean and the median of these multiply censored data are within the standard set for them.

```
Paired Prentice-Wilcoxon test
(NonPar test for equality of paired left-censored data)

        Ho:  distribution of June = Std
vs      Ha:  greater than

Test Statistic: −3.713
        p value: 1.000
```

9.8 SUMMARY OF TWO-SAMPLE TESTS FOR CENSORED DATA

Two-sample tests corresponding to the t-test, rank-sum and sign tests are available for use with censored data. The decision whether to use a parametric or nonparametric test is made in the same way as for uncensored data, judging how closely the data, or their transformed values, follow a normal distribution. Censored t-tests are available in survival analysis by solving for regression parameters with maximum likelihood. The slope coefficient estimates the difference between the means of the two groups, and testing this coefficient determines whether the difference is significant. For nonparametric tests, the generalized Wilcoxon test expands on the uncensored Wilcoxon rank-sum (or Mann–Whitney) test, adjusting the ranks to incorporate information contained in censored observations. Wilcoxon tests can be used on both singly and multiply censored data. However, for a single reporting limit the standard rank-sum or sign test could also be used. Therefore standard nonparametric tests can be used for data with one reporting limit.

Paired tests analogous to the paired t-test and signed-rank tests are available for multiply censored data. They can also be used to test whether one set of multiply censored data exceed a legal or other standard. The sign test provides the simplest binary approach, while a MLE confidence interval on the mean and the paired Prentice–Wilcoxon test allow data with multiple reporting limits to be analyzed.

EXERCISES

9-1 Eppinger et al. (2003) measured metals concentrations in stream sediments at 82 sites in New Mexico in 1996. After wildfires occurred throughout the region in 2000, each site was resampled to determine if concentrations had changed following the fires. Several mechanisms were proposed for why this might be so. Data for lead are found in SedPb.xls. Test to determine whether lead concentrations, some of which are recorded as below a single reporting limit of 4 µg/L, have changed pre- and postfire. Note that the data are paired by sampling location.

9-2 Yamaguchi et al. (2003) measured concentrations of the pesticide lindane in fish and eels collected at several sites in the United Kingdom. One site was below Swindon, an active industrial area draining to the Ray River, a tributary to the Thames. To avoid differences due to types of fish, data presented here are for only one species (Roach) at two sites, Swindon and a site further downstream on the Thames River, in the file Roach.xls. There was one reporting limit, at 0.08 µg/kg. Test whether lindane concentrations are the same or different at the two sites, using both the parametric "t-test" performed with censored regression and a nonparametric Wilcoxon score test.

9-3 Squillace et al. (1999) related VOC concentrations in groundwater throughout the United States to population density. The data for one compound, chloroform, in the state of California is presented in ChlfrmCA.xls. Observations are grouped by whether they are from urban areas (population density > 386 people per acre) or rural areas (population density <386 people per acre). Determine if the mean/ median concentration of chloroform is higher in the urban areas than in the rural areas. Use both a parametric and nonparametric test. There are two reporting limits.

10 Comparing Three or More Groups

Scientists must at times evaluate environmental data that are classified into more than two groups. Are the means, or medians, or probabilities of observing a detected value the same within each group, or is at least one different? What can be said when there are censored observations below several reporting limits scattered throughout the groups? For uncensored data, comparisons among groups are made using the parametric analysis of variance (AOV or ANOVA) and the nonparametric Kruskal–Wallis (KW) test. For data with censored observations, the methods surveyed in Chapter 9 for differentiating between two groups—binary, ordinal, maximum likelihood, and score test methods—can be extended to three or more groups and employed here. Differences between means can be tested using maximum likelihood, using software for censored regression. Nonparametric Wilcoxon score tests look for differences in the empirical distribution functions (or survival distributions) among the groups.

As the primary example for this chapter, concentrations of trichloroethylene (TCE) in shallow ground waters of Long Island, NY were reported by Eckhardt et al. (1989). The waters sampled were from wells surrounded by one of three land-use types: low-, medium-, or high-density residential lands. Sources of TCE were expected in each land-use type due to the past use of TCE as a septic system cleaner, and as a solvent in a variety of residential and light industrial uses. At issue is whether the occurrence of TCE is similar in waters under the three land-use types, or whether concentrations appear to differ in at least one land use. Because samples were sent to different laboratories, and because precision changed over time, four reporting limits at 1, 2, 4, and 5 µg/L are found in the data. Boxplots of the data are shown in Figure 10.1, where the boxes for all three land-use groups lie below the highest reporting limit. The percentage of values above the highest reporting limit of 5 µg/L in each of the three groups is listed in Table 10.1.

These data can be found in TCE.xls. Concentrations or reporting limits are stored in the column named TCEConc. The column named BDL=1 indicates a censored observation with a value of 1, and a detected value with a 0. The land-use groups are listed in the column named Density.

Statistics for Censored Environmental Data Using Minitab® and R, Second Edition.
Dennis R. Helsel.
© 2012 John Wiley & Sons, Inc. Published 2012 by John Wiley & Sons, Inc.

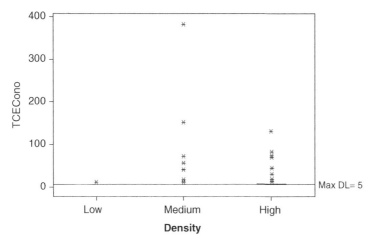

FIGURE 10.1   TCE concentrations in ground water under three land-use types (Eckhardt et al., 1989).

10.1   SUBSTITUTION DOES NOT WORK—INVASIVE DATA

Substituting the reporting limit for all censored observations can quickly be accomplished by ignoring the indicator variable, and using only the values stored in the concentration column. An ANOVA run directly on the concentration column produces an ANOVA with substitution of the reporting limits. This produces a high bias for the means of each group. As already shown in previous chapters, substitution in the case of multiple reporting limits can either artificially produce patterns not seen in the data themselves or obscure patterns that should be detected. ANOVA on the column of TCE concentrations produces the following results:

```
One-way Analysis of Variance

Analysis of Variance for TCEConc
Source      DF          SS          MS          F           P
Density      2          1120        560         0.60        0.547
Error      244        225876        926
Total      246        226996
```

ANOVA compares a ratio of the mean square (MS) for the signal (Density) to the MS of the background noise (Error). This ratio is the F statistic for the test,

TABLE 10.1   Percent of TCE Concentrations in Long Island Groundwater Greater than 5 μg/L

Land-Use Density	Low	Medium	High
Percent above 5 μg/L	0	9	20

Data from Eckhardt et al. (1989).

$560/926 = 0.60$. The expected value for F when there is no difference among the group means is approximately 1. If the F statistic is much larger than 1, the null hypothesis of no difference among means can be rejected. For the TCE data with substitution, the F statistic of 0.60 is not large. The magnitude of difference in means it represents can be expected to occur about 55% of the time when the null hypothesis is true ($p = 0.547$). This is insufficient evidence to reject the null hypothesis, and therefore mean TCE concentrations for the three groups are considered to be similar by this test. This result could also be due, however, to either the non-normality of the data (censored data sets are often quite skewed) or to the inaccuracy of fabricating the pattern of concentrations by substitution. Substituting a constant for all nondetects declares that you know that all these values are exactly the same, an invasive pattern that is unlikely to have been in the original data.

Substitution of either one-half the reporting limit or zero also results in insignificant F-tests. It should be noted that the simplest nonparametric test for these data, a contingency table test presented in the next section, finds significant differences in the proportions of data above 5 µg/L among the three land-use groups. Substitution's failure to see any differences emphasizes again the weaknesses of the method. There is no need to use substitution. There are better ways.

10.2 NONPARAMETRIC METHODS AFTER CENSORING AT THE HIGHEST REPORTING LIMIT

10.2.1 Binary Methods: Contingency Tables

If the censored response variable is collapsed into two values, below and above the maximum reporting limit, contingency tables will test whether the proportion of values in those two categories changes among groups. Collapsing data into two response categories loses information, but (depending on where the highest reporting limit is located) the information that remains may be the major component of what is available. Contingency tables are easily understood, and easily illustrated with a simple bar chart. The proportions above and below the highest reporting limit are unambiguous, and the test results definitive. The test is available in all commercial statistical software. This is the simplest test to perform with a censored response variable, but it has less power than score tests. The mechanics of the method are found in many statistics textbooks, including Conover (1999).

For the TCE data, observations greater than or equal to the highest reporting limit of 5 µg/L are assigned a unique value, either numeric or text, and those below assigned a second value. The test determines whether the percentage of response values in each category is similar across all groups. Figure 10.2 shows the observed percentages of TCE concentrations at or above 5 µg/L for the three density groups.

A contingency table test of the TCE data can be produced by invoking the Minitab® macro "detects" by typing

```
 > %detects 'TCECONC' 'BDL_1' 'Density'
```

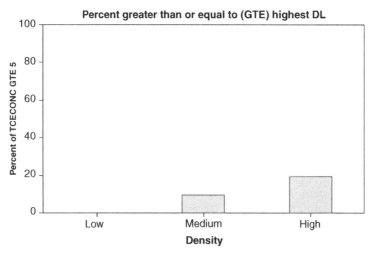

FIGURE 10.2   Percent of TCE concentrations at or above the highest reporting limit of 5 μg/L for three land-use categories (data from Eckhardt et al., 1989).

where TCECONC is the column of data plus reporting limits, BDL_1 is the indicator variable for censoring, and Density is the group assignment. This produces the output:

```
Test whether % GTE 5 is the same in all groups
Pearson Chi-Square = 9.238, DF = 2, P-Value = 0.010
Likelihood Ratio Chi-Square = 11.697, DF = 2, P-Value = 0.003
```

The two chi-square tests use alternate equations for comparing the observed number of counts to the expected number of counts in each cell of the table. Expected counts are the numbers expected when the null hypothesis is true. If the observed counts are similar to expected, the test statistic is small and the null hypothesis is not rejected. If the observed counts differ from what is expected, the test statistic is large and the null hypothesis is rejected. Both tests indicate that the proportions of data at or above 5 μg/L differ among the three land-use groups, with a p-value of no more than 0.01. The chi-square test detected differences that were obscured when an analysis of variance on substituted data was performed. If a "quick and dirty" test is required, this test of proportions is just as quick, and far less dirty, than substitution.

10.2.2   Ordinal Methods: Kruskal–Wallis Test

A second approach to analyzing these data without a distributional assumption is the KW test. The KW test determines whether the distribution functions (edfs) of three or more groups of data are similar, or if at least one is different. The test is applied to censored data by setting all observations below the highest reporting limit to the same value. When ranked, these observations become tied at the lowest rank. Data above the highest reporting limit are ranked using the same method as for uncensored data. If survival analysis software is not available, the KW test provides a valid nonparametric alternative. However, score tests in survival analysis software will provide greater power for multiply censored data, without recensoring data to the highest reporting limit.

For the TCE data, all observations below the highest reporting limit of 5 µg/L are assigned the same value, any value less than 5. All <1, <2, and so on, as well as all detected values of 1, 2, 3, and 4 are assigned an identical low number to represent that they are less than 5. The Minitab macro censKW assigns a -1 to these lowest values. In this way, all values censored below the highest reporting limit are given tied ranks. A negative number is used to reinforce that this is an assigned value, rather than an actual measurement. The KW p-value of 0.01 shows that even with approximately 80% censoring, differences among the edfs of the three groups can be discerned. As mentioned in Chapter 9, a large proportion of censoring is not in itself a reason to avoid performing a rank-based test like the KW test. If the proportion of censoring differs significantly among the groups, as it does here, that difference can be discerned by the test.

```
Kruskal-Wallis Test on TCEConc-
```

Density-	N	Median	Ave Rank	Z
Low	25	-1.000	109.0	-1.11
Medium	130	-1.000	120.4	-0.84
High	92	-1.000	133.2	1.56
Overall	247		124.0	

H $=2.95$	DF $=2$	P $=0.228$	
H $=9.17$	DF $=2$	P $=0.010$	(adjusted for ties)

```
Use the tie adjustment. All values below the max dl were set
as tied at -1.
```

Note the estimated median of -1 for all three groups results from the assignment of that value to all censored observations. The -1 values are to be interpreted as censored values, so the median is <5.

10.2.3 The Kruskal–Wallis Test in R

The Kruskal–Wallis test is available in R with the kruskal.test command. But first, the TCE data must be recensored so that all values below 5 are set to a low number such as -1. This can be done relatively easily in R and gives insight into how data can be managed using R.

The TCE data set is one of the data sets that is packaged with NADA for R. So after loading NADA for R, the data set can be activated using

```
>     data(TCE)
>     attach(TCE)
```

and variable names plus the first 6 rows of data can be seen using the head command:

```
> head(TCE)
       Density     LowEq1      MedEq1      HiEq1       TCEConc     TCECen
1      Low         1           0           0           1           TRUE
2      Low         1           0           0           1           TRUE
3      Low         1           0           0           1           TRUE
4      Low         1           0           0           1           TRUE
5      Low         1           0           0           1           TRUE
6      Low         1           0           0           1           TRUE
```

It takes only two steps to recensor all values below 5 to equal -1. In the first step, all TCEConc data below 5 are set to -1. In the second, all censored <5 values are set equal to -1. Note the use of brackets to define a subset of a column's data. Read "TCE5[TCEConc<5]" as "TCE5 values for rows where TCEConc is less than 5." Also note that a declaration of equality other than numeric is defined using $==$.

```
> TCE5=TCEConc
> TCE5[TCEConc<5]=-1
> TCE5[TCECen==TRUE]=-1
```

Now the Kruskal–Wallis test is computed on the recensored TCE5 variable.

```
> kruskal.test(TCE5,Density)

    Kruskal-Wallis rank sum test

data: TCE5 and Density
Kruskal-Wallis chi-squared = 9.1725, df = 2, p-value = 0.01019
```

The p-value reported is the one corrected for ties, showing that the group edfs are significantly different.

10.3   MAXIMUM LIKELIHOOD ESTIMATION

Maximum likelihood estimation (MLE) can be used with censored data to perform hypothesis tests similar to analysis of variance. As with other parametric methods, the results will be valid if the data used for the test closely follow the distribution assumed by the MLE. When the data do not match the distribution assumed by MLE, hypothesis tests may have low power to discern differences present between groups.

In order to use MLE, the data are usually coded as interval-censored data in the Interval Endpoints format described in Chapter 4. Censored observations have a lower bound of either 0 or the method detection limit (if it is detected but not quantified). The upper bound is usually the reporting limit. Hypothesis tests will then recognize that data cannot go negative. Uncensored quantified observations have the same value in each of the two endpoint columns. Alternatively, for skewed distributions the data (or their reporting limits) can be log-transformed and then flipped. In addition to accounting for skewness, this transformation allows software for right-censored data to be used to conduct the test.

10.3.1   Likelihood-Ratio Tests

Parametric hypothesis tests for differences among groups in the means of censored data can be accomplished using censored regression methods available in survival analysis software. A likelihood-ratio test equivalent in purpose to ANOVA's F-test determines whether there is a significant increase in the log-likelihood of data classified by group in comparison to the log-likelihood for data when unclassified (the null likelihood). In other words, does the classification by group explain a significant portion of the variation observed in the data? If so, the means of the groups are not all similar, and classification reflects that the location of at least one group is not identical to the others. When the log-likelihood statistic with classification is significantly greater (less negative) than that for no classification, the null hypothesis of no difference among means is rejected. The test statistic is the "-2 log-likelihood"

$$-2 \log\text{-likelihood} = -2(L_{\text{null}} - L_{\text{groups}}) \qquad (10.1)$$

where L is the log-likelihood of each situation. Because these are logs of the likelihood, subtracting one from the other is equivalent to a ratio of likelihoods, hence the name "likelihood-ratio test."

Some software packages compute both the log-likelihood L and the -2 log-likelihood, comparing the latter to a chi-square distribution with $k-1$ degrees of freedom, where k is the number of groups. It is a one-step process. Other packages require separate commands to print the model and null likelihoods, requiring you to compute the test statistic using equation (10.1), and its p-value. However this is not an onerous task. Before the TCE data are used as an example, the first step in computing a likelihood-ratio test is to represent the data in the interval endpoints format.

10.3.2   Representing Interval-Censored Data

TCE concentrations have a lower bound of zero, as do most environmental data. Some statistical software allows entry of low and high endpoints for censored data, called "interval censored" or "arbitrary censored" data. Setting the lower bound at zero is important in accurately representing the possible values of the data and for returning correct values for coefficients and tests. If no lower bound is set (the lower end is a missing value), values are assumed to be able to go as low as minus infinity. This produces a low bias for estimates of the mean, high bias for standard deviation, and incorrect test results in parametric hypothesis tests.

Data in the indicator format are often assumed to be right-censored with no upper bound. This is not an issue with nonparametric methods—the highest observation receives the highest rank. However, flipping left-censored environmental data and running MLE survival analysis procedures with no upper bound can produce incorrect results. If a normal distribution is assumed, a lower bound *must* be represented. This can be done with software that allows interval censoring, such as Minitab, R, and SAS. When using a lognormal distribution, however, the lower end of the logarithms can go to minus infinity because this translates into a lower bound of zero in original units. If available software only allows right-censored data, compute the logarithms of data, flip the logarithms by subtracting from any number larger than the maximum log value, and run the procedure on these right-censored values (y_{flip} in equation (10.2).

$$y_{flip} = C - \ln(x) \qquad (10.2)$$

where C is any number larger than the maximum of $\ln(x)$. The result when retransformed back into units of x will be correct for lognormal data with a lower bound of zero, the latter represented by the plus infinity value for y_{flip}. As most environmental data more closely follow a lognormal rather than normal distribution, this procedure should work well for many data sets if interval-censored software is not available.

Using Minitab's `Regression with Life Data` procedure, interval-censored data can be directly entered as the response variable in a censored regression (Figure 10.3). For the TCE data, the low end of the interval is the variable TCE0, where all censored observations are represented by a value of 0. The high end of the interval is the variable TCEConc, with censored observations set to the reporting limit. The format is the "interval endpoints" format of Chapter 4. As shown later,

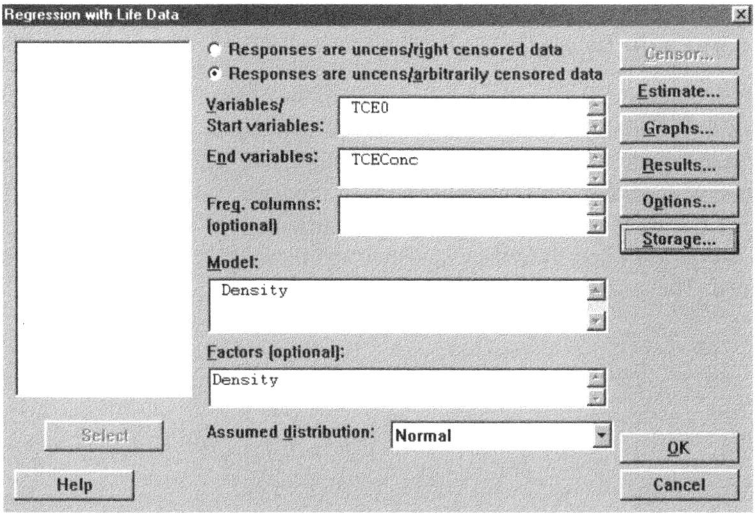

FIGURE 10.3 Entry window for censored regression in Minitab. Interval ("arbitrary") censoring results from use of the Start and End variables.

censored regression software will produce a parametric test for differences among the means of these grouped data.

10.3.3   Representing Data Groups with Regression Software

Minitab and several other survival analysis packages allow explanatory variables in a regression to be specified as a "factor" or "grouped variable" or a "category." These variables have a single value for every observation in a group and so define the group identity for each observation. These may be numeric or text variables—in the TCE example the group identifier is the text variable Density. Explanatory variables entered into the Model dialog box are designated as factors in Minitab by also entering them into the Factors dialog box, as shown in Figure 10.3.

Specification of variables as factors in regression can be accomplished by hand if the software does not have an option to do so automatically. To do this, the group identifier is recoded into binary (0/1) variables, using the same process as for analysis of covariance with regression software (Helsel and Hirsch, 2002, Chapter 11). Membership in one of k groups is represented by $(k - 1)$ binary variables, each with values either as 0 or 1. Membership is usually indicated as the value 1. It takes two binary variables to represent the same information contained in the three groups designated by the variable Density. One variable is named MedEq1, and has a value of 1 for every observation in the medium density group. Observations from the low- or high-density groups have a value of 0 for the MedEq1 variable. Similarly, the variable HiEq1 has the value 1 for all data from the high-density group, and zeros otherwise. The low-density group is defined as the "baseline" situation, with values for both MedEq1 and HiEq1 equal to 0. Thankfully, most statistics software will do this for you by declaring Density as a factor.

10.3.4   The TCE Example

With TCE concentrations in the interval zero to the reporting limits represented by TCE0 and TCEConc, Density groups entered as a factor (Figure 10.3), and the assumed distribution set to a normal distribution, Minitab's MLE life-regression procedure will estimate an overall likelihood ratio used in determining whether the means of the three groups are significantly different. In essence it produces an ANOVA-type test where differences and their significance are estimated by maximum likelihood. The output of the procedure is

```
                       Standard                        95.0% Normal CI
Predictor    Coef      Error       Z       P       Lower       Upper
Intercept    1.01988   6.08187     0.17    0.867   -10.9003    12.9401
Density
  High       6.74525   6.85866     0.98    0.325    -6.69748   20.1879
Medium       7.00662   6.64111     1.06    0.291    -6.00971   20.0229
Scale       30.4048    1.36799                      27.8383    33.2078

Log-Likelihood = -1069.163
```

To determine whether significant differences in mean TCE are found between Density groups, the log-likelihood for no grouping (the null log-likelihood) is obtained by running the

```
Stat > Reliability/Survival > Distribution Analysis
(Arbitrary Censoring) > Parametric Distribution Analysis
```

procedure, setting the assumed distribution to the same as for the grouping results, here a normal distribution. The output from the procedure estimates the mean and standard deviation for the entire TCEConc column without breaking it into groups, and produces the null log-likelihood:

Parameter	Estimate	Standard Error	95.0% Normal CI Lower	Upper
Mean	7.21985	1.93959	3.41830	11.0213
StDev	30.4761	1.37120	27.9037	33.2857

Log-Likelihood=−1069.741

To test for whether there are significant differences in mean TCE among the three Density groups, compute the likelihood-ratio test statistic (equation 10.1)

$$-2\log\text{-likelihood} = -2(L_{null} - L_{groups}) = 2(-1069.741 - [-1069.163]) = 1.156$$

This statistic is compared to a chi-square distribution with $(k-1)=2$ degrees of freedom. The number of degrees of freedom is the number of binary variables added to the null equation, and so equals $(k-1)$. The resulting p-value is 0.56, indicating that no significant differences are found among the three group means.

If there had been no designation for a grouping variable (factor) available in Minitab, the two binary variables MedEq1 and HiEq1 described previously could have been entered as explanatory variables in the model dialog box (Figure 10.4). The resulting output shown below is identical to that produced when the factor designation was used. Internally, the software designates all but one of the values for the factor variable as a 0/1 binary variable and enters those as explanatory variables in the regression model.

Predictor	Coef	Standard Error	Z	P	95.0% Normal CI Lower	Upper
Intercept	1.01988	6.08187	0.17	0.867	−10.9003	12.9401
MedEq1	7.00662	6.64111	1.06	0.291	−6.00971	20.0229
HiEq1	6.74525	6.85866	0.98	0.325	−6.69748	20.1879
Scale	30.4048	1.36799			27.8383	33.2078

Log-Likelihood=−1069.163

10.3.5   Importance of the Assumed Distribution

The validity of results for a parametric method, whether involving censored data or not, depends on the adherence of the observed data to the assumed distribution. One of the common procedures for evaluating adherence to the assumed distribution is a

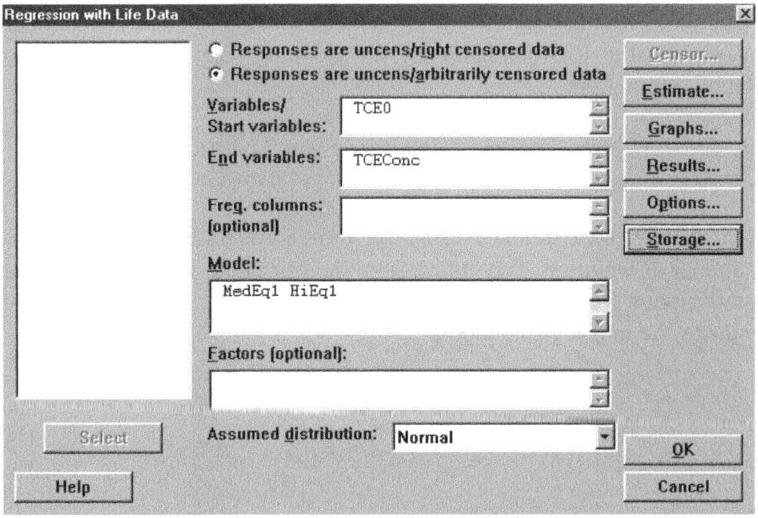

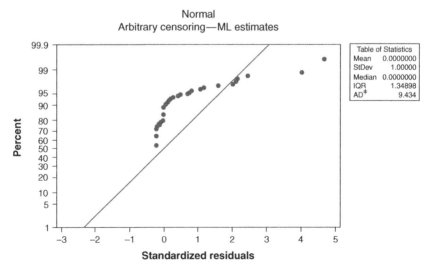

FIGURE 10.4 Entry window for censored regression using 0/1 binary variables instead of a factor.

probability plot of the residuals. Figure 10.5 is a probability plot of residuals from the censored regression for the TCE data. The departure of the residuals from the straight line indicates that it is unlikely they arose from a normal distribution. When data depart from the assumed distribution, the power of parametric procedures is low and findings of no difference between group means can result. Therefore, the TCE data should be tested again, this time with a lognormal distribution.

FIGURE 10.5 Probability plot of residuals from the MLE test for differences in TCE group means. The data do not appear linear, indicating the residuals do not follow a normal distribution.

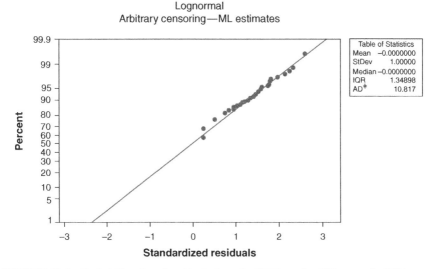

FIGURE 10.6   Probability plot of residuals from the MLE test for differences in TCE group means, assuming a lognormal distribution. The linear pattern indicates that a lognormal distribution is a reasonable assumption.

Testing for differences between the mean logarithms of the three TCE groups can be quickly accomplished in Minitab by choosing the lognormal distribution in the dialog box. First, the probability plot of residuals (Figure 10.6) is much more linear than Figure 10.5. The data are more appropriately analyzed in log units.

Second, the overall test determines whether there are any difference among the three mean logarithms (geometric means). To obtain the overall test we must compute the -2 log-likelihood test. The null log-likelihood has a value of -316.40. The output from the three-group model is below.

```
Standard                     95.0% Normal CI
Predictor   Coef      Error      Z       P       Lower       Upper
Intercept   -3.78195  1.15396   -3.28    0.001   -6.04369    -1.52021
Density
  High       3.05966  1.13781    2.69    0.007    0.829594    5.28974
Medium       1.40339  1.11891    1.25    0.210   -0.789640    3.59642
Scale        2.85286  0.317678                    2.29349     3.54867

Log-Likelihood = -308.700
```

The $-2\log\text{-likelihood} = -2(L_{\text{null}} - L_{\text{groups}}) = 2(-316.40 - [-308.70]) = 15.40$, which when compared to a chi-square distribution with 2 degrees of freedom, corresponds to a p-value for the test of 0.0005. The groups do differ in their geometric means. Now, which groups differ from the others?

Some of the relevant individual comparisons are printed in the three-group output. The test identified with the high row is the test for whether the high geometric mean (median) differs from the low (base, not shown) group geometric mean. The p-value of 0.007 shows that these two groups are significantly different. The test identified with the medium row is the test for whether the medium group geometric mean is significantly different than the low group geometric mean. With a p-value of 0.210, it is not. The final pairwise significance test (does the high geometric mean differ from the medium geometric mean?) is not printed on the output, but by rearranging the terms and rerunning, the resulting p-value of 0.003 shows that these groups are significantly different as well.

10.3.6   Censored Regression "ANOVA" Using NADA for R

The cenmle command computes a censored regression, estimating slopes by MLE. Using this command, the overall chi-square test of significance is reported, as well as some of the individual group mean comparisons.

```
> tcemle = cenmle(TCEConc, TCECen, Density, dist="lognormal")
> tcemle
                   Value Std. Error          z           p
(Intercept)       -0.722      0.416      -1.73   8.28e-02
DensityLow        -3.060      1.138      -2.69   7.17e-03
DensityMedium     -1.656      0.553      -2.99   2.76e-03
Log(scale)         1.048      0.111       9.41   4.76e-21
Scale = 2.85

Log Normal distribution
Loglik(model) = -308.7  Loglik(intercept only) = -316.4
Loglik-r: 0.2459125

Chisq = 15.41 on 2 degrees of freedom, p = 0.00045
Number of Newton-Raphson Iterations: 4
n = 247
```

The chi-square test statistic is the -2 log-likelihood, twice the difference of the log-likelihoods for the density factor model minus the null (intercept-only) model. The resulting p-value of 0.0004 leads us to reject the null hypothesis that the mean log of concentrations in all three groups are equal. The cenboxplot command produces a picture of the differences, given that the box portions below the highest reporting limit are estimated using the ROS method (Figure 10.7)

```
> cenboxplot(TCEConc, TCECen, Density)
```

10.3.7   Censored Regression "ANOVA" with Other Software Packages

How would this "censored ANOVA" test be computed if the software only allowed input of right-censored data, and could not construct the 0/1 factor indicators from grouping variables? The same results would be obtained as above, assuming a lognormal distribution, using these steps:

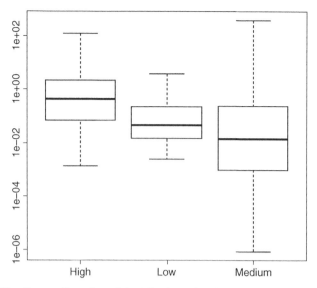

FIGURE 10.7 Censored boxplots of the TCE data using the cenboxplot command of NADA for R.

1. Take logarithms of the TCE data.
2. Flip the logs by subtracting from a large constant. This produces a right-censored data set whose maximum at infinity will map back to a zero lower bound for TCE.
3. Create the two binary variables MedEq1 and HiEq1. Use these as the two explanatory variables in the regression equation. The entry window for the setup using Minitab is given in Figure 10.8.
4. From the output (shown below) compare the log-likelihood of the model to the null log-likelihood. The test statistic value is the same 15.40 found for the more automated procedure above. Also note that the slopes for the two explanatory variables are just (-1) times the values found above for the same variables. The sign is now negative because the data have been flipped.

```
                     Standard                        95.0% Normal CI
Predictor   Coef     Error      Z        P       Lower       Upper
Intercept   13.7819  1.15396    11.94    0.000   11.5201     16.0436
MedEq1      -1.40339 1.11891    -1.25    0.210   -3.59642    0.789640
HiEq1       -3.05966 1.13781    -2.69    0.007   -5.28974    -0.829594
Scale       2.85286  0.317678                    2.29349     3.54867

Log-Likelihood=-197.761
```

$$-2 \text{ log-likelihood} = -2(L_{\text{null}} - L_{\text{groups}}) = 2(-205.46 - [-197.76]) = 15.40$$

Either the automated grouping or the manual creation of binary variables results in the same likelihood-ratio test and p-value, the same individual slope coefficients

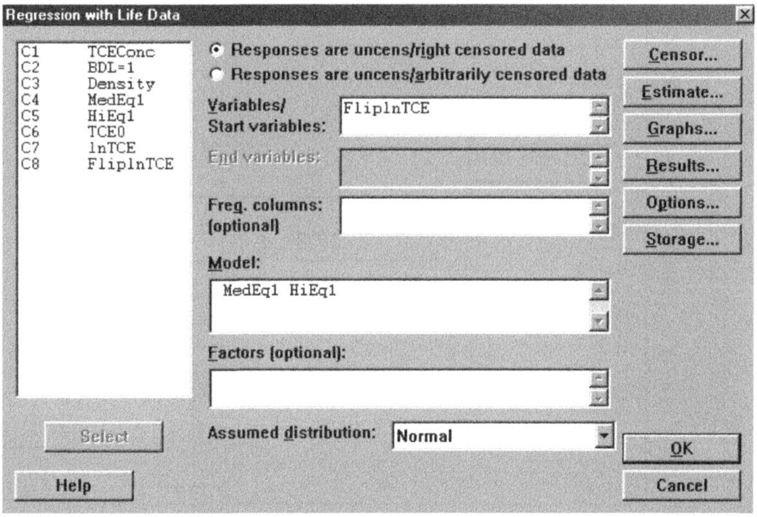

FIGURE 10.8   Entry window for right-censored flipped logarithms of TCE data.

(multiplied by -1 if the data were flipped), and the same tests of significance on the individual pairwise comparisons between groups. Pairwise comparisons are the way in which multiple comparisons can be computed using MLE software for censored data.

10.3.8   Multiple Comparison Tests

If an overall test is found to be significant, the next question is often "which groups differ from the others?" A series of individual comparisons between group means can be performed to answer this question. To determine the entire pattern of k group means requires $g = k(k-1)/2$ comparisons. For three groups, the $g = 3$ comparisons are between groups 1 and 2, groups 1 and 3, and groups 2 and 3. The end result might be something like "groups 1 and 2 are not significantly different, but both are lower than the mean of group 3." If an overall error rate of 5% is desired, so that there is no more than a 5% chance of making one error in the ordering of the means, each group comparison must be made at an individual error rate smaller than 5%. One commonly used formula for determining the individual error rate is Bonferroni's formula in equation (10.3):

$$\text{individual error rate} = \frac{\alpha}{g} \tag{10.3}$$

where α is the desired overall error rate (often 5%, 0.05) and g is the number of comparisons between means to be made. Any p-values for tests of differences between two means that are below the Bonferroni-adjusted level are considered significantly different at a 5% error rate for the overall pattern.

For the test of logarithms of the TCE data with an overall rate of $\alpha = 0.05$, the p-values for individual comparisons must be less than $0.05/3 = 0.017$. Both of the tests found significant at the 0.05 level are still significant when $\alpha = 0.017$ because their p-values are less than this. For the tests found to be significant, the slope coefficient describes by how much they differ. The difference between mean (logarithms) of the high and low groups is 3.06. To express this in the original units of concentration, the mean logarithm (geometric mean) for the high-density group is typically $e^{3.06} = 21.3$ times higher than the geometric mean for the low-density group.

Bonferroni's procedure for multiple comparisons is a "conservative" method, because the individual α-levels may be lower than necessary in order to achieve an overall 5% error rate. An alternative to Bonferroni's multiple comparison procedure is Tukey's honest significant difference test (Zar, 1999). Tukey's test uses the sample sizes in each group to adjust the distances by which two means must be separated in order to call them significantly different. The test statistic q for comparing the means of any two groups (groups a and b) is shown in equation (10.4):

$$q = \frac{\overline{x}_b - \overline{x}_a}{\sqrt{(s^2/2)((1/n_\alpha) + (1/n_b))}} \tag{10.4}$$

where s is the scale coefficient printed in the regression output from maximum likelihood regression. The calculated statistic q is compared to a table of $q_{\alpha,\nu,k}$, the Studentized range statistic, which is a function of α (the overall error rate), ν (the error degrees of freedom $= n - k$), and k (the total number of groups to be compared). However, this test is not currently setup to function under MLE software for survival analysis, so it must be computed by hand.

10.4 NONPARAMETRIC METHOD—THE GENERALIZED WILCOXON TEST

Nonparametric tests do not assume the data follow any particular distribution, and therefore no transformations are required prior to computing the tests. With many commercial statistics packages, nonparametric censored methods allow only right-censored data to be entered. If this is true, the data must be flipped to become right-censored before using these methods. The resulting p-values will be identical to those that would be computed if left-censored data were allowed. Flipping data for nonparametric tests merely changes the order of ranking from high to low instead of from low to high. Flipping censored data was previously discussed in Chapter 2.

10.4.1 The Generalized Wilcoxon Score Test

The generalized Wilcoxon score test for three or more groups is an extension of the score test for two groups presented in Chapter 9. Like the KW test, score tests determine whether distributions (edfs) of groups are the same, or if at least one is different. Unlike the KW test, score tests extract more information from the data by assigning estimated percentiles (or scores) to uncensored observations falling between multiple censoring thresholds. When data have multiple reporting limits, a score test will have more power (ability to see differences between edfs) than the KW test because no additional censoring is required.

Using Minitab, the TCE data were first flipped by subtracting each concentration from 400 to produce a right-censored variable "FlipTCE." The Gehan version of the generalized Wilcoxon test is computed on FlipTCE using the

```
Reliability/Survival > Distribution Analysis (Right Censoring) >
Nonparametric Distribution Analysis
```

command.

```
Comparison of Survival Curves

Test Statistics

Method        Chi-Square      DF      P-Value
Log-Rank      16.2794         2       0.000
Wilcoxon      16.0761         2       0.000
```

The generalized Wilcoxon (Gehan) test statistic is 16.076, with a corresponding p-value of <0.001, indicating that the distributions of concentrations differ significantly among the three groups. The lower p-value of the score test as compared to the KW test ($p = 0.01$) illustrates the greater power of the score test, as it uses the information below the highest reporting limit of $5\,\mu$g/L more efficiently.

Flipping data is tedious, so a Minitab macro called "gw" will perform the generalized Wilcoxon test after invisibly flipping the data. The command and resulting output for the TCE data are given below. The interim output, Kaplan–Meier statistics on the data in flipped units, is not repeated here. The final resulting p-value is seen as identical to the one above that required flipping the data by hand.

```
> %gw 'TCECONC' 'BDL_1' 'Density'
Method        Chi-Square      DF      P-Value
Log-Rank      16.2795         2       0.000
Wilcoxon      16.0762         2       0.000
```

Another benefit of the macro is that the edfs for each group are plotted (Figure 10.9). These are plots of 1 minus the survival functions, or the "cumulative failure" probabilities of the data. Following a line across the graph at the 90th percentile clearly shows that for this upper end of the distributions, the high-density group has larger concentrations than the other two groups. This reflects the larger proportion of detected concentrations originally seen in the boxplots of Figure 10.1. The spread between the curves of the three edfs is the significant difference identified by the Wilcoxon score test.

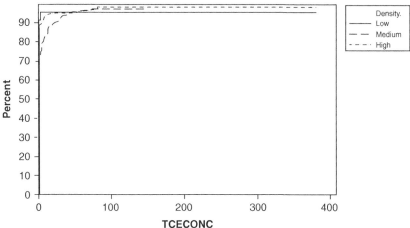

Turnbull method
Arbitrary censoring

FIGURE 10.9   Cumulative failure plot (edf) of TCE concentrations for three land-use categories. Differences appear substantial between the 80th and 99th percentiles.

10.4.2   The Wilcoxon Score Test Using NADA for R

The cenfit command reports descriptive statistics estimated using the Kaplan–Meier procedure for data in each factor group NA means 'below the RL'.

```
> cenfit(TCEConc, TCECen, Density)
                    n      n.cen     median     mean        sd
Density=High       92      58        NA         7.778019    19.4895231
Density=Low        25      23        NA         1.166667    0.6364688
Density=Medium     130     113       NA         7.867264    38.5791132
```

Plotting the cenfit object produces the left-censored edf plot for each group (Figure 10.10):

```
> plot(cenfit(TCEConc, TCECen, Density),xlab="TCE concentration")
> legend(20,0.4,legend=c("High","Low","Medium"),lty=c(1:3))
```

The Peto-Prentice or HF-1 version of the generalized Wilcoxon test is computed using the cendiff command:

```
> cendiff(TCEConc, TCECen, Density)
                    N      Observed    Expected    (O-E)^2/E    (O-E)^2/V
Density=High       92      30.45       18.2        8.26         15.65
Density=Low        25      1.73        5.7         2.76         3.62
Density=Medium     130     15.47       23.8        2.89         6.76

Chisq=16.3 on 2 degrees of freedom, p=0.000295
```

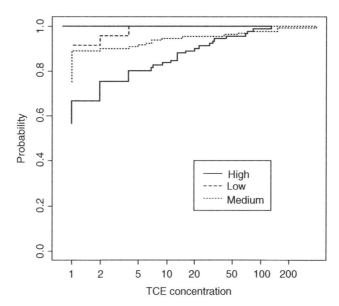

FIGURE 10.10   Plot of edfs for TCE concentrations (log scale) for three land-use categories. Differences are substantial between the 80th and 99th percentiles.

The HF-1 test statistic is 16.3 while the Gehan version was 16.08, both strongly significant. This nonparametric test handling multiple reporting limits without substitution of any kind has determined that these data with 70–80% censored values show significant differences between the edfs of the three groups. The high-density group tends to produce higher concentrations and higher probabilities of detected concentrations than the other two groups.

10.4.3   Multiple Comparison Tests

Multiple comparison procedures are a logical next step following rejection of the null hypothesis of similarity by the Wilcoxon score test. To date, the only approach available with commercial software for nonparametric multiple comparisons for censored data is to perform a series of two-group score tests between each pair of groups. If the p-value is less than the Bonferroni individual comparison level obtained using equation (10.3), the two groups can be declared to have different distribution functions at the chosen overall error rate.

For the $k = 3$ groups of TCE data there are $g = 3$ pairwise comparisons. Using Bonferroni's method, each pairwise comparison must have a p-value below $0.05/g = 0.017$ for the group edfs to be considered different at the overall error rate of 5%. Computing each of these pairwise comparisons for the TCE data using the gw macro in Minitab results in

Low versus Medium

Method	Chi-Square	DF	P-Value
Wilcoxon	0.68890	1	0.407

Low versus High

Method	Chi-Square	DF	P-Value
Wilcoxon	7.09906	1	0.008

Medium versus High

Method	Chi-Square	DF	P-Value
Wilcoxon	11.5275	1	0.001

Both the low- and medium-density areas have significantly different TCE concentrations than the high-density area, using nonparametric score tests. The low- and medium-density areas are not different from each other. These are the same results assuming a lognormal distribution using the parametric MLE test.

Other methods for nonparametric multiple comparisons exist, but are not coded into survival analysis software. These include the Dwass et al. test (Hollander and Wolfe, 1999, p. 240), a test computed using ranks of the data. For censored data the Wilcoxon scores can be used rather than the ranks. A second nonparametric multiple comparison test is the slippage test (Conover, 1968), a test that counts the number of observations in a group that are greater than the highest observation in the next lowest group. If more observations than expected exceed the top of the next group, that group has "slipped" significantly lower and a difference is declared. Because the test statistic is based primarily on observations at the high end of each group, left censoring is not often an issue when computing the slippage test.

To illustrate the slippage test, the $k = 3$ residential-density groups of TCE concentrations are ordered based on the magnitude of their maximum observations (see Table 10.2). By this criteria the medium-density group is "higher" than the high-density group. Starting with the highest group, the number of observations r_i is counted that exceed the maximum of the next lowest group. For example, three TCE concentrations in the medium-density group exceed 130 µg/L, the highest concentration in the high-density group.

TABLE 10.2   Computation of the Slippage Test for the TCE Data

Density	Medium	High	Low
Maximum TCE	382	130	<5 or 4
# exceedances r_i	3	18	–
Sample size, n	130	92	25
Prob $(edf_i = edf_j)$	0.275	0.016	–

The probability that the distribution of the ith group Gp_i is identical to the distribution of the next lowest group Gp_{i+1}, is (Conover, 1968)

$$\text{Prob}\,(Gp_i > Gp_{i+1}) = \left[\frac{n_i - 1}{\left(\sum_{j=i}^{k} n_j\right) - 1}\right]^{r-1} \tag{10.5}$$

This is equivalent to the p-value for the individual comparison test. For the comparison between the $(i=1)$ medium-density group and the next lowest (high-density) group,

$$\text{Prob}\,(Gp_1 > Gp_2) = \left[\frac{129}{(130 + 92 + 25) - 1}\right]^{2} = 0.275$$

and between the $(i=2)$ high-density group and the $(i=3)$ low-density group,

$$\text{Prob}\,(Gp_2 > Gp_3) = \left[\frac{91}{(92 + 25) - 1}\right]^{17} = 0.016$$

The p-values get smaller as the test statistic r, the number of exceedances, gets larger. When r equals 1, the p-value resulting from equation 10.5 equals 1.

Individual comparisons are declared different if their p-values are less than $\alpha/(k-1)$, where k is the number of groups and α is the desired overall error rate. For $\alpha = 0.05$ and $k = 3$, the p-values are compared to an individual error rate of 0.05/2, or 0.025. Therefore, the slippage test concludes that there is a significant difference between the high- and low-density groups, but not between the medium and high-density groups.

The results of different multiple comparison tests do not always agree. The slippage test and individual Wilcoxon tests did not agree as to which groups differed from others. The slippage test focuses more on the high end of each group than does the Wilcoxon test, and is insensitive to differences occurring in the central and lower portions of each group. The results of the slippage test seem to agree with the visual impression of the boxplots of Figure 10.1, which only show the high-end values above the highest reporting limit. The choice of which test to use should be based on which characteristic of each group is the most appropriate to distinguish.

Perhaps in future releases, software will provide more in the way of nonparametric multiple comparison procedures for censored data.

10.4.4 Interval-Censored Score Tests: Testing Data That Include (DL to RL) Values

The interval-censored form of the generalized Wilcoxon test coded into the "interval" contributed package for R (Fay and Shaw, 2010) will perform equally as well for three and more groups of data as it did in Chapter 9 for two groups. Data with censored values reported as (0 to DL) for nondetects and (DL to QL) for data between the detection and quantitation limits can be directly input to these procedures.

```
> TCE$TCElo=with(TCE, TCEConc*(1-TCECen))
> TCE$Lt=with(TCE, TCE$TCEConc<1000)
> TCE$Rt=with(TCE, !TCECen)
> TCEtest=ictest(TCElo,TCEConc,Density,rho=1,Lin=TCE$Lt,Rin=TCE$Rt)
> TCEtest

Asymptotic Wilcoxon k-sample test (permutation form)
data: {TCElo,TCEConc} by Density
Chi Square=16.4905, p-value=0.0002625
alternative hypothesis: survival distributions not equal

              n       Score Statistic*
Low          25       3.966279
Medium      130       8.291785
High         92      -12.258064
* like Obs-Exp, positive implies earlier failures than expected
```

Test results are the same as the Harrington–Fleming rho $= 1$ form of the test in the previous section. The ictest command adds the important ability to incorporate data at nonzero low ends of the censoring interval (though this data set does not contain any of those). No substitution is required; no numbers need be estimated between the detection and quantitation limits. Censored data are designated as either below the detection limit or in between the two limits. The resulting test is completely without a distributional assumption. While becoming familiar with the software takes adjustment if you are not used to using R and its contributed packages, the result is well worth the time. This test strongly finds a difference, as opposed to the "no difference" result from substituting one-half of each limit and running analysis of variance shown at the beginning of this chapter. The last sentence tells the direction of difference, believe it or not—"positive implies earlier failures than expected" refers to the score statistic for each group. The two groups with positive scores have "earlier failures," or lower detected values, than expected if all showed the same distribution of concentrations. The high group's negative score means that it has "later failures" or higher detects than what is expected on average if all groups are the same.

The contributed package "interval" with its ictest command is a major advance in handling censored data that has occurred since the first edition of this book was published in 2005.

10.5 SUMMARY

Tests for differences in the distributions of three or more groups of censored environmental data can be carried out in several ways. Simpler tests consider all values below the highest reporting limit to be identical. The binary contingency table test and the ordinal Kruskal–Wallis test can then be computed on the recensored values. While losing some information, these simpler tests avoid all of the problems due to the invasive patterns introduced by substitution methods.

Maximum likelihood estimation provides parametric tests of hypotheses without substitution. When the assumption of a distributional shape such as the lognormal can be verified, MLE methods perform well. These MLE analogs of analysis of variance are directly conducted on left-censored data using methods for "arbitrary" or interval-censored data, where censored observations are considered to be within two ends of an interval. Group differences can be tested using either a "factor" designation for the group variable or with binary variables representing group membership. Tests for whether regression slopes are different from zero are in this case parametric tests for differences in group means of censored data.

Nonparametric score tests extract the maximum information for determining group differences from multiply censored data without assuming a distributional shape. The generalized Wilcoxon test, an extension of the Kruskal–Wallis test to multiply-censored data, efficiently detects differences in the distributions of groups without assuming normality. With most commercial software, data must still be flipped into a right-censored format prior to testing. However, with more advanced software packages such as the "interval" package in R, nonparametric methods can be computed directly using interval-censored data.

Reasonable methods for plotting multiply-censored data are available, starting with survival function plots, which plot from right to left the edfs of censored data. Cumulative failure plots will reverse this, plotting the edfs in the expected direction. These plots show the results of the Wilcoxon score test, much as censored boxplots do for the Kruskal–Wallis test and bar graphs of percentages above the highest reporting limit do for contingency table analyses.

EXERCISES

10-1 Golden et al. (2003) measured concentrations of lead in the blood and in several organs of herons in Virginia, in order to relate those concentrations to levels found in feathers. The objective was to determine whether feathers were a sensitive indicator of exposure to lead. If so, feathers could be collected in the future so that the birds would not need to be sacrificed in order for their exposure to lead to be evaluated. The herons received different doses of lead—exposure was categorized into one of four groups: a control group receiving no additional lead, and groups receiving 0.01, 0.05, and 0.25 mg lead per g of body weight. Determine whether the lead found in feathers of these birds differed among the four exposure groups at an $\alpha = 0.05$ level. If so, run a multiple comparison test to determine which groups differ from the others. Use the methods of this chapter (not substitution!) to test for differences, using the data found in Golden.xls.

10-2 Brumbaugh et al. (2001) measured mercury concentrations in fish of approximately the same trophic level across the United States, as well as characteristics for the watersheds they lived in. The data are found in HgFish.xls. The variable

"LandUse" reflects the dominant land use within the watershed, and includes categories of Ag/Forested (or "A/F"), Ag, Mining, Urban, and Background. Test to see if mercury concentrations in fish (variable "Hg") differ among the five land-use categories. The mercury concentrations have been censored at three reporting limits, as indicated with a value of 1 for the variable "HgBDL1." Note that the A/F land-use includes watersheds containing the largest proportion of wetlands.

10-3 Yamaguchi et al. (2003) measured concentrations of PCBs in fish collected at four sites draining to the Thames River, UK. Three sites are below Swindon, an active industrial area draining to the Ray River, a tributary to the Thames. The fourth site, Burford, is on the Windrush tributary and not downstream of Swindon. Test whether PCB concentrations are the same or different in fish at the four sites, using both parametric (censored regression) and nonparametric (Wilcoxon score) tests. The data are found in Thames.xls.

11 Correlation

How strong is the association between two variables? As X increases, how likely is it that Y consistently increases, or decreases? The strength of the association or dependence between two variables, how predictably they covary, is measured by a correlation coefficient. Correlation coefficients can take on values between -1 and $+1$. At zero, no correlation is observed. As the coefficient moves away from 0 in either direction, evidence for correlation increases. At values of $+1$ or -1 perfect dependence is observed, with the sign denoting whether the two variables move in the same direction ($+$) or in opposite directions ($-$). One of the most common uses for correlation coefficients in environmental studies has been for investigation of trends. When X represents time, the test for significance of the correlation between Y and X is a test for a trend in Y. Do the observations consistently increase or decrease over the time period of collection?

11.1 TYPES OF CORRELATION COEFFICIENTS

The traditional (parametric) correlation coefficient is Pearson's r. Pearson's coefficient measures the linear correlation between Y and X. As seen in equation (11.1), Pearson's r involves computing both the mean and standard deviation for both X and Y, as well as a measure of the distance each observation is from its mean. All three items are difficult to calculate with censored data. Software for censored data does not attempt to compute Pearson's r.

$$\text{Pearson's } r = \frac{1}{n-1} \sum_{i=1}^{n} \left(\frac{x_i - \bar{x}}{s_x} \right) \left(\frac{y_i - \bar{y}}{s_y} \right) \tag{11.1}$$

Spearman's rho (ρ) is a nonparametric correlation coefficient that is computed by calculating Pearson's r on the ranks of the original data. Where those ranks can be computed unequivocally, as when there is only one reporting limit, Spearman's rho provides a feasible alternative to Pearson's r for censored data. For multiple reporting limits, however, computation of Spearman's rho involves deciding how to rank observations such as a <1, a detected 3 and a <5. ρ is not used where there are multiple reporting limits.

Statistics for Censored Environmental Data Using Minitab® and R, Second Edition.
Dennis R. Helsel.
© 2012 John Wiley & Sons, Inc. Published 2012 by John Wiley & Sons, Inc.

Kendall's tau (τ) is another nonparametric correlation coefficient that is commonly used in tests for trend, and can easily be applied to censored data. No computation of means, or distances from means, is required. τ is computed for a data set of n (X, Y) observations as the number of concordant pairs of observations (N_c) minus the number of discordant pairs (N_d), divided by the number of total pairs, or

$$\text{Kendall's } \tau = \frac{N_c - N_d}{n(n-1)/2} \tag{11.2}$$

One great advantage of Kendall's tau is that it can be adapted for use with multiple reporting limits.

Finally, correlation coefficients can be computed for binomial data, data whose values are categorized as a 0 or 1. These coefficients are applied to data analyzed by contingency tables (see Chapter 10 for more on contingency tables). For censored data, if all values below a single reporting limit are assigned a 0, and all uncensored observations assigned a 1, these coefficients indicate the correlation of the percent detections to a grouping variable. Does the frequency of detection change from one group to the next? Two coefficients useful in this situation are again Kendall's tau, and a measure called the phi (ϕ) coefficient.

In parallel to the discussions in other chapters, binary and ordinal methods for computing correlation coefficients are first demonstrated, followed by maximum likelihood and nonparametric survival analysis methods.

11.2   NONPARAMETRIC METHODS AFTER CENSORING AT THE HIGHEST REPORTING LIMIT

11.2.1   Binary Methods—The Phi Coefficient

For singly censored data, or data recensored to the highest reporting limit, observations can be classified into two categories, either greater than or equal to (GTE), or less than the reporting limit. The phi coefficient is a correlation coefficient for paired binomial data—in fact it is identical to Pearson's r computed on data represented as one number per class (Conover, 1999, p. 234). When observations have a high proportion of censoring, the preponderance of information contained in the data is represented by the proportions in each category, rather than by numerical values of individual observations. Binomial methods efficiently capture this proportion information. ϕ may be the easiest measure of association to explain when data are severely censored, though it is not required in that situation—ρ and τ may also be computed for severely censored data. For example, Kolpin et al. (2002) computed ρ for the relation between censored/uncensored observations and values of single explanatory variables.

To compute ϕ, consider variables X and Y, each consisting of observations classed as either high or low. Four combinations of the two variables are possible, as shown below.

X	Y
High	High
High	Low
Low	High
Low	Low

ϕ will have the largest positive value when both variables are high together, and both low together. Large negative values result from consistent classifications into opposite categories. A mix of conditions will produce ϕ values close to 0, as with any correlation coefficient. Counts of occurrences of the four combinations may be visualized as a table (Table 11.1). The order of the columns and rows of the table should be such that cells a and d represent positive correlation, and cells b and c negative correlation.

Table 11.1 contains the counts of pairs of low (<0.01) and high (detected, GTE 0.01) values for the 10 observations in the atrazine data of Table 11.2. There are six pairs where atrazine was detected in both June and September, and two where both were censored observations. There are two pairs where atrazine was below the reporting limit in June, but above in September (even though one was just barely above, detected at 0.01). There are no cases where atrazine was detected in June but not in September.

If $r_1 =$ the sum of counts in row 1 of Table 1.1, r_2 the sum of counts for row 2, c_1 the sum of counts for column 1 and c_2 the sum of counts for column 2, then the phi coefficient is computed as

$$\phi = \frac{ad - bc}{\sqrt{r_1 r_2 c_1 c_2}} \tag{11.3}$$

For Table 11.1 classes,

$$\phi = \frac{(2 \times 6) - (0 \times 2)}{\sqrt{2 \times 8 \times 4 \times 6}} = \frac{12}{\sqrt{384}} = 0.61$$

The test of whether ϕ is significantly different from 0 is computed by multiplying ϕ by the square root of n, where n is the number of paired observations, and comparing

TABLE 11.1   Two Variables Classified into Two Categories Each

		June	
		Low	High
September	Low	$a = 2$	$b = 0$
	High	$c = 2$	$d = 6$

A 2×2 contingency table.

TABLE 11.2   A Subset of the Atrazine Concentrations Reported by Junk et al. (1980)

June	September	Rank of June	Rank of September
0.38	2.66	10	10
0.04	0.63	8	8
<0.01	0.59	2.5	7
0.03	0.05	6.5	5
0.03	0.84	6.5	9
0.05	0.58	9	6
0.02	0.02	5	4
<0.01	0.01	2.5	3
<0.01	<0.01	2.5	1.5
<0.01	<0.01	2.5	1.5

Used as an example of data having one reporting limit.

the product to a standard normal distribution (Conover, 1999). If we are looking for correlation in only one direction, a one-sided test is performed. When both positive and negative correlations are of interest, a two-sided test is performed. For Table 11.1, the test statistic is $\phi\sqrt{N} = 0.61\sqrt{10} = 1.93$. For the case where only a positive correlation between June and September concentrations is of interest, the one-sided $p = 0.027$ and we conclude that there is an association, a positive correlation. If both positive or negative correlations were of scientific interest, the more general two-sided p-value equals twice this, or $p = 0.054$, right on the edge of the default 0.05 criteria for significance.

The p-value for ϕ is generally larger than (less significant than) the p-values for ρ and τ. This is because less information is used, and less required, to compute ϕ. In particular, the rank ordering of values above the reporting limit is ignored by ϕ. All values above the reporting limit are only considered equivalent; they are in the same category. ϕ is appropriate for data below and above a single threshold for each variable. For multiple reporting limits, either use counts above and below the highest reporting limit, or instead compute Kendall's tau. There is little reason to prefer ϕ over τ, and τ will make use of the information in the ranks above the reporting limits.

11.2.2   Ordinal Methods for Correlation—Rho and Tau

Nonparametric ordinal methods for correlation are widely used in the environmental sciences. Spearman's rho and Kendall's tau correlation coefficients both meet this definition. Both can easily be computed for data with one reporting limit. Kendall's tau is also applicable to data with multiple reporting limits. However, standard commercial software for Kendall's tau is not coded to compare values when multiple limits are present. This unfortunately requires the data to be recensored below the highest reporting limit unless routines such as those available in NADA for R and the Minitab® ckend macro are used. Both coefficients will be first discussed in the context of one reporting limit, or recensored data at the one highest reporting limit.

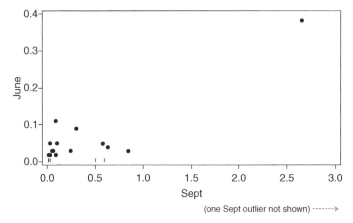

FIGURE 11.1   Censored scatterplot of the atrazine subset data of Table 11.2. Censored observations shown as dashed lines.

11.2.3   Spearman's Rho

Spearman's rho is a rank-transform measure of the monotonic association between Y and X (Helsel and Hirsch, 2002). A positive monotonic correlation means that as X increases, Y consistently increases, although the pattern may or may not be linear. Thus, ρ (and τ) are more general measures of correlation than is Pearson's r, which measures the strength of linear associations.

ρ is calculated by ranking each variable separately, and then computing Pearson's correlation coefficient on the ranks. The ranks when data have only one reporting limit are uniquely defined. To illustrate the computation of ρ, 10 paired observations from the atrazine data of Junk et al. (1980) were listed in Table 11.2 (also see Figure 11.1). Their ranks were reported in the two right-hand columns. Values less than the reporting limit are considered ties and given their average rank; the four June observations at <0.01 are each given the rank of 2.5, the average of ranks 1 through 4. The two censored September observations are each given the rank of 1.5; they are tied and are the two lowest values for that group.

Computing Pearson's r on the two columns of ranks produces a Spearman's rho of 0.74 for the atrazine data, as shown in the following output from Minitab:

```
Pearson correlation of rankJune and rankSept = 0.743
P-Value = 0.014
```

The R command for computing Spearman's rho correlation is cor.test:

```
> cor.test(June, September, method="spearman")

        Spearman's rank correlation rho
data: AtraSubset$June and AtraSubset$September
S = 42.4866, p-value = 0.01390
```

```
alternative hypothesis: true rho is not equal to 0
sample estimates:
       rho
0.7425056
```

The small p-value (0.014) and positive slope are evidence that concentrations in June are correlated with concentrations in September. Wells with high concentrations in June are generally also high in September (Figure 11.1). This correlation may or may not be linear, but it is monotonic.

When Y and X are both censored and have one reporting limit each (though not necessarily the same reporting limit), ρ can easily be computed. This was the case with the data in Table 11.2. When X or Y have multiple reporting limits, however, the data must be recensored at the highest reporting limit in order to compute Spearman's rho. This is because it is difficult to judge the relative rank of a <10 as opposed to a 5, or a <1. Though some adaptations of ρ have been suggested in this situation, such as using Kaplan-Meier scores in the place of ranks, no standard software implements adaptations of ρ, and its characteristics are not well described in the literature.

11.2.4   Kendall's Tau for Data with One Reporting Limit

The standard implementation in software of Kendall's tau also operates on data censored at one reporting limit. As seen in equation (11.2), τ is computed using the number of concordant and discordant pairs of observations. Concordant pairs are those pairs of observations where Y increases as X increases, or where there is a positive slope. Discordant pairs are those where X and Y are going in opposite directions, or where there is a negative slope. Pairs of observations with ties in either X or Y are assigned a 0. Numerous ties can occur with censored data, as when Y values of <0.01 and <0.01 are compared. Unlike Pearson's r, τ measures relations that are curved as well as linear. In other words, the rate of change between points may change, resulting in a curved relationship, or the rate of change may stay constant, producing a straight line. Whether straight or curved, τ detects monotonic (one consistent direction) relationships. And if Y increases only about half the time as X increases, and decreases half the time, τ is close to zero and no significant monotonic correlation between the two variables is found.

Kendall's tau for the Figure 11.1 data is computed using the R command cor.test:

```
> cor.test(June, September, method="kendall")

       Kendall's rank correlation tau
data: AtraSubset$June and AtraSubset$September
z = 2.4299, p-value = 0.01510
alternative hypothesis: true tau is not equal to 0
sample estimates:
      tau
0.6358508
```

The p-value of 0.015 for τ indicates a significant monotonic correlation between June and September atrazine concentrations, as did Spearman's rho. The lower value for τ (0.64) than for ρ does not indicate a lower sensitivity of τ; as Helsel and Hirsch (2002) note, τ is measured on a different scale than ρ. Like centigrade and Fahrenheit, the same phenomenon may be represented by different numbers on different scales. Kendall's tau is generally 0.15 or so lower than ρ and Pearson's r for the same strength of correlation (Helsel and Hirsch, 2002).

11.3 MAXIMUM LIKELIHOOD CORRELATION COEFFICIENT

One measure of the quality of a regression equation is the coefficient of determination (r^2). Modern textbooks on statistics caution against overdependence on this statistic, noting that poor regression models can have high r^2, and vice versa. However, r^2 is a reasonable measure as long as the data follow a straight-line pattern, the residuals are close to a normal distribution, and no single point overly affects the position of the regression line (Ryan, 1997). Maximum likelihood methods are capable of producing several measures similar in concept to r^2, the most popular being the likelihood r^2. Computation of the likelihood r^2 (called the generalized r^2 by Allison, 1995) is very different from the process used by standard least-squares regression with uncensored data. The process is similar to that for likelihood testing of group differences in Chapters 9 and 10. The log-likelihood statistic is determined for the regression of Y versus one or more X variables (the "full model") computed using maximum likelihood. A second log-likelihood statistic is determined for the "null model," using no X variables. For the null model, the variation of Y is only noise around the mean of Y. The difference between these two likelihood statistics measures the explanatory power of the X variable. The magnitude of the difference is multiplied by -2 to produce the "-2 log-likelihood" or G_0^2. G_0^2 measures the increase in the likelihood of producing the observed pattern of Y when the relationship with the X variables is taken into consideration. For large values of G_0^2 the null hypothesis of no linear relationship between Y and the X variables is rejected, and the slope and correlation of the relationship are determined to be nonzero. In order to evaluate its significance, G_0^2 is compared to a chi-square distribution with degrees of freedom equal to the number of X variables in the full model. For correlation, only one X variable is usually considered at a time.

The value of G_0^2 is used to compute the likelihood r^2:

$$\text{likelihood } r^2 = 1 - \exp\left(-\frac{G_0^2}{n}\right) \tag{11.4}$$

where n is the number of (X, Y) paired observations. For large values of G_0^2, the value of r^2 will be close to 1. The likelihood-r correlation coefficient is the square root of the likelihood r^2, with algebraic sign identical to that for the slope of the relationship.

$$\text{likelihood } r = \text{sgn(slope)} \sqrt{1 - \exp\left(-\frac{G_0^2}{n}\right)} \tag{11.5}$$

This correlation coefficient expresses the strength of the relationship between Y and X as measured by maximum likelihood.

As an example of data with multiple reporting limits, Hughes and Millard (1988) presented dissolved iron concentrations collected during summers from the Brazos River, Texas:

Dissolved iron (Y):	20	<10	<10	<10	<10	7	3	<3	<3
Time (years) (X):	1977	1978	1979	1980	1981	1982	1983	1984	1985

Do summer dissolved iron concentrations exhibit a trend during this period? The data are plotted in Figure 11.2.

Censored regression by maximum likelihood assuming a normal distribution of residuals in Minitab is computed (for further information, see Chapter 12) using

```
Stat > Reliability/Survival > Regression with Life Data
```

A log-likelihood of -13.184 was produced and the slope was negative with a Wald's p-value of <0.001. The Wald's test is not considered as accurate as the likelihood-ratio test for measuring the response of a censored variable. The log-likelihood of the null model was -16.858 by fitting a normal distribution to the data without an explanatory variable using

```
Stat>Reliability/Survival > Distribution Analysis (Arbitrary
Censoring) > Parametric Distribution Analysis
```

The resulting $G_0^2 = 7.35$, the slope is negative, and from equation (11.5) the likelihood $r^2 = 0.56$. The likelihood r is therefore $= -0.75$. The test of the null hypothesis that $r = 0$ is identical to the test for whether the slope equals 0, just as in ordinary linear regression. Therefore we reject the null hypothesis that X and Y are uncorrelated, and conclude that there is a significant linear correlation.

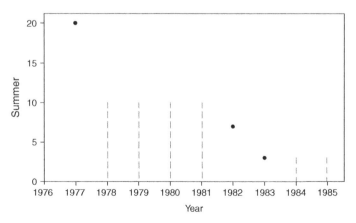

FIGURE 11.2 Censored scatterplot of summer dissolved iron concentrations from Hughes and Millard (1988). Censored observations are shown as dashed lines.

In NADA for R the cenreg command will perform a regression of a censored Y variable versus an uncensored X variable, reporting the likelihood-r correlation and the p-value for the likelihood-ratio test of significance, rather than using the Wald's test of Minitab. The format of the arguments is cenreg (Y, Y censoring indicator $\sim X$) where the tilde ($\sim$) denotes a model of Y versus an explanatory variable X. The data set DFe is included with NADA for R package. The names command lists the names of the stored variables.

```
> data(DFe)
> attach(DFe)
> names(DFe)
[1]"Year"        "YearCen"        "Summer"        "SummerCen"
> DFeReg = cenreg(Cen(Summer, SummerCen)~Year, dist="gaussian")
> DFeReg
                 Value        Std. Error      z        p
(Intercept)      3426.07      859.278         3.99     6.69e-05
Year             -1.73        0.434           -3.98    6.90e-05
Log(scale)       1.13         0.315           3.60     3.15e-04
Scale = 3.11


Gaussian distribution
Loglik(model) = -13.2      Loglik(intercept only) = -16.9
Loglik-r: 0.747004
Chisq= 7.35 on 1 degrees of freedom, p= 0.0067
```

The p-value of 0.0067 for the significance test for slope and correlation coefficient differs from the Wald's test result given by Minitab. NADA for R provides the p-value for the likelihood-ratio test, comparing the -2 log-likelihood G_0^2 to a chi-squared distribution with 1 degree of freedom, as there is one X variable. The correlation of Y and X variables is significant.

If a lognormal distribution were found to be a better fit to the data, the command to test the correlation of Y in log units versus X is just as above except that the distribution is set to "lognormal." Just as with Pearson's r, the changed units of Y used will change the value of the likelihood-r coefficient and its p-value. The likelihood-r correlation of the natural log of iron concentration versus year is found to be 0.737, with a p-value of 0.0079. The correlation is again strongly significant.

```
> DFeReg = cenreg(Cen(Summer, SummerCen)~Year, dist="lognormal")
> DFeReg
                 Value        Std. Error    z        p
(Intercept)      507.472      106.3237      4.77     1.82e-06
Year             -0.255       0.0537        -4.76    1.97e-06
Log(scale)       -1.118       0.4106        -2.72    6.48e-03
Scale = 0.327


Log Normal distribution
Loglik(model) = -9.3      Loglik(intercept only) = -12.8
Loglik-r: 0.7371631
Chisq= 7.06 on 1 degrees of freedom, p= 0.0079
```

For this example, the value of the likelihood r (-0.75) for summer iron concentrations versus year happens to be between the values for Pearson's r computed by substitution of the maximum (reporting limit) and minimum (zero) values possible for censored data. This is not always the case, due to the vagaries of substitution methods' computation of standard deviations. Do not depend on substitution of zero and the reporting limit to bracket the possible values for correlation! The likelihood-r correlation and associated significance test have the advantage that they can be computed for data subject to multiple censoring limits. Their validity will depend on whether there are enough data to determine that the relationship is in fact linear, and whether the variation around the line is normally distributed. These are the same constraints required of Pearson's r.

11.4 NONPARAMETRIC CORRELATION COEFFICIENT—KENDALL'S TAU

Kendall's tau (τ_a, equation (11.2)) can also be applied to multiply censored data. τ is computed by comparing all pairs of observations, counting the number of positive slopes minus the number of negative slopes, and dividing by the total number of comparisons between pairs of observations. Kendall (1955) gives a complete description of the coefficient. Brown et al. (1974) adapted τ for use with censored data in heart transplant studies. Kendall's tau has several advantages over the likelihood-r correlation coefficient.

1. τ can use a censored explanatory variable.
2. It does not require a normal distribution.
3. It will have the same value and significance test before and after a power transformation such as the logarithm.

To compute Kendall's tau, the first Y observation is compared to all subsequent observations. Concordant observations are those where Y increases (a positive slope, as X is increasing). Assign a $+$ to those comparisons. Discordant observations are those where Y decreases as X increases, a negative slope. Assign a $-$ to those observations. Comparisons to the first observation (20) of the dissolved iron data are shown below their respective observation. All are decreasing.

```
Dissolved Iron (Y):     20   <10   <10   <10   <10   7   3   <3   <3
   sign of difference:          -     -     -     -    -   -    -    -
```

Multiple reporting limits have posed no problem for these comparisons. Going from a 20 down to a <10 is a decrease, as is going down to a <3.

Next, the second Y observation is compared to all subsequent values. Consider each observation in its interval endpoints format. If the intervals overlap, the data are tied and assigned a zero difference. Here none of the comparisons is clearly an increase or decrease. It is impossible to determine whether a (0–10) is higher or lower than a detected 7, or a (0–3). All of the comparisons with the second observation are given zeros, neither concordant nor discordant.

```
Dissolved Iron (Y): 20    < 10   <10     <10     <10     7     3     <3    <3
    as intervals    (20,20) (0,10) (0,10) (0,10) (0,10) (7,7) (3,3) (0,3) (0,3)
      sign of difference:          0       0       0     0     0     0     0
```

The scores for all $9 \times (8/2) = 36$ comparisons are shown below:

```
Dissolved Iron (Y):     20   <10   <10   <10   <10   7    3    <3   <3
   sign of difference:        -     -     -     -    -    -    -    -
                                    0     0     0    0    0    0    0
                                    0     0     0    0    0    0    0
                                          0     0    0    0    0    0
                                          0     0    0    0    0    0
                                                     -    -    -
                                                     -    -    -
                                                               0
```

There are 0 concordant pairs and 13 discordant pairs, so the numerator for $\tau = -13$. The denominator is 36 comparisons. Kendall's tau for the dissolved iron data is therefore

$$\text{Kendall's } \tau = \frac{0 - 13}{(9(8))/2} = -0.361$$

The test for significance of Kendall's tau is computed using the numerator $S = N_c - N_d$, divided by its standard error. The square of the standard error is the variance of S, which for the case of no ties equals

$$\text{var}[S] = \frac{N(N-1)(2N+5)}{18}$$

The test statistic Z is compared to values of the standard normal distribution.

$$Z = \frac{S - \text{sgn}(S)}{\sqrt{(N(N-1)(2N+5))/18}} \qquad (11.6)$$

where sgn is the algebraic sign function. The numerator uses a continuity correction of sgn(S) to better calculate p-values from a smooth normal distribution function, adding or subtracting a value of 1 from S (Kendall, 1955).

With many ties resulting from comparisons among censored values, a correction is required for the variance of S. Software that performs this adjustment is crucial for censored data, because censored observations produce many tied comparisons. For censored data, the tie correction is more complex than if all ties resulted from uncensored observations. This is because when there are four observations tied at 10, all comparisons between the four observations are ties, so there are $4(3)/2 = 6$ ties among the four observations. Tie corrections in commercial software usually compute the number of ties as (obs (obs − 1))/2 for a set of tied observations. But there are fewer ties among a set of four observations that include both censored and uncensored values, such as <1, 4, 7,<10. In this case there are three ties: [<1, <10], [4, <10], and [7, <10]. The other three comparisons are known to be pluses: [<1, 4], [<1, 7], and [4, 7].

The ckend macro for Minitab computes Kendall's tau and its significance test as described above, adjusting both τ and Z for ties originating from multiply censored data. The output from the ckend macro for the dissolved iron data is

```
S              -13.0000
tau            -0.361111
taub           -0.600925
z              -1.50787
pval            0.131587
```

For these data, τ is not significantly different from zero ($p = 0.13$). There is not sufficient evidence of monotonic correlation to declare that there is a trend in concentration.

Kendall (1955) proposed an alternate correlation coefficient when tied values are considered as "no information" that has been given the name Kendall's τ_b (equation (11.7)). Note that this is not the usual situation in environmental science, when an early <1 tied with a later <1 (or a detected 7 with a later detected 7) indicates a situation with no change. However in some disciplines, a tie is noninformative. In that case, τ_b is computed by dropping out the tied observations from both the numerator and denominator.

$$Kendall's\ \tau_b = \frac{N_c - N_d}{\sqrt{((N(N-1)/2) - \#\text{ties}_X)((N(N-1)/2) - \text{ties}_Y)}} \qquad (11.7)$$

where ties_X is the number of ties in the X variable and ties_Y is the number of ties in the Y variable. An alternate way to state τ_b is equation (11.8):

$$Kendall's\ \tau_b = \frac{N_c - N_d}{N_c + N_d} \qquad (11.8)$$

For the dissolved iron data, there are no ties among the values of X, but 23 ties in the comparisons between Y observations, including the comparisons that were unclear due to censoring. Kendall's τ_b for these data is therefore

$$Kendall's\ \tau_b = \frac{0 - 13}{\sqrt{(((9(8))/2) - 0)(((9(8))/2) - 23)}} = -0.60$$

τ_b is analogous to the tie adjustment in other nonparametric procedures that delete zero observations, as in the sign test (Chapter 9). As was true for the sign test, dropping zero observations artificially and often incorrectly makes the statistic too significant. It slights the information that favors the null hypothesis. For the case of censored data, a tie does not mean zero information, but in fact is evidence for the null hypothesis of no correlation. Therefore, Kendall's original coefficient (τ_a, equation (11.2), called simply τ in this book) should generally be used for censored data rather than τ_b.

11.5 INTERVAL-CENSORED SCORE TESTS: TESTING CORRELATION WITH (DL TO RL) VALUES

The interval-censored form of the generalized Wilcoxon test is one of a class of procedures called linear-rank tests (Klein and Moeschberger, 2003) in survival analysis. Different scores such as varying the Harrington–Fleming rho statistic from zero to one change the type of test conducted, but all of them have the same general goal. The goal is to discern whether the proportion of detected values, and proportion of data above cut points of detected values, change with an explanatory variable, or whether that variable has no effect. In essence this tests whether the distribution function of the Y variable changes as X changes. In Chapters 9 and 10, the explanatory variable was a group assignment. But linear-rank tests also allow an ordinal variable, a variable with a finite number of ordered classes, to serve as the explanatory variable. In this situation the test determines whether the edf changes as the explanatory variable increases or decreases, and so serves much the same function as does a test for significance of correlation. As nonparametric tests, linear-rank tests will not provide an equation analogous to the censored regression or ATS lines presented in Chapter 12, but they do provide a test of association between an interval-censored variable such as concentration, and an explanatory variable whose values can be ordered or ranked in value. Kendall's tau correlation can serve the same purpose, but standard correlation software is not set up to allow input of data in the interval format. Software for interval-censored data such as the "interval" contributed package for R is designed for this purpose, allowing data coded as (0 to DL) and other values (DL to QL) to be combined with uncensored numbers to determine whether concentration changes with changing values of the explanatory variable.

To illustrate the procedure, the TCE data from Eckhardt et al. (1989) used extensively in Chapter 12 is tested for correlation with population density, one of the explanatory variables considered in the next chapter. The data set is called TCEReg. Values for population density go from 1 to 19 as a series of integers, and so are ordered categories. After reading in the data, the lower end of the interval is set to 0 for all censored values, creating the new variable TCElo. The lt and rt variables define inclusion of the value at the low end of the censoring interval, and exclusion of the value at the upper end of the interval, just as in Chapters 9 and 10. The linear-rank test is computed using the ictest command, and the Wilcoxon score used. The Surv function represents the interval-censored Y values by declaring TCElo as the low end, TCEConc as the high end, and type="interval2" to designate interval censoring. This joint-Y variable is modeled (as represented by the tilde "$\sim$") using the PopDensity variable as the X variable.

```
> data(TCEReg)
> attach(TCEReg)
> head(TCEReg)
> TCEReg$TCElo=with(TCEReg, TCEConc*(1-TCECen))
> TCEReg$Lt=with(TCEReg, TCEReg$TCEConc<1000)
> TCEReg$Rt=with(TCEReg, !TCECen)
> mk=ictest(Surv(TCEReg$TCElo,TCEConc, type="interval2")
+ ~Pop Density, scores="wmw", lin=TCEReg$lt, rin=TCEReg$rt)
> mk

     Asymptotic Wilcoxon trend test(permutation form)
data: Surv(TCEReg$TCElo, TCEConc, type = "interval2") by PopDensity
Z = -4.4698, p-value = 7.83e-06
alternative hypothesis: survival distributions not equal

     n              Score Statistic*
[1,] 247                -100.8901
* negative so larger covariate values give later failures than expected
```

Because the explanatory variable is ordinal, this is a test for trend of Y with X. The Z-test statistic is a comparison of the observed percent exceedances above various cut points compared to what is expected if the null hypothesis is true, and nothing changes as X changes. The resulting p-value is very low (0.000007). Therefore, the TCE concentrations do evidence a trend with PopDensity. The final sentence tells the direction—"larger covariate values give later failures than expected." In other words, for larger values of the explanatory variable (PopDensity, the covariate), "failures" or detected values are "later" or larger than expected if the null hypothesis had been true. So concentrations increase with increasing PopDensity (not the clearest statement to the eyes of an environmental scientist, right?).

Finally, the dissolved iron data of Millard and Deveral is slightly altered to contain two intervals, as if the detection limit had been 3 µg/L and the reporting limit 10 µg/L. The change is that the <10 values now become (3, 10) interval values between the detection and reporting limits. The ictest procedure is run to illustrate a nonparametric trend test for data coded (0, DL) and (DL, RL) as well as detected values.

```
> data(DFe)
> attach(DFe)
> DFe$FElo=with(DFe, Summer*(1-SummerCen))
> DFe$Lt=with(DFe, DFe$Summer<1000)
> DFe$Rt=with(DFe,!SummerCen)
> DFe$FElo[2]=3
> DFe$FElo[3]=3
> DFe$FElo[4]=3
> DFe$FElo[5]=3
> DFe
```

```
        Year     YearCen Summer    SummerCen   FElo    Lt      Rt
1       1977     FALSE   20        FALSE       20      TRUE    TRUE
2       1978     FALSE   10        TRUE        3       TRUE    FALSE
3       1979     FALSE   10        TRUE        3       TRUE    FALSE
4       1980     FALSE   10        TRUE        3       TRUE    FALSE
5       1981     FALSE   10        TRUE        3       TRUE    FALSE
6       1982     FALSE   7         FALSE       7       TRUE    TRUE
7       1983     FALSE   3         FALSE       3       TRUE    TRUE
8       1984     FALSE   3         TRUE        0       TRUE    FALSE
9       1985     FALSE   3         TRUE        0       TRUE    FALSE
> mk=ictest(Surv(DFe$FElo,Summer, type="interval2")~Year, scores="wmw",
+ lin=DFe$Lt, rin=DFe$Rt)
> mk
        Exact Wilcoxon trend test(permutation form)
data: Surv(DFe$FElo, Summer, type = "interval2") by Year
p value = 0.004
alternative hypothesis: survival distributions not equal


        n       Score Statistic*
[1,] 9          10.66667
* positive so larger covariate values give earlier failures than expected
p-value estimated from 999 Monte Carlo replications
99 percent confidence interval on p-value:
 1.003509e-05 1.482735e-02
```

The p-value for the trend test is 0.004, showing that dissolved iron concentrations do change with year. Note that the change in lower limit of the <10s to (3, 10) moves these data to a position higher than the two (0, 3) values, so adds strength to the signal over what was tested with Kendall's tau. The "larger covariate values give earlier failures than expected" should be interpreted as "as Year increases, summer iron concentrations decrease over what is expected if there were no trend." A downtrend is found.

11.6   SUMMARY: A COMPARISON AMONG METHODS

Maximum likelihood estimation and the likelihood-r correlation coefficient are limited to data where only the Y variable is censored. Correlation where both the X and Y variables are censored can be accounted for directly with Kendall's tau, and can be accounted for with Spearman's rho and the phi coefficient if there is only one reporting limit per variable, and with the interval-censored Wilcoxon test for trend if there is one or no reporting limit for the explanatory variable. For the atrazine data in Table 11.2 censored at one reporting limit (both June and September data were censored so the likelihood-r coefficient is not applicable), Table 11.3 compares the possible correlation coefficients.

Spearman's rho and Kendall's tau both measure monotonic correlation, and give similar results. Note the similarity of their p-values. The two coefficients are measured on different scales, with ρ on the same scale as the traditional Pearson's r coefficient. τ is expected to be about 0.15 smaller than ρ for the same strength of correlation

TABLE 11.3   **Correlation Coefficients for Table 11.2 Atrazine Data**

	Correlation Coefficient	Two-Sided p-Value
Likelihood r	NA	NA
Spearman's rho	0.74	0.014
Kendall's tau	0.58	0.018
phi	0.61	0.054

NA: not applicable.

(Helsel and Hirsch, 2002). Kendall's tau might be preferred due to its ability to be used for multiply censored data.

ϕ measures the association between variables after reducing the data into two categories. This is a heavy price to pay in information content unless the data are strongly censored. The loss of information from classifying all observation sat or above 0.01 in this data set as merely "GTE 0.01" produces the higher p-value for ϕ as compared to τ or ρ.

For data with multiple reporting limits, options are somewhat more limited. Spearman's rho and phi cannot be computed unless the data are recensored at the highest reporting limit, which greatly reduces the information content of the data. If only one variable is censored the likelihood r can be computed by MLE. Kendall's tau can be computed for multiply censored data if the software is coded for that situation. The interval-censored Wilcoxon test is the most adaptable, allowing multiple censoring levels for the Y variable. For the multiply censored dissolved iron data the likelihood r, Kendall's tau and interval-censored Wilcoxon test results are listed in Table 11.4. Note that the interval-censored Wilcoxon test was performed on the original values of (0, 10) below, unlike the illustrative example given previously.

The difference in p-values between the first two coefficients is due to the "information" contained in the assumptions of linearity and normality used by the likelihood-r coefficient. The choice of which to use is a choice of whether to use a "data only" method (the nonparametric Kendall's tau) or a "data plus model" method, the parametric likelihood-r coefficient. If the data are linear with a normal distribution, the four <10s at the early part of the record must be close to the higher end of their range. If so, a strong correlation results. This is the "information" generated by the model that is used by the parametric (model-based) likelihood-r coefficient. Kendall's tau makes no assumption of linearity or normality of residuals. It allows the uncertainty of the positions of these four values to remain. It calls a <10 tied with a <3, for example, when stating what is known (or more importantly, not

TABLE 11.4   **Correlation Coefficients for the Summer Dissolved Iron Data**

	Correlation Coefficient	Two-Sided p-Value
Likelihood r	-0.75	<0.001
Kendall's tau	-0.36	0.132
icens Wilcoxon	5.33 (score statistic)	0.122

known) about the data. The interval-censored Wilcoxon test also does not differentiate between the (0 to 10) and (0 to 3) values. The likelihood-r coefficient should be used only if the assumptions of linearity and normality of residuals can be assumed to be true. Based only on the information in this small data set, there is insufficient information to make that assumption; the likelihood-r coefficient gives a strongly significant test statistic even though there are only three detected observations to base the answer upon. The information in the assumed model is being too heavily counted by likelihood r. This effect of the model will not be as large if the data set is larger, but small data sets often occur in environmental studies, so scientists should be aware of the possible effects of making an unverifiable model assumption.

11.7 FOR FURTHER STUDY

Hughes and Millard (1988) computed the significance test for Kendall's tau in a different way than described here, determining all possible permutations of ranks allowed by the censoring scheme and computing τ and its significance test using the average of the possible ranks for each observation. Computations for this "expected rank statistic" method are quite complex, as thousands of permutations are possible for even a moderate sized data set. The values for S and τ will be the same as for the procedures to compute Kendall's tau outlined in this chapter, because τ is defined as the average of possible values for all permutations (Kendall, 1955). The test of significance for the permutation procedure will differ, however, from the Brown et al. (1974) tie correction method used by the Ckend macro. For the summer iron data, Hughes and Millard report a test statistic Z of -2.27 with the corresponding p-value of 0.012, declaring a trend for the period. This is greater evidence for a trend (smaller p-value) than found by the Brown et al. procedure. They state that their expected rank method recovers some of the information lost by the Brown et al. procedure when comparing a <DL to a small detected value or to another censored value. Further investigation of the two methods and their applications to environmental data is warranted.

Oakes (1982) presents the Brown et al. procedure in more detail and applies it to two example data sets. Oakes' results differ from those of the Ckend macro in that he did not use a continuity correction, and did not account for ties in both X and Y. When ties in X occurred, as happened in the example data he presents, he randomly chose an ordering for the tied observations, giving them untied ranks. Therefore his results are not as accurate as those presented here using the Ckend macro, which incorporates ties in both X and Y.

Isobe et al. (1986) first applied the Brown et al. procedure for censored correlation to left-censored data in the field of astronomy. They argued for routine adoption of these methods in the field of astronomy, much as this book does for the field of environmental sciences.

Akritas and Siebert (1996) derived a test for partial correlation using a partial Kendall's tau. This allows the correlation of X and Y (when one or both include censored values) to be adjusted to account for the influence of one or more covariates.

Libiseller and Grimvall (2002) applied the partial τ correlation concept to trend detection of environmental data. Effects of covariates were removed in a mechanism analogous to multiple regression, and so relevant to the situations in the next chapter.

EXERCISES

11-1 Golden et al. (2003) measured concentrations of lead in the blood and in several organs of herons in Virginia, in order to relate those concentrations to levels found in feathers. The objective was to determine whether feathers were a sensitive indicator of exposure to lead. If so, feathers could be collected in the future so that the birds would not need to be sacrificed in order for their exposure to lead to be evaluated. Compute a correlation coefficient to determine whether lead concentrations in feathers are associated with concentrations in blood. Note that both have censored values. The data are found in Golden.xls.

11-2 Brumbaugh et al. (2001) measured mercury concentrations in fish of approximately the same age and trophic level across the United States. Even so, the size of fish varied due to differences in age and species. Determine whether there is a significant correlation between mercury concentrations ("Hg") and fish length. The data are found in HgFish.xls.

11-3 Yamaguchi et al. (2003) measured concentrations of dieldrin and lindane in fish collected at four sites draining to the Thames River, United Kingdom. Determine whether concentrations of the two contaminants in fish are correlated. Note that both concentrations contain censored values. The data are found in Thames.xls.

12 Regression and Trends

One of the most frequently used techniques in statistics is linear regression—relating a response variable to one or more explanatory variables by use of a linear model. Estimates of slopes and intercept are computed by least squares, with partial t-tests determining if slope estimates differ significantly from zero. Those that do are worth including in the model. The least-squares procedure produces optimal estimates of slope and intercept if the residuals, the distances in the y-direction between observations and the fitted line, follow a normal distribution, have constant variance across the range of x values, and the data are linear. The resulting regression line is a conditional mean of y given x; the parametric regression line can be thought of as a "linear mean." If one of the explanatory variables is a measure of time, the test for significance of the slope of that variable is a test for (temporal) trend.

A nonparametric analog to linear regression, commonly used in trend analysis of environmental data, is the Theil–Sen line and slope estimator (Helsel and Hirsch, 2002). The slope and intercept for the Theil–Sen line are estimated by an entirely different method than least squares. The slope is related to the nonparametric Kendall's tau correlation coefficient—the slope is the ratio $\Delta y/\Delta x$ that, if subtracted from the response variable y, would produce a set of residuals having a Kendall's tau correlation coefficient of zero. The significance test for the Theil–Sen slope is the same test as that for the Kendall's tau correlation coefficient. The Theil–Sen line is a "linear median" not strongly influenced by the presence of outliers. When the explanatory variable is a measure of time, the significance test for the slope is a test for (temporal) trend.

There are several methods for incorporating censored observations into linear models. Maximum likelihood estimates of slope and intercept produce a parametric regression model without resorting to substitution. MLE methods provide a best-fit line for data with one or more reporting limits, assuming the residuals follow the chosen distribution. For a nonparametric approach, a variation of the Thiel-Sen line can be computed for censored data. One advantage of this line is that it can be computed when values for both x and y are censored. A third approach—binary logistic regression—is performed after the y variable is classified as either detect or nondetect, analogous to the contingency table process for testing group differences.

Statistics for Censored Environmental Data Using Minitab® and R, Second Edition.
Dennis R. Helsel.
© 2012 John Wiley & Sons, Inc. Published 2012 by John Wiley & Sons, Inc.

Logistic regression is used to model how the proportions of uncensored and censored observations change as a function of one or more explanatory variables. Each of these methods is examined in turn.

12.1 WHY NOT SUBSTITUTE?

By now there should be no doubt that substitution methods are inadequate. The example below demonstrates this in the context of regression.

Consider again the summer dissolved iron (DFe) data presented in Table 5 of Hughes and Millard (1988):

DFe:	20	<10	<10	<10	<10	7	3	<3	<3
Year:	1977	1978	1979	1980	1981	1982	1983	1984	1985

To determine whether there is a trend in dissolved iron over time, a regression of DFe (Y variable) versus Year (X variable) can be computed and the slope tested to determine if it is significantly different from zero. One analyst might set censored observations to the value of their reporting limits, while another sets all censored observations to 0. The results for both are given below.

censored observations = dl
```
The regression equation is
DFe = 3508 - 1.77 YEAR

Predictor    Coef      SE Coef    T          P
Constant     3508.2    662.4      5.30       0.001
Year         -1.7667   0.3344     -5.28      0.001

S = 2.590        R-Sq = 80.0%         R-Sq(adj) = 77.1%
```

censored observations = 0
```
The regression equation is
DFeZero = 2215 - 1.12 YEAR

Predictor    Coef      SE Coef    T          P
Constant     2215      1627       1.36       0.215
Year         -1.1167   0.8211     -1.36      0.216

S = 6.360        R-Sq = 20.9%         R-Sq(adj) = 9.6%
```

When substituting the reporting limit for all censored observations, the slope of DFe versus Year (-1.77) appears significantly different from zero with a t-test statistic of -5.28 and a p-value of 0.001. A significant trend of decreasing dissolved iron is declared. However when zeros are substituted for censored observations, the slope for Year (-1.12) is not significantly different from zero $(p = 0.216)$, so no trend is found. The values for the intercept change by about one-third. The r-squared statistic changes from 80 to 21%. Neither equation is necessarily correct. Even if both had produced the same result of nonsignificance, the choice to substitute values somewhere in-between these two could produce a significant test result. Clearly substitution produces inadequate information on which to base any decision.

In contrast, survival analysis software can be used to compute regression equations for left-censored data. The result is a unique solution, with a defensible test for whether the slope coefficient differs from zero given the assumptions of the method. Assuming a normal distribution for the residuals, MLE produces the estimates for slope and intercept below. These estimates are defensible as the best-fit parameters, given censored data and the assumptions of normality and linearity. The slope for Year (-1.73) has a p-value of essentially zero; a downtrend in summer iron concentrations occurs over this time period if a linear trend is assumed.

Coefficients estimated by mle

```
DFe = 3426 - 1.73 YEAR
Estimation Method: Maximum Likelihood
Distribution: Normal

Regression Table
                        Standard                      95.0% Normal CI
Predictor    Coef       Error       Z        P        Lower      Upper
Intercept    3426.1     859.3       3.99     0.000    1741.9     5110.2
Year         -1.7260    0.4337      -3.98    0.000    -2.5760    -0.8760
Scale        3.1083     0.9785                        1.6771     5.7607

Log-Likelihood = -13.184
```

Chung (1990) compared substitution versus MLE for regression of trace elements in geochemical exploration. He found that MLE provided slope coefficients quite close to the true values, while none of the substitutions $(0, 0.1, 0.2, \ldots$ to 1 times the reporting limit) worked nearly as well. Thompson and Nelson (2003) compared the bias and precision of MLE regression results to that of substituting one-half the reporting limit and found that substitution produced biased estimates of slope, as well as producing confidence intervals that were too small. Their study used simulations with consistently defined limits, and therefore the errors they found were likely not as large as would be found in practice for environmental data, given the inconsistencies among laboratories in the determination of reporting limits (see Chapter 3). A more realistic setting would be to first censor at several higher reporting limits for some proportion of the data, and then substitute fabricated numbers based on those limits. Errors from substitution in this more realistic scenario would likely be considerably larger than those found in their study. Yet even with their smaller errors, their study strongly advocated the method of maximum likelihood over substitution for performing regression with censored data.

12.2   NONPARAMETRIC METHODS AFTER CENSORING AT THE HIGHEST REPORTING LIMIT

12.2.1   Binary Method—Logistic Regression

Standard regression methods use a continuous variable such as concentration as their response variable. With logistic regression, the response variable is a probability π, such as the probability of the concentration being greater than or equal to (GTE) the reporting limit. The probability of falling below the reporting limit is then $1 - \pi$ (the Greek letter π is used here to avoid using the letter p and so avoid confusing the probability of detection with a p-value). Logistic regression models the probability π as a function of the effects of one or more explanatory variables. Observations for the response variable are recorded as belonging either in one category or another, either below or GTE a reporting limit. The regression equation predicts the probability of falling into one category.

Logistic regression can be written in a form that appears much like least-squares regression:

$$\ln\left(\frac{\pi}{1 - \pi}\right) = b_0 + b_j X_j \tag{12.1}$$

Logistic regression relates this logistic (or logit) transformation of the Y variable (the left-hand side of equation 12.1) to a linear function of the X variable(s). The right-hand side of equation 12.1 looks similar to a least-squares multiple regression equation, where X_j represents a vector of $j = $ one or more explanatory variables, and the b_j are the fitted slope coefficients. The left-hand side of the equation is called the logit or logistic transform, the natural logarithm of the odds $\pi/(1 - \pi)$ for the occurrence of an event. If $\pi = 0.8$ or four-fifths, then the odds for the event are 0.8/0.2, or 4 to 1.

The logistic transform in equation 12.1 can be solved for the probability π, producing equation 12.2.

$$\pi = \frac{\exp(b_0 + b_j X_j)}{1 + \exp(b_0 + b_j X_j)} \tag{12.2}$$

For a unit increase in the jth explanatory variable X_j, the odds $\pi/(1 - \pi)$ increase by a multiplicative factor of e^{b_j}. When π is plotted versus an explanatory variable as in Figure 12.1, the result is an S-shaped curve. The S-curve is flexible, becoming almost linear at the central portion near $\pi = 0.5$ while changing value much more slowly near the extremes of $\pi = 0$ or 1. At the inflection point of $\pi = 0.5$ where the probability of detection equals 50%, the logit function on the left-hand side of equation 12.1 equals $\ln(0.5/(1 - 0.5))$, or 0.

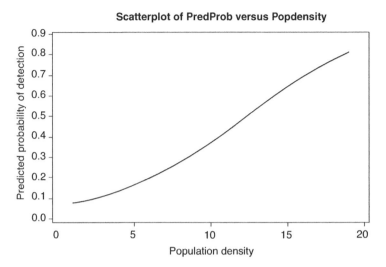

FIGURE 12.1   Logistic regression model of TCE detection. Note the S shape of the function.

The S-curve that best fits observed frequencies of detection is determined by maximum likelihood. Equation 12.3 is the log-likelihood equation for $i = 1$ to n observations where π is the probability of a detect.

$$L = \sum_{i=1}^{n} (Y_i \ln[1 - \pi] + (1 - Y_i)\ln[\pi]) \tag{12.3}$$

The Y value for each observation is either a 1 (nondetect) or a 0 (detect). For a nondetect, the right-hand side of the expression inside the summation sign is zero and the left-hand side becomes $\ln(1 - \pi)$. This will be a maximum of $\ln(1)$ for $\pi = 0$, which fits the observed value the best. For a detected observation $Y = 0$, the left-hand side becomes zero and the right-hand side becomes $\ln(\pi)$. This will fit the observed data best at $\ln(1)$ when $\pi = 1$. Maximum likelihood is a matching process, aiming for a perfect match of a detect with the probability of a detect $= 1$, and the match of a nondetect with the probability of a detect $= 0$. Maximizing the log-likelihood function is a search for values of the intercept b_0 and the slopes b_j that produce a value for L closest to the perfect match of $\ln(1) = 0$. For a less than perfect match, L will be negative. The solution maximizes the match between the estimated probabilities, which are a function of the j explanatory variables X_j, and the observed data.

A single log-likelihood value tells the investigator very little. As a sum of numbers for each observation, the log-likelihood value is primarily a function of the sample size n for any given data set. Instead, what is informative are comparisons between two or more models for the same data set. The two models are compared using a likelihood-ratio test, to determine which model is preferred—which explanatory variables should be included in the logistic regression equation. The test for determining the overall significance of a logistic regression model, similar in concept to the overall F-test in least-squares regression, is the overall likelihood-ratio test. This test determines whether the entire model is an improvement over using no model at all, that is over the "null model" where the slope coefficients for all explanatory variables equal 0. For the null model, the best estimate of π is the average proportion of detections, regardless of the value of X. The overall test statistic is

$$
\begin{aligned}
G_0^2 &= 2[L(\beta) - L(0)] \\
&= [-2\,L(0)] - [-2\,L(\beta)]
\end{aligned}
\tag{12.4}
$$

where β represents the population slope coefficients being estimated, $L(\beta)$ is the log-likelihood of the tested model and $L(0)$ is the log-likelihood of the null model. The test statistic G_0^2 is compared to a chi-square distribution with k degrees of freedom, where k is the number of explanatory variables in the model, to determine a p-value for the overall test.

Note that while Minitab® and most other software computes π as the probability of observing a value of 1, SAS considers π as the probability of observing a 0. The results in one direction can be directly converted to the other, yet confusion results from not knowing which direction is being used. Hypothesis tests results (log-likelihood statistics and p-values) will be identical, but the slope coefficients themselves will have the opposite sign if the modeled event is switched from a 1 (detect) to a 0 (nondetect). Estimates for π will equal $1 - \pi$ for the other direction. Estimated odds ratios will be the inverse of the other (3 to 1 versus 1 to 3). If your software is computing logistic regression in the opposite direction from what you expect, the easiest remedy is to reverse the binary 0/1 assignment of the response variable and run the procedure again.

12.2.2 Example: TCE Concentrations in Groundwater

Is the probability of detecting TCE in the shallow groundwaters of Long Island, NY related to population density and/or depth to the water table? The 247 observations from Eckhardt et al. (1989) are found in TCEReg.xls. Logistic regression is run in Minitab using the command

```
Stat > Regression > Binary Logistic Regression
```

The original data have several reporting limits, so a binary response variable GTE5 was calculated to record whether or not each observation was GTE (1) or below (0) the maximum reporting limit of 5 μg/L. The overall likelihood-ratio test determines whether the two variables Popden and Depth together significantly effect the probabilities of TCE concentrations being equal to or above 5. If not, the best estimate of π for all depths and population densities is the average exceedance probability of 30/247 or 12.1%. The results are printed below.

Binary Logistic Regression: GTE5 versus PopDensity, Depth
```
Link Function: Logit

Variable     Value          Count
GTE5         1              30 (Event)
             0              217
             Total          247
```

```
Logistic Regression Table
                                                     Odds        95% CI
Predictor    Coef        SE Coef      Z       P      Ratio   Lower   Upper
Constant     -2.81967    0.535295    -5.27   0.000
PopDensity   0.156350    0.0514576    3.04   0.002   1.17    1.06    1.29
Depth        -0.0013185  0.0017043   -0.77   0.439   1.00    1.00    1.00

Log-Likelihood = -84.961
Test that all slopes are zero: G = 12.768, DF = 2, P-Value = 0.002
```

The (Event) designation for the value of 1 on the output shows that Minitab is computing the probabilities of observing a 1. The log-likelihood $L(\beta)$ of this model equals -84.961. G_0^2 is reported as 12.768, which is significant at a very low p-value of 0.002. Therefore, the two variables Popden and Depth together significantly affect the probabilities of TCE being at or above 5 µg/L in groundwater.

Using R, the logistic regression model is estimated using the generalized linear model (glm) command. GLM.2 is the null model. GLM.3 is the two variable, Depth + PopDensity model. The -2 log-likelihood test to compare the two models is computed using an ANOVA command:

```
GLM.2<-glm(formula = GTE5 ~ 1, family = binomial(logit))
GLM.3<-glm(formula = GTE5 ~ Depth + PopDensity, family = binomial
  (logit))
> anova(GLM.2, GLM.3, test="Chisq")
```

resulting in

```
Coefficients:
             Estimate   Std. Error  z value  Pr(>|z|)
(Intercept) -2.819673   0.535304    -5.267   1.38e-07 ***
Depth       -0.001319   0.001704    -0.774   0.43918
PopDensity   0.156350   0.051458     3.038   0.00238 **
---
Signif. codes: 0 '***' 0.001 '**' 0.01 '*' 0.05 '.' 0.1 ' ' 1

    Null deviance: 182.69 on 246 degrees of freedom
Residual deviance: 169.92 on 244 degrees of freedom
AIC: 175.92

Analysis of Deviance Table
Model 1: GTE5 ~ 1
Model 2: GTE5 ~ Depth + PopDensity
```

	Resid. Df	Resid. Dev	Df	Deviance	P(>|Chi|)
1	246	182.69			
2	244	169.92	2	12.768	0.001689 **

The other tests for significance that are important in logistic regression are partial tests, which determine whether the effect of each independent variable on π is significant. Partial tests are used for model building, determining which explanatory variables to keep in the model and which to ignore. Both Minitab and R print out the results of partial Wald's tests. Wald's tests compute an estimate of the slope for each explanatory variable, and an estimate of the standard error of that slope. Test statistics are commonly printed out in computer software in a similar location to partial t-tests in linear regression. The Wald's ratio of slope to its standard error is approximately normally distributed. Minitab and R report a partial Wald's test for Popden that is significant at a p-value of 0.00238. The partial test for Depth is not significant ($p = 0.439$). Therefore, a model with Popden as the sole explanatory variable would be preferable to the two-variable model above. Hosmer and Lemeshow (2000) state that Wald's estimates of standard error are unreliable for all but large sample sizes, with test statistics that are often too small, resulting in p-values larger than they actually should be. They recommend using partial likelihood-ratio tests instead of Wald's tests.

A partial likelihood-ratio test for each β coefficient can be obtained by running the MLE procedure twice, one model with and one without the explanatory variable to be tested. The difference in log-likelihoods measures how the fit to the data is improved by the use of that variable. A p-value for the test is computed by comparing twice the difference in log-likelihoods to a chi-square distribution with one degree of freedom (equation 12.5):

$$\begin{aligned} G^2_{partial} &= 2[L(\beta_{with}) - L(\beta_{without})] \\ &= [-2L(\beta_{without})] - [-2L(\beta_{with})] \end{aligned} \tag{12.5}$$

If the p-value is less than the significance level α, the null hypothesis that the coefficient β equals 0 is rejected, and the explanatory variable provides a significant improvement in model fit. If the p-value is large and the null hypothesis is not rejected, the variable is dropped from the list of useful predictor variables.

The log-likelihood test for whether Depth significantly affects the exceedance probability of TCE can be computed by hand if a logistic regression model is run using Popden as the only explanatory variable, and its log-likelihood statistic recorded.

Binary Logistic Regression: GTE5 versus PopDensity
Link Function: Logit

Response Information
Variable	Value	Count	
GTE5	1	30	(Event)
	0	217	
	Total	247	

```
Logistic Regression Table
                                                   Odds       95% CI
Predictor    Coef       SE Coef     Z      P       Ratio  Lower  Upper
Constant     -3.10111   0.419878   -7.39   0.000
PopDensity   0.170198   0.0493273   3.45   0.001    1.19   1.08   1.31

Log-Likelihood = -85.296
Test that all slopes are zero: G = 12.099, DF = 1, P-Value = 0.001
```

The log-likelihood without Depth as an explanatory variable is -85.296. Comparing this to the log-likelihood for the model with both Popden and Density, the partial likelihood-ratio test statistic for Depth is

$$G_{\text{partial}}^2 = 2(-84.961 - [-85.296]) = 0.67$$

From Minitab's

```
Calc > Probability Distributions > chi-square
```

command, the probability of exceeding 0.67 in a chi-square distribution with 1 degree of freedom is 0.413. Because this p-value is much larger than an alpha of 0.05, the null hypothesis of no effect is not rejected; Depth is adding little to the model and can be dropped. Similarly, to compute a log-likelihood test of the effect of Popden, drop Popden from the two-variable model. With Depth as the only explanatory variable, a log-likelihood of -89.765 results. Subtracting from the log-likelihood of -84.961 when both variables were in the model, and multiplying by 2, produces a test statistic of 9.608 and a p-value of 0.002. The effect of Popden is significant, and should remain in the model. Note that for this relatively large data set of 247 observations the results of the Wald tests and log-likelihood tests, as measured by their p-values, are essentially the same.

12.2.3   Likelihood r-Squared

An overall measure of the 'strength' of the regression relationship for logistic regression is the likelihood r-squared (equation 11.4, repeated here as equation 12.6):

$$\text{likelihood } r^2 = 1 - \exp\left(\frac{G_0^2}{n}\right) \tag{12.6}$$

where G_0^2 is the overall likelihood-ratio test statistic for the comparison between the selected model and the null model. For the TCE exceedance data with Popden as the sole explanatory variable (our preferred model), $G_0^2 = 12.099$ and the likelihood r-squared equals

$$\text{likelihood } r^2 = 1 - \exp\left(\frac{12.099}{247}\right) = 0.048$$

Though this model is better than no model, there is still much variability left to explain in these data.

12.2.4   Confidence Interval for the Slope

A confidence interval associated with the slope of an explanatory variable can be computed using the Wald's normal approximation for large samples (equation 12.7). Ryan (1997, p. 272) comments that the Wald's confidence intervals "should be used cautiously, especially if the sample size is not large."

$$\text{CI on } \beta : \beta \pm z_{1-\alpha/2} \cdot se_\beta \tag{12.7}$$

For the TCE exceedance data a 95% confidence interval on the slope for Popden is

$$0.170 \pm 1.96 \times 0.049 = [0.074, 0.266]$$

12.2.5   Rate of Change of the Odds—the Odds Ratio

In the logistic regression equation the slope b_1 describes the change in the log of the odds (logit) per unit change in x_1. Exponentiating, $\exp(b_1)$ is the amount by which the odds changes for a unit change in x_1. For TCE exceedances, the odds increases for a unit change in Popden by a multiplicative factor of $\exp(0.170) = 1.19$, where 0.170 is the estimated slope for Popden. For example, at a population density of 5 the probability of exceeding 5 µg/L TCE is 9%. The odds is 0.09/0.91, or 0.10. For the same conditions but at a population density of 6, the odds will increase to 0.10 times 1.19, or 0.119. Using equation 12.2, this equals a probability of exceedance π of 0.119/1.119, or 10.7%.

Though $\exp(\beta)$ is the best point estimate for the rate of change of the odds (usually called the "odds ratio") as a function of the explanatory variable, a 95% confidence interval for that change also can be calculated. If the Wald's estimates of standard deviation are acceptable, the confidence interval on the slope can be exponentiated into a confidence interval around the odds ratio:

$$\text{CI on the change in the odds}[\pi_L, \pi_H] = \exp[\beta \pm z_{1-\alpha/2} \cdot se_\beta] \tag{12.8}$$

For the TCE data, the 95% confidence interval on the odds ratio for PopDensity, as a function of a unit change in population density, is

$$\exp[\beta \pm z_{1-\alpha/2} \cdot se_\beta] = \exp[0.170 - 1.96 \times 0.049, 0.170 + 1.96 \times 0.049]$$

$$= \exp[0.074, 0.266] = [1.08, 1.31]$$

With 95% confidence, the rate of change of the odds is between 1.08 and 1.31 for a unit change in population density. This is how the confidence interval printed out by Minitab on the line for each explanatory variable in logistic regression is computed.

12.2.6   Confidence Interval for the Proportion π

Of great interest is the estimated or predicted probability of detection π as a function of reasonable values for the explanatory variables: the S-shaped logistic surface. This is provided by equation 12.2. Also of interest is a confidence interval around the fitted surface. How precisely is the estimate of π known? An estimated probability of exceedance of 40%, or 0.4, may lead to different consequences if it is known with relative precision, say 0.4 ± 0.05, than if it can be estimated only crudely, say as 0.4 ± 0.5. In the latter case, the commitment of scarce resources or expensive solutions would probably not be justified.

First the point on the logistic regression surface π is estimated. The tenth observation in the TCE data set has a population density of 11. The estimated probability of detection π for this observation based on the model with Popden as the sole explanatory variable is

$$\pi = \frac{\exp(-3.10 + 0.170 \times 11)}{1 + \exp(-3.10 + 0.170 \times 11)} = \frac{\exp(-1.23)}{1 + \exp(-1.23)} = 0.226$$

Overall, 22.6% of the samples collected for that population density are expected to have TCE concentrations greater than $5 \, \mu g/L$.

An approximate $(1 - \alpha)\%$ confidence interval around π based on Wald's statistics, can be computed as equation 12.9 (Ryan, 1997):

$$[\pi_L, \pi_U] = \frac{\exp(b_0 + b_i X_i \pm z_{1 - \alpha/2} s \sqrt{h_{ii}})}{1 + \exp(b_0 + b_i X_i \pm z_{1 - \alpha/2} s \sqrt{h_{ii}})} \tag{12.9}$$

where h_{ii} is the leverage statistic for the ith observation (a statistic that is large for outlying observations), and s is an estimate of the error around the estimated proportions. Ryan (1997) notes that this equation is only close to a true $(1 - \alpha)\%$ interval when sample sizes are "very large." I would propose that "very large" be at least 100 observations total, with no fewer than 20 in either category. Fortunately, the TCE data meet this criteria.

The leverage statistic is a multivariate measure of the influence each individual observation has on the final regression model. Leverage measures by how much the estimates of slope and intercept would change when an observation is deleted from the data set. Outlying observations away from the main "cloud" of data have higher leverage, and deleting those observations changes the estimates of slope and intercept more so than observations closer into the bulk of the data. Minitab will compute and save h_{ii} as the leverage (Hi), if requested after clicking on the "Storage" button. Other software packages do much the same.

However, the value for s is a matrix, and software programs do not print it out. To work around this, the standard error of the logit $s\sqrt{h_{ii}}$, which changes based on the values of the explanatory variable(s), can be computed using the variance–covariance matrix. This matrix is output by logistic regression software. The standard error of the logit can be computed using equation 12.10 (based on Hosmer and Lemeshow, 2000, p. 19):

$$s\sqrt{h_{ii}} = \sqrt{\text{Var}(b_0) + x^2\text{Var}(b_1) + 2x\text{Cov}(b_0, b_1)} \qquad (12.10)$$

where the variance of the intercept b_0 and slope b_1 are the squares of their respective standard errors provided in the table of partial statistics. Equation 12.10 describes the standard error of the logit $\text{SE}[\ln(\pi/(1-\pi))]$, or equivalently, the standard error of the linear estimates $\text{SE}[b_0 + b_j X_j]$. Variance estimates for each parameter, along with the covariance between all pairs of parameters $\text{Cov}(b_0, b_1)$ can be stored as an option within Minitab.

For the TCE–Popden model, the variance-covariance matrix is stored as matrix M1 and printed:

```
Matrix XPWX5
  0.183850    -0.0188711
 -0.018871     0.0025017
```

The value 0.183850 is the variance of b_0, and equals the square of the standard error 0.428778 reported in the table of partial statistics. The value 0.0025017 is the variance of b_1, and equals the square of the standard error 0.0500166 reported in the table of partial statistics. The value -0.0188711 is the covariance between b_0 and b_1. The estimated standard error of the logit is therefore (following equation 12.18):

$$s\sqrt{h_{ii}} = \sqrt{0.183850 + x^2 \times 0.0025017 - 2x \times 0.0188711}$$

The probability of exceedance π for a population density (x) of 11 was previously computed to be 0.226. A 95% confidence interval around this estimate is computed as

$$s\sqrt{h_{ii}} = \sqrt{0.183850 + (11)^2 \times 0.0025017 - 22 \times 0.0188711} = \sqrt{0.0714} = 0.267$$

$$\pi = \frac{\exp(-3.10 + 0.170 \times 11 \pm 1.96 \times 0.267)}{1 + \exp(-3.10 + 0.170 \times 11 \pm 1.96 \times 0.267)} = \frac{\exp(-1.23 \pm 0.524)}{1 + \exp(-1.23 \pm 0.524)}$$

$$= \left[\frac{0.173}{1.173}, \frac{0.4936}{1.4936}\right] = [0.147, 0.330].$$

At a population density of 11, a 95% confidence interval for the probability of TCE concentrations exceeding 5 μg/L is between 14.7 and 33%, with the best single estimate of the probability at 22.6%. This confidence interval will be accurate when there are good estimates of the variances and covariance of the parameters, resulting in a good approximation to the true standard error. This is more likely for large sample sizes, such as for the TCE data.

For multiple explanatory variables the process is similar to the above. There will be multiple covariance estimates between pairs of parameters. For a further discussion of how confidence intervals on π might be computed using the variance–covariance matrix output by statistical software, see Chapter 2 of Hosmer and Lemeshow (2000).

As an alternative for computing confidence intervals with smaller samples, repeated logistic regressions can be performed to produce bootstrapped estimates for π. This will obviously be computer intensive, but the $100 \times \alpha/2$th and $100(1 - \alpha/2)$th percentiles of the bootstrapped estimates of π will form the ends of an $\alpha\%$ confidence interval for the desired proportion. This bootstrapped interval does not require an assumption of normality for Wald's statistics that cannot be adequately tested by the smaller sample sizes common to most environmental studies.

12.2.7   Ordinal Methods for One Reporting Limit

A nonparametric alternative to the estimate of slope in linear regression is the median of all possible slopes between pairs of data points (Helsel and Hirsch, 2002). Suggested first by Theil (1950), Sen (1968) placed the estimator in the context of being a Hodges–Lehman estimator, determining its confidence interval and establishing it as a robust alternative to the least-squares slope of ordinary linear regression. Hirsch and Slack (1984) used the Theil–Sen method to estimate the trend slope for the Seasonal–Kendall test, which has become one of the most popular tests in environmental studies for determining changes over time. The Theil–Sen slope and confidence interval are related to Kendall's tau correlation coefficient (for more on tau, see Chapter 11). The test for whether the Theil–Sen slope is significantly different from zero is also the test for whether Kendall's tau is significantly different from zero. Neither relies on an assumption of normality for their validity. If the trend $\Delta y/\Delta x$ measured by the Theil–Sen slope is subtracted from the y variable, the correlation between the residuals and the x variable will have a tau coefficient of zero.

With one reporting limit the Kendall's tau correlation coefficient available in commercial software will produce a unique and accurate result. The Theil–Sen line will not. Both the slope and intercept are affected by any value chosen to represent censored observations, making the Theil–Sen method invalid for censored data unless there are very few censored observations. Helsel (1990) stated that for as many as 15% censored values, the standard Theil–Sen approach would be fine. While a guess at the time, it is not true. Depending on the arrangement of the data, 15% censored values could effect the computation of a median slope. Thankfully, since then a version of the Theil–Sen line called the Akritas–Theil–Sen (ATS) procedure was developed that solves for the slope in a different manner, one that does not require a substituted or estimated value for censored observations. It is described more fully later in this chapter. ATS and not Theil–Sen should be the method used to obtain a "linear median" with censored data, and is valid for one or multiple censoring limits. ATS is available in the NADA Minitab macros and NADA for R routines on the Practical Stats web site, but is not yet found as part of commercial statistics software. In this instance of the Theil–Sen ordinal nonparametric method for censored regression, keeping things simple is just not possible.

12.3   MAXIMUM LIKELIHOOD ESTIMATION

Maximum likelihood methods are used in a variety of disciplines to compute regression models. In reliability analysis, they are called "failure time models" (Meeker and Escobar, 1998). The response variable in that case is the time until a product fails, that is, until a light bulb burns out. In medical statistics, "accelerated failure time models" are used to predict the time until the recurrence of a disease, or until death (Collett, 2003). In economics, "censored maximum likelihood models" are used to predict the time until an event such as an interest rate change occurs, but have also been used with a response variable other than time. For example, Chay and Honore (1998) modeled incomes using MLE regression, where records were right-censored at tax-category ceilings. For the specific case of left-censored data that can include true zeros and whose residuals follow a normal distribution, MLE is sometimes called "Tobit analysis" after the economist Tobin (1958). All of these methods are fundamentally identical. Terminology, as always, can be confusing.

For regression of right-censored failure time models, uncensored observations are those where the subject's length of time, such as the time until death, is known exactly. The event has occurred and is recorded. What is known as a "death" or "failure" in that literature is a "detect" in environmental applications. Censored data are observations where the event has not yet happened by the time the experiment is finished. For these observations the time to occurrence is known only to be greater than some value. Regression analysis determines the significance of the effects of explanatory variables on the time until occurrence of the event - for example, does life expectancy increase with the amount of daily exercise?

In environmental science MLE can be used to model the response of left-censored variables other than time, such as concentrations or streamflows. For example, Slyman et al. (1994) modeled the concentrations of the trace element tin in minnows as a function of exposure time to wastewaters. Twenty percent of tin concentrations were below reporting limits. Liu et al. (1996) modeled atrazine concentrations in near-surface aquifers using Tobit regression. Kroll and Stedinger (1999) estimated low-flow quantiles in streams using censored regression. To be sure, there are environmental studies where the response variable is time, such as time until death of a sentinel species organism. Chaloupka et al. (2004) modeled time to failure of transmitters on sea turtles, for example (the turtles outlived the transmitters). These models use standard right-censored failure-time analysis. However, the primary interest in this book is in using MLE for left-censored variables where something other than time is modeled.

MLE regression software incorporates left-censored observations using an interval endpoints or "interval censored" or "arbitrary censored" format. In order to correctly model variables that do not go negative such as concentration, the lower bound of zero must be specified. In left-censored survival analysis applications the left boundary is negative infinity. This would produce biased parameter estimates for variables that actually could not go below zero. Of course the lower bound of concentration may be a value higher than zero, as in the data "flagged" as being between detection and quantitation limits. For these data, the lower bound would be the detection limit.

As a parametric method, hypothesis tests using MLE require that the data follow an assumed distribution. For skewed environmental data where most variables have values spanning two or more orders of magnitude, a lognormal distribution is most often assumed. The lognormal distribution has the additional benefit that estimated values cannot go negative, a problem that is often encountered when assuming a normal distribution for low-level contaminants. Parameter estimates for the slopes and intercept of a linear regression are computed by MLE (also see Chapter 2). A likelihood function is written as

$$L = \prod p[e_i]^{1-\delta_i} \cdot F[e_i]\delta_i \qquad (12.11)$$

where δ_i is the indicator of 1 for a censored observation and 0 for a detected observation, $F[e_i]$ is the cumulative distribution function of the residuals, equaling $\text{Prob}(e_i \leq t)$ for limit t, $p[e_i]$ is the probability density function of the residuals, and e_i are the residuals from the regression equation

$$e_i = y_i - \sum \beta_j x_{ij} \qquad (12.12)$$

where the βs are the coefficients for the j explanatory variables.

The derivative of L with respect to the βs is set to 0. Solving these equations produces the coefficients β_j with the highest likelihood of matching the observed data, both censored and uncensored. The optimization is something like varying the regression surface (a line for one β, a plane for two βs, and a higher dimensional surface in more than two dimensions) as it slices through the data until it finds the position with the smallest residual error.

12.3.1   Testing Relative Merits of MLE Models

Statistics software usually provide Wald's partial tests for the significance of each coefficient in MLE regression. Wald's tests estimate each coefficient along with an estimate of their standard error. The ratio of the coefficient to its standard error approximately follows a normal distribution, at least asymptotically (for large sample sizes) and when the distribution of the regression residuals follows the assumed distribution. Likelihood-ratio tests have been recommended instead of Wald's tests for MLE survival analysis procedures, due to the uncertainty in how quickly the Wald's ratios converge on their true value for small data sets that only approximately fit the assumed distribution. Likelihood-ratio tests still require that data follow the assumed distribution, but unlike Wald's tests, their validity does not depend on whether the coefficients themselves follow a normal distribution.

The test for determining the overall significance of an MLE regression model, similar in concept to the overall F-test in least-squares regression, is the overall likelihood-ratio test. This test determines whether the entire model being tested is an improvement over using no model at all, that is, over the "null model" where all βs equal 0. The overall test statistic is

$$
\begin{aligned}
G_0^2 &= 2[L(\beta) - L(0)] \\
&= [-2L(0)] - [-2L(\beta)]
\end{aligned}
\tag{12.13}
$$

where $L(0)$ represents the log-likelihood of the null model. Statistical software that does not report G_0^2 will report either the log-likelihood, $L(\beta)$ or the "-2 log-likelihood," $-2L(\beta)$. For example, Minitab prints the log-likelihood $L(\beta)$ for each regression model, as well as G_0^2. When it is not reported, G_0^2 can be computed by first obtaining the log-likelihood for the null model (sometimes done by having no explanatory variables except the number 1). The test statistic G_0^2 is compared to a chi-square distribution with k degrees of freedom, where k is the number of explanatory variables in the model, to determine a p-value for the overall test. As with least-squares regression, the overall test is not of much help in deciding which of the individual explanatory variables should be retained in the final MLE regression model.

Partial likelihood tests for the β coefficients of each explanatory variable in the model are alternatives to the partial Wald's tests. They are similar in concept to the partial t-tests of least-squares regression (though the statistic used is compared to a chi-squared rather than t distribution). To evaluate the effect of parameter B, for example, a model of all proposed variables $A + B + C$ is compared to a model without the variable to be evaluated, so $A + C$. The difference in log-likelihoods of the two models is a measure of the effect of variable B. The test statistic is twice the difference in log-likelihoods between the two models (equation 12.14), measuring how the fit to the data is improved by the use of that variable:

$$
\begin{aligned}
G_{\text{partial}}^2 &= 2[L(\beta_{\text{with}}) - L(\beta_{\text{without}})] \\
&= [-2L(\beta_{\text{without}})] - [-2L(\beta_{\text{with}})]
\end{aligned}
\tag{12.14}
$$

A p-value is obtained by comparing the test statistic to a chi-square distribution with one degree of freedom (a difference of one explanatory variable between the two models). If the p-value is less than the significance level α, the null hypothesis that the coefficient β equals 0 is rejected, and including the explanatory variable provides a significant improvement in model fit. If the p-value is large and the null hypothesis is not rejected, then the variable can be dropped from the list of useful predictor variables.

Partial likelihood or Wald's tests allow comparisons between nested models, sets of variables where the more complex equation ("with") contains all the variables of the simpler model ("without"). To compare models that are not nested, say a model with variables $A + B + C$ against a model containing variables $B + D$, other criteria are needed. Akaike's Information Criterion (AIC) and its corrected version (AICc) are two of the common tools for deciding among models (Harrell, 2001). Like Mallow's Cp and adjusted r-squared for least-squares regression, these statistics perform a cost-benefit analysis on the addition of variables to an equation. Adding a variable introduces a cost, which may or may not be offset by a reduction in the residual noise unexplained by the model. AIC for MLE models has the formula

$$AIC = -2L - 2p \qquad (12.15)$$

where $-2L$ is the -2 log-likelihood, expressing the attained error level, and $2p$ is twice the number of estimated parameters ($p = \#$ of explanatory variables plus 1 for the intercept), representing the cost. The best model is the one with the lowest AIC for a given data set. AIC is said to favor more complicated models than necessary for smaller data sets (Harrell, 2001). This complaint led to the corrected version, AICc

$$AICc = -2L - 2p\left(1 + \frac{p+1}{n-p+1}\right) \qquad (12.16)$$

which tends to favor smaller models when there are smaller amounts of data.

12.3.2   Example: MLE Regression of TCE in Groundwater Data

Are TCE concentrations in groundwater a function of population density, land use, or depth to the water surface? The data set TCEReg contains information on TCE concentrations in ground waters of Long Island, NY, along with data on three explanatory variables (Eckhardt et al., 1989). Of interest is determining whether any of the explanatory variables significantly affect TCE concentrations, and if so, estimating their slopes.

The relationship between TCE concentration and the three explanatory variables—%IndLU (percent industrial land use), Depth (depth to water), and Popden (population density surrounding the well location)—is measured without substitution of values for censored data by MLE regression. The column TCEConc contains the concentrations and reporting limits for TCE. There are multiple reporting limits. The column BDL0 is the indicator column for TCE, having a value of 0 for data below reporting limits (hence the name) and a 1 for uncensored observations. MLE software for left-censored or arbitrarily censored data requires a start and an end column, the interval endpoints format of Chapter 3. The TCEConc column is the end column, containing the upper limit value of the reporting limit for censored observations. A start column must be created by multiplying TCEConc times BDL0. The result is stored in the column TCE0. In TCE0 all censored observations are represented by a value of 0. All uncensored observations have the same TCE concentration in both the start and end columns. A few example entries for these data are given below. As an example, the last entry is a detected concentration of 1 μg/L, the smallest possible detected concentration for the time period when the reporting limit was 1.

BDL0	TCEConc	LU	Popden	%IndLU	Depth	TCE0
0	1.000	9	9	10	103	0.000
0	1.000	9	11	0	32	0.000
0	1.000	8	3	4	142	0.000
1	32.00	9	11	3	69	32.000
0	1.000	9	11	0	32	0.000
1	13.00	9	14	7	89	13.000
0	1.000	5	6	1	23	0.000
1	150.0	8	6	6	177	150.000
1	1.000	8	3	4	207	1.000

The maximum likelihood equation is solved by using the

```
Stat > Reliability/Survival > Regression with Life Data
```

command of Minitab. With start variable TCE0 and end variable TCEConc, and the three explanatory variables entered into the Model window, the output from MLE regression assuming a normal distribution is

```
Estimation Method: Maximum Likelihood
Distribution: Normal

Regression Table
Predictor   Coef        Error       Z      P       Lower       Upper
Intercept   6.31118     4.55867     1.38   0.166   -2.62366    15.2459
Popden      0.373907    0.538686    0.69   0.488   -0.681899   1.42970
%IND LU     0.0791126   0.439033    0.18   0.857   -0.781377   0.939602
Depth       -0.0091394  0.0129492   -0.71  0.480   -0.0345193  0.0162406
Scale       30.3921     1.36742                    27.8267     33.1939

Log-Likelihood = -1069.059
```

The probability plot of residuals for the regression (Figure 12.2) indicates that a transformation is necessary. The residuals do not match the straight line representing the normal distribution.

A lognormal distribution is assumed for the residuals and the procedure run again. The residuals are essentially linear (Figure 12.3), so that a lognormal distribution is suitable for MLE regression of the TCE data. The output from the procedure is given below.

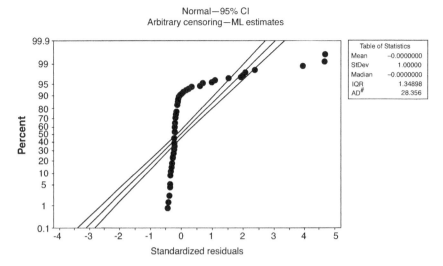

FIGURE 12.2   Probability plot of residuals for MLE regression of the TCE data, assuming a normal distribution.

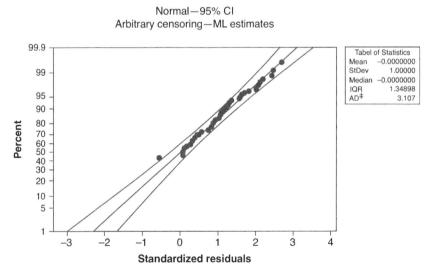

FIGURE 12.3   Probability plot of residuals for MLE regression of the TCE data, assuming a lognormal distribution.

```
Estimation Method: Maximum Likelihood
Distribution: Lognormal

Regression Table
                            Standard                      95.0% Normal CI
Predictor    Coef          Error       Z        P         Lower       Upper
Intercept    -2.88026      0.823547    -3.50    0.000     -4.49438    -1.26613
Popden       0.250904      0.0745204   3.37     0.001     0.104846    0.396961
%IND LU      0.0406455     0.0526390   0.77     0.440     -0.0625251  0.143816
Depth        -0.0043726    0.0023329   -1.87    0.061     -0.0089450  0.0001998
Scale        2.81166       0.311129                       2.26345     3.49264

Log-Likelihood=-302.931
```

To compute the overall test for whether this three-variable model predicts TCE concentrations better than simply the mean concentration (the null model), the log-likelihood for no explanatory variables is calculated using the

```
Stat > Reliability/Survival > Distribution Analysis
(Arbitrary censoring) > Parametric Distribution Analysis
```

command. Assuming a lognormal distribution, the null model produces a log-like-lihood of -316.404:

```
Estimation Method: Maximum Likelihood
Distribution: Lognormal

Parameter Estimates
                        Standard    95.0% Normal CI
Parameter    Estimate   Error       Lower        Upper
Location     -1.77893   0.415959    -2.59420     -0.963676
Scale        2.93033    0.327988    2.35311      3.64913

Log-Likelihood=-316.404
```

Using equation 12.13, the test statistic G_0^2 is computed for the overall test as

$$G_0^2 = 2[L(\beta) - L(0)] = 2[-302.931 - (-316.404)] = 26.95$$

Comparing 26.95 to a table of the chi-square distribution with $k = 3$ degrees of freedom (three explanatory variables), the resulting p-value equals <0.001, less than the alpha of 0.05. So the three-variable model is considered better than no model at all.

The Wald's tests (Z statistics) for coefficients of the three explanatory variables previously indicated that Popden is a significant predictor of TCE concentration ($p = 0.001$), that %IndLU is not ($p = 0.440$), and that Depth is on the edge with a

p-value of 0.06. Since the Wald's tests are believed by many to be only approximate, a likelihood-ratio test can be conducted for a variable such as Depth whose Wald's test conclusion might be in doubt. To do this, the lognormal MLE regression is again computed, this time without Depth as an explanatory variable. The output for this two-variable regression model is below.

```
Distribution: Lognormal
Relationship with accelerating variable(s): Linear, Linear

Regression Table
                       Standard                        95.0% Normal CI
Predictor   Coef       Error       Z       P       Lower        Upper
Intercept   -3.82240   0.767808   -4.98   0.000   -5.32727    -2.31752
Popden      0.300775   0.0740682   4.06   0.000    0.155604    0.445946
%IndLU      0.0376751  0.0529982   0.71   0.477   -0.0661996   0.141550
Scale       2.83152    0.314002                    2.27838     3.51896

Log-Likelihood = -305.035
```

Following equation 12.14, the partial log-likelihood test for Depth is computed as

$$G^2_{partial} = 2[L(\beta_{with}) - L(\beta_{without})] = 2[-302.931 - (-305.035)] = 4.208$$

with an associated p-value from a chi-square distribution with 1 degree of freedom of 0.04. This p-value is smaller than for the Wald's test (and the log-likelihood test is to be preferred), so that Depth is considered a significant variable and should be retained in the model. Based on the partial log-likelihood tests, the best regression model for these data has Popden and Depth as explanatory variables. The final model for explaining TCE concentrations is

$$\ln TCE = -2.79 + 0.260 \times Popden - 0.004 \times Depth$$

```
Distribution: Lognormal
Relationship with accelerating variable(s): Linear

Regression Table
                       Standard                        95.0% Normal CI
Predictor   Coef        Error       Z       P       Lower        Upper
Intercept   -2.79066    0.810181   -3.44   0.001   -4.37859    -1.20273
Popden      0.259589    0.0740544   3.51   0.000    0.114445    0.404733
DEPTH       -0.0043407  0.0023406  -1.85   0.064   -0.0089282   0.0002468
Scale       2.81474     0.311546                    2.26582     3.49665

Log-Likelihood = -303.227
```

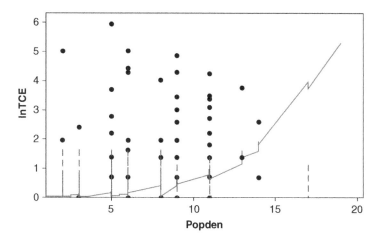

FIGURE 12.4 Log of TCE concentration plotted as a function of population density (Popden). Censored observations plotted as vertical dashed lines. The lognormal MLE two-variable regression model is shown as the jagged curve.

The MLE lognormal regression model is pictured as a jagged curve in Figure 12.4. Most of the data are censored observations, and plot on top of one another as the dashed vertical lines. The curve has responded to the predominance of censored data at low population densities. The jags on the curve result from wells of different depths at the same Popden value. An additional variable, which would improve this model is one that would explain the high concentrations seen at lower population densities.

The cenreg procedure performs the same test in NADA for R. Partial tests are Wald's tests.

```
> tcemle2= with(TCEReg,cenreg(Cen(TCEConc,TCECen)
   ~PopDensity+Depth,dist="lognormal"))
> tcemle2
              Value      Std. Error    z        p
(Intercept)  -2.79067    0.81018     -3.44    5.72e-04
PopDensity    0.25959    0.07405      3.51    4.56e-04
Depth        -0.00434    0.00234     -1.85    6.37e-02
Log(scale)    1.03487    0.11068      9.35    8.78e-21

Scale=2.81

Log Normal distribution
Loglik(model)=-303.2  Loglik(intercept only)=-316.4
Loglik-r:  0.318126

Chisq= 26.35 on 2 degrees of freedom, p= 1.9e-06
Number of Newton-Raphson Iterations: 4
n=247
```

To determine which MLE regression to use, AIC and/or AICc is easily computed from the $-2 \times$ log-likelihood statistic to compare the two-variable model with the following PopDensity-only model:

```
> tcemle1 = with(TCEReg, cenreg(Cen(TCEConc, TCECen)
  ~PopDensity,dist="lognormal"))
> tcemle1
                  Value      Std. Error     z         p
(Intercept)      -3.734      0.7493       -4.98      6.23e-07
PopDensity        0.309      0.0736        4.20      2.71e-05
Log(scale)        1.042      0.1109        9.39      5.88e-21


Scale = 2.83


Log Normal distribution
Loglik(model) = -305.3  Loglik(intercept only) = -316.4
Loglik-r:   0.2934194


Chisq = 22.24 on 1 degrees of freedom, p = 2.4e-06
Number of Newton-Raphson Iterations: 4
n = 247
```

AIC for the two-variable model $= -2 \times (-303.2) - 2 \times 3 = 606.4 - 6 = 600.4$
AIC for the one-variable model $= -2 \times (-305.3) - 2 \times 2 = 610.6 - 4 = 606.6$
AICc for the two-variable model $= -2 \times (-303.2) - 2 \times 3 \times [1 + 4/243]$
$\quad = 606.4 - 6.1 = 600.3$
AICc for the one-variable model $= -2 \times (-305.3) - 2 \times 2 \times [1 + 3/244]$
$\quad = 610.6 - 4.05 = 606.55$

By either criterion, choose the two-variable PopDensity + Depth model as it has smaller AIC values. The correction makes little difference with a data set this large.

12.4 AKRITAS–THEIL–SEN NONPARAMETRIC REGRESSION

For censored data, a Theil–Sen type of slope estimate can be calculated as the slope that when subtracted from the Y variable most closely produces a tau correlation coefficient of zero between the residual and the X variable. Developed by Akritas et al. (1995), the Akritas–Theil–Sen slope estimate is implemented in the ATS Minitab macro and the cenken routine in NADA for R. A major advantage of the ATS slope estimator over the MLE estimator is that ATS can be computed for doubly censored data, data where X and Y are both censored. The more traditional estimators of slope for censored data, including Buckley–James regression (see next section) and MLE methods, allow only the Y variable to be censored.

A major disadvantage of the ATS estimator is that the procedure has not yet been generalized to multiple X variables, and so for now is only available in the context of one explanatory variable. Dietz and Killeen (1981) described a multivariate application of the Theil–Sen slope to test for trend in uncensored data. Libiseller and Grimvall (2002) used the partial Kendall's tau correlation concept to separately evaluate the effects of multiple explanatory variables, in essence a multivariate Theil–Sen procedure. Akritas and Siebert (1996) derived Kendall's tau for partial correlation with censored data, leading to possibilities of implementing an ATS approach for multivariate censored data. Wahlin and Grimvall (2010) extended the partial Kendall's tau of Libiseller and Grimvall (2002) to the case of interval-censored observations. This last article in essence presents a multivariate ATS computation for censored (and uncensored) data. To date, it has been implemented only in the authors' visual Basic code but has great promise for extending the methods of this section to a multiple regression context.

Akritas et al. (1995) found that their method had lower bias and standard error than several alternatives, including a weighted least-squares approach and a median of pairwise slopes method. A later study by Wilcox (1998) showed that the Akritas–Theil–Sen slope had a "substantial advantage" in bias and precision over Buckley–James regression, the most commonly used nonparametric regression method for censored data (see Section 12.5). The Akritas–Theil–Sen method has as much utility for trend analysis and other regression models for censored data as does the original Theil–Sen slope for uncensored data.

To compute the Akritas–Theil–Sen slope estimator, set an initial estimate for the slope, subtract this from the Y variable to produce the Y residuals, and then determine Kendall's S statistic between the residuals and the X variable. Next, conduct an iterative search to find the slope that will produce an S of zero. Because the distribution function of the test statistic S is a step function, there may be more than one slope that will produce a value of zero S. Therefore, the final Akritas–Theil–Sen slope is considered to be the one halfway between the maximum and minimum slopes that produce a value of zero for S.

12.4.1   Example: TCE in Groundwater

The nonparametric ATS slope estimate of TCE concentration as a function of population density (Popden) is computed using the macro ats.mac. The same four arguments used for the ckend macro are required: X, Y, and censoring indicators for X and Y (Figure 12.5).

```
%ats c4 c2 c7 c1; cens 0.
```

The subcommand cens is used because the indicator for censored data is a 0 in this data set, rather than the default of 1. The subcommand allows any designator, numeric or text, to be used to specify which observations are censored.

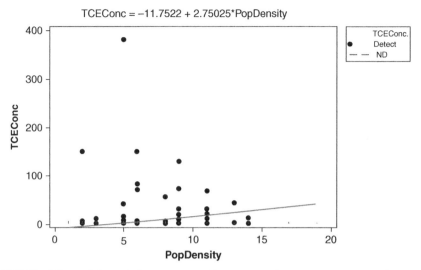

FIGURE 12.5   ATS line relating censored TCE concentrations to population density (Pop-Density). Censored observations plotted as vertical dashed lines.

A-T-S line
```
stau                4431.00
tau                 0.145848

TCEConc = −11.7522 + 2.75025*PopDensity
Slope               2.75025
Intercept           −11.7522
```

As a nonparametric test, ATS makes no assumption about the distribution of the residuals of the data. Yet computing a slope implies that the data follow a linear pattern. If the data do not follow a linear pattern, either the Y or X variables should be transformed to produce one before a slope is computed. Otherwise the statement that a single slope describes the change in value for data in the original units is not correct. If the data are curved, the rate of change $\Delta Y/\Delta X$ changes as X increases. A straight line, whether by MLE or ATS, implies a constant rate of change in those units. Logarithms of TCE were previously used with MLE to improve the linear relationship with population density. After taking the natural logs of the TCE concentrations, the ATS macro for ln TCE versus population density produces (Figure 12.6):

A-T-S line
```
stau                4431.00
tau                 0.145848
lnTCE = −2.83228 + 0.383507*PopDensity
Slope               0.383507
Intercept           −2.83228
```

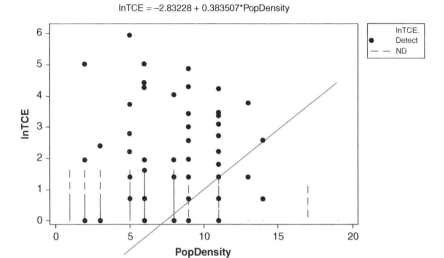

FIGURE 12.6 ATS line relating natural log of TCE concentrations to population density (PopDensity). Censored observations plotted as vertical dashed lines are easier to see here than in Figure 12.5.

The equation indicates that the natural log of TCE concentrations increase by 0.384 units for every unit increase in population density. This translates into an average increase of 47% per unit increase in population density, using the formula

$$\text{percent change in } Y \text{ per year} = \left(e^{b_1} - 1\right) \times 100 \qquad (12.17)$$

where b_1 is the slope in natural log units.

The ATS procedure is implemented in NADA for R with the cenken command. With cenken, the last argument of a censoring indicator for the X variable is optional. If omitted, all values of the X variable are considered detected. For the untransformed TCE data the results are

```
> data(TCEReg)
> attach(TCEReg)
> ats=cenken(TCEConc,TCECen,PopDensity)
> ats
slope
[1] 2.75025

intercept
[1] -11.75225

tau
[1] 0.1458477

p
[1] 0.0003007718
```

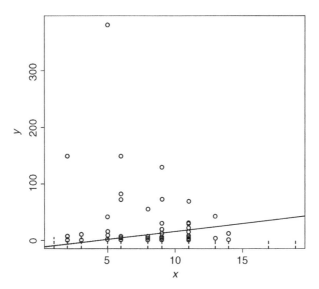

FIGURE 12.7 ATS line relating censored TCE concentrations to population density (Pop-Density) using the cenxyplot and cenken commands in NADA for R.

A censored scatterplot with superimposed ATS line can be drawn after typing three R commands. The first "with" command draws the scatterplot in a graphics device. Note that the order of arguments for the cenxyplot command are (X, Xcen, Y, Ycen), a different order than in the cenken command. The second "with" command computes the cenken line and saves it as the object "reg" (or any name you choose). Then the lines command adds the cenken line to the scatterplot. The result is Figure 12.7.

```
> with(TCEReg, cenxyplot(PopDensity,PopCen,TCEConc,TCECen))
> reg=with(TCEReg, cenken(TCEConc,TCECen,PopDensity))
> lines(reg)
```

Previously we instead settled on using the log TCE model. Taking logarithms of TCE and solving using the commands below, the result is

```
> lnTCE=log(TCEConc)
> atslog=cenken(lnTCE,TCECen,PopDensity)
> atslog
slope
[1] 0.3835066

intercept
[1] -1.150520

tau
[1] 0.1458477

p
[1] 0.0003007718
```

12.4.2   Nonparametric Estimates of Intercept

Several possible nonparametric estimates for an intercept to accompany the Theil–Sen slope have been evaluated for uncensored data (Dietz,1987, 1989; Hollander and Wolfe, 1999, Section 9.4). However, their use has not been explicitly evaluated for the case of censored data. Dietz (1987, 1989) found that the median residual (equation 12.18) was a relatively efficient measure of the intercept for a linear equation based on the Theil–Sen slope b_{TS}.

$$\hat{b}_0 = \text{median}[Y_i - b_{TS} \cdot X_i] \quad \text{for } i = 1, \ldots, n \tag{12.18}$$

A second estimator had slightly lower mean squared error under specific circumstances. It was the median of all pairwise (Walsh) averages of residuals, in the Hodges–Lehmann class of estimators (equation 12.19):

$$\hat{b}_{HL} = \text{median}\left(\frac{[Y_i - b_{TS} \cdot X_i] + [Y_j - b_{TS} \cdot X_j]}{2}\right) \quad \text{for } i, j = 1, \ldots, n \text{ and } j \neq i \tag{12.19}$$

A third estimator was attributed to Conover (1999). It had higher mean square error but is simpler to compute. Remember from basic statistics that the least-squares regression line goes through the point $(\overline{X}, \overline{Y})$. The Theil–Sen line is often placed through the median of X (X_{med}) and median of Y (Y_{med}) by using the intercept in equation 12.20:

$$\hat{a} = Y_{med} - b_{TS} \cdot X_{med} \tag{12.20}$$

where b_{TS} is the Theil–Sen slope estimator. This is the form of the trend line in the Seasonal–Kendall trend analysis process of Hirsch and Slack (1984). For censored data, it would require that the median of both X and Y variables be computed by Kaplan–Meier (see Chapter 6) or another method appropriate for censored data. Additional information on these three estimates of intercept using uncensored data is found in Hettmansperger et al. (1997).

The median residual intercept of equation 12.18 was implemented in the ATS Minitab macro and cenken R routine, as it is more efficient than equation 12.20 and simpler to compute for censored data than is equation 12.19. Note that though Akritas et al. (1995) derived the slope estimate, their paper looked only at slope estimates and did not evaluate any corresponding estimates for intercept. Until a study of intercept terms is conducted specifically for censored data, the uncensored results favoring equations 12.18 or 12.19 is all that is available. Censored Y observations produce interval-censored residuals, and so to solve the equivalent of equation 12.18; the median residual is computed using a Turnbull estimate (see Chapter 6).

The result of using any of these intercept estimates, along with the ATS slope estimate, is a line less strongly affected by outliers than is regression that assumes a normal error distribution, such as MLE regression. The ATS line predicts the conditional median of Y (the median of Y for any given X), rather than the conditional

TABLE 12.1 Comparison of slopes and intercepts for straight lines fit to the natural logarithms of TCE concentrations

Method	Slope	Intercept	p-Value
MLE (lognormal)	0.309	−3.73	<0.000
ATS (log of TCE; p from ckend macro)	0.383	−2.83	<0.000

The single explanatory variable is population density.

mean of Y provided by MLE methods. If the residuals follow a normal distribution, the ATS line and the MLE line should be quite similar. A comparison of MLE and ATS results for fitting a straight line to the logarithms of the TCE data is listed in Table 12.1, and plotted in Figure 12.8.

The ATS line has a higher slope than MLE, and with a log model the differences increase with increasing population density. Remember that most of the data are censored observations, with the proportion of censored observations decreasing as population density increases. The lines are sensitive to those data, keeping the estimated values low for low densities, as a result of using those proportions in computing the final result.

12.5 ADDITIONAL METHODS FOR CENSORED REGRESSION

Other methods have also been used for regression models with censored data. Three examples are Cox proportional hazards models, Buckley–James regression, and quantile regression. The first two methods are regularly applied to censored data in the field of survival analysis. Proportional hazards is a standard tool for the

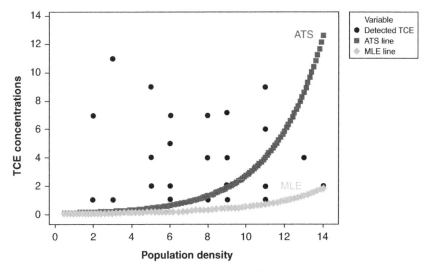

FIGURE 12.8 ATS and MLE models for log of TCE concentrations versus population density. Data are from Eckhardt et al. (1989).

biostatistician, and is discussed in essentially all textbooks on survival analysis. Buckley–James regression is a more specialized procedure found in some software and textbooks. Quantile regression is a newer method used for analysis of uncensored data, and appears in current statistical journals, but not yet in many statistics textbooks.

Cox proportional hazards is a semiparametric method used to estimate the proportional effect of one or more explanatory variables X_k on the censored Y variable, without modeling the underlying relationship between Y and the Xs. As such it does not result in a regression equation–there is no intercept and no functional form that will predict Y from the Xs. Instead, a rate increase is modeled, the change in the "hazard function" $h(Y)$, as a function of the explanatory variables X:

$$h(Y) = h_0(Y)e^{\beta_i X_i} \tag{12.21}$$

A hazard function is the ratio of the probability density function $f(Y)$ to the survival function $S(Y)$. It is the "'approximate' probability of an individual of age X experiencing the event in the next instant" (Klein and Moeschberger, 2003). Translated into environmental applications, it is the approximate probability of a concentration falling below a reporting limit as concentration decreases a minute amount. The concept of a hazard function has not translated well to environmental studies, perhaps leading to the lack of its use there.

However, the frequent use of proportional hazards models in medical studies may warrant a new look at these models for environmental applications. The focus of interest would be in the slope coefficient β. The β slope coefficient in equation 12.19 is the source of statements reported to the general public in a form something like "the medication resulted in a threefold decrease in risk of heart attack." By keeping the exact form of the h_0 base function unspecified, no functional form of the relationship between X and Y need be assumed. The risk of heart attack for individuals may be high, or low, and may or may not be linear with respect to diet and exercise. None of that matters, because it is all contained in the unspecified base risk term h_0. What is of interest is that, given whatever risk there may be, that risk proportionally changes as X changes. All terms but the effect of X are "blocked out" by putting them into the base risk h_0.

Proportional hazards might be of use in environmental studies with censored data for performing procedures similar in purpose to analysis of covariance. With analysis of covariance, all effects on concentration other than the one to be studied are adjusted for. If X were a measure of exposure, such as industrial activity or proportion of soils with high metals content, the rate of risk for increasing the resultant concentration Y as X changes could be determined without assuming an underlying model for other effects that might be occurring. Proportional hazards could also be useful as a screening tool for determining which of several X variables has a significant effect on the distribution of concentrations Y, in a mode not unlike stepwise regression is now often (though perhaps inappropriately) used. The reader is referred to textbooks on survival analysis such as Lee and Wang (2003), Klein and Moeschberger (2003), Kalbfleisch and Prentice (2002), or Collett (2003) for a much more thorough treatment of hazard functions and proportional hazard models.

Buckley and James (1979) presented a method to construct a linear regression model for right-censored response variables (the x variables cannot be censored) without assumption of a normal distribution. The method is an iterative procedure to find a slope β that fits a weighted combination of uncensored values for Y and the censored survival function for right-censored data. Ireson and Rao (1985) compared this method to the Theil–Sen slope, finding that Buckley–James regression always had larger confidence intervals than the Theil–Sen method, and so comparatively lacked precision. Buckley–James also suffers from a possible failure for the algorithm to converge to an answer. Wilcox (1998) showed that a modified form of Buckley–James regression had much larger bias than did Theil–Sen regression. From these studies there appears little reason to prefer Buckley–James regression over the Theil–Sen methods described in this chapter.

Quantile regression (Cade and Noon, 2003) estimates the conditional quantiles or percentiles of a distribution as a linear function of explanatory variables. The impetus in ecology seems to be for fitting data with heterogeneous errors—residuals that show a pattern of increasing variance. For such data the slope of Y with X may be moderate for central data, such as near the mean or median, but larger slopes would better characterize the data when higher quantiles are of interest. The objective is something like "How do concentrations that are exceeded only 25 and 10% of the time change as a function of X?" Multiple equations for simultaneously fitting different quantiles are usually constructed.

Quantile regression has not often been applied to censored data. Slope coefficients are normally calculated for uncensored data using weighted absolute deviation. These equations would have to be solved by maximum likelihood in order to correctly incorporate censored observations. In any case, quantile regression is not yet found in most commercial statistical software, and routine application to censored data will require further study.

EXERCISES

12-1 Atrazine concentrations were measured in streams across the Midwestern United States (Mueller et al., 1997). Data are found in recon.xls. Measured at each site were the following explanatory variables:

Name	Description
Area	Basin size
Applic	Atrazine application rate, estimated from statewide estimates
Corn%	Percent of land area of watershed planted in corn
Soilgp	Soil hydrologic group, a measure of soil permeability found in STATSGO
Temp	Annual average temperature (a north–south indicator)
Precip	Annual average precipitation (mostly an east–west indicator)
Dyplant	Days since planting (and therefore since last atrazine application)
Pctl	Percentile of streamflow (standardizes across streams of varying size)
Atraconc	Atrazine concentration, in μg/L

Using censored parametric regression, build a multiple regression model to relate atrazine concentrations to the variables in the list above. Determine what units are the best to be working in before settling on a final model. Find the explanatory variables that are all significant at $\alpha = 0.05$.

12-2 Brumbaugh et al. (2001) measured mercury concentrations in fish of approximately the same age and trophic level across the United States. Determine a regression equation for the dependence of mercury ("Hg") on one or more of the possible explanatory variables listed below. Transformation of the explanatory variables may be required. All explanatory variables included in the model should have a p-value of 0.05 or less. The data are found in HgFish.xls.

Name	Description
WatMeHg	Methyl mercury concentrations in stream water
WatTotHg	Total mercury concentrations in stream water
SedMeHg	Methyl mercury concentrations in stream sediments
SedTotHg	Total mercury concentrations in stream sediments
WatDOC	Dissolved organic carbon concentrations in stream water
SedLOI	Loss of ignition (a measure of organic carbon content) in stream sediment
SedAVS	Sediment acid–volatile sulfides
% wetland	Percent of the basin occupied by wetlands

12-3 Using the data in recon.xls collected by Mueller et al. (1997), compute a logistic regression equation for predicting the probability of observing an atrazine concentration above 1 mg/L. The variable GE_1 has a value equal to 1 for all atrazine concentrations greater than 1 mg/L, and 0 otherwise. Candidate explanatory variables are the same as those considered in Exercise 12-1.

13 Multivariate Methods for Censored Data

Environmental studies usually involve measurement of more than one contaminant, and more than one explanatory or causative agent. A suite of trace elements is measured on soils at both background and possibly affected areas. A suite of organics is measured in waters across the state to see if patterns are related to population and land use. Biological community structure, including counts of many types of indicator organisms, is combined with chemical and physical measures at a number of watersheds to better understand the effects of those measures on the patterns of community health. Looking only at each chemical or type of organism individually comes nowhere close to providing the information available in such data sets. Multivariate statistical methods provide insight and clarity into the patterns and relationships among numerous measures. Similar patterns among contaminants indicate that those chemicals are routinely found together. Similar patterns among contaminants and explanatory variables provide insight into possible causes. Grouping of variables into "factors" aids in understanding of the processes operating in the system. Groupings of locations based on similar patterns in chemistry and community structure indicate how geology or elevation or other characteristics interact with anthropogenic influences, or might be used to decrease costs by sampling only a subset of the sites that are similar to many others.

While terminology and concepts present one hurdle to using multivariate methods (training in the methods has not been sufficient in many undergraduate programs in environmental science), another serious hurdle is how to employ them when chemical data include censored observations. Documented deficiencies when substituting fabricated values for nondetects in the univariate case (see Chapter 6) are reflected in multivariate findings. Hopke et al. (2001) compared imputation procedures to substitution prior to running multivariate methods. They found that substituting zero or RL produced a significant bias. Substituting one-half or 1/(square root of 2) times the RL underestimated the variance for those variables. Both effects caused problems with later interpretations. Farnham et al. (2002) found with a simulation study that with as little as 20% censoring, slopes of principal components were not correctly computed. Substitution produced problems for principal components analysis (PCA) and cluster methods with percent censoring of 30% and higher.

Statistics for Censored Environmental Data Using Minitab® and R, Second Edition.
Dennis R. Helsel.

Their recommendation was to not use variables with 30% or more censoring. The level at which problems occur is likely lower, as their censoring simulation did not mimic most of the common sources of variation in reporting limits found in field studies such as changing sample amounts, changing lipid content, and so on. As noted by Reimann et al. (2002), the results of PCA and factor analysis can change radically depending on which variables are included and excluded. Excluding variables just due to a specific level of censoring will have important but unrecognized negative consequences on the findings. Aruga (1997) found that estimating values for non-detects from a PCA on only detected observations, dropping the nondetects, gave "unacceptable results" with less than 5% censoring.

Scientists have struggled with what to do with censored data when applying multivariate methods. Grünfeld (2005) simply deleted the nondetect data, with unknown consequences. O'Connell et al. (2010) used methods from this book for their univariate procedures, but fell back into substitution prior to performing PCA. Stetzenbach et al. (1999) stated that when running PCA "... if the distribution cannot be determined from the data (as in the case in this study), a substitution method is the only available alternative," which as you will see in this chapter is not correct. They then justified substitution based on a USEPA guidance document that recommended it for univariate methods. Griffith et al. (2002) substituted one-half the RL before performing correspondence analysis and redundancy analysis. After collecting volumes of chemical, physical and biological data using state of the art methods, they employed a back of the envelope interpretation procedure whose effect on their conclusions is unknown. There are better ways.

Rather than using flawed methods, use multivariate tools that capture the information in the pattern of detected values along with the pattern in the frequency of data below each reporting limit. The same classes of procedures as in other chapters are available: binary methods, ordinal nonparametric methods, and methods that operate on scores or percentiles while adjusting for frequencies of censored observations.

13.1 A BRIEF OVERVIEW OF MULTIVARIATE PROCEDURES

Multivariate procedures can be classified into one of two groups depending on their objective. The first class of "interdependent" procedures treats all variables as equal in function, discerning patterns of covariance. Methods in this class include PCA, exploratory factor analysis, cluster analysis, correspondence analysis, and multi-dimensional scaling (MDS). The patterns of interest may be between variables (R mode analysis) or between samples/sites (Q mode analysis). There generally are no hypothesis tests involved with "interdependent" procedures, but only an exploration of the inherent patterns of the data. When there are no hypothesis tests there are no required assumptions of a normal distribution. PCA, factor analysis, and classical MDS are however linear procedures, modeling linear relationships among variables. Transformations are sometimes required to make relationships within the data more linear so that a linear model fits well, or to reduce the effect of one or a few outliers on the outcome.

In the second class of multivariate procedures, some variables are response variables and some explanatory variables, much like in multiple regression or analysis of variance but operating on multiple y (response) variables. These can be called "analysis of dependence" procedures. Example explanatory variables include group assignments, continuous physical variables such as light intensity or pollution index, or time or spatial sequences. These are used to model and/or explain observed patterns in the response variables. Discriminant function analysis (DFA), canonical correlation, multivariate analysis of variance (MANOVA), partial least squares, analysis of similarity (ANOSIM), and tests for seriation (a nonparametric trend test) are some of the procedures in this class. Some procedures such as MANOVA are parametric, requiring an assumption of multivariate normality. This assumption is difficult to check when there are censored observations, though the issue has been addressed (Tempelman and Akritas, 1996). Other procedures such as ANOSIM are nonparametric, employing permutation tests to obtain p-values rather than requiring multivariate normality. Nonparametric multivariate procedures have more utility for censored data, operating on ranks of resemblances rather than absolute magnitudes. Permutation tests are computer intensive, so were envisioned before but not routinely implemented until computing power was sufficient, coming into common use in the 1990s. Multivariate permutation procedures are found in the commercial software Primer, as well as in R.

This rich collection of multivariate procedures cannot be fully described and illustrated in one chapter, including this one. The objective of this chapter will be to focus on how to include censored observations into representative examples of multivariate procedures. For a more comprehensive description of multivariate procedures, see Everitt and Dunn (2001). Implementation of multivariate procedures in environmental sciences can be found in Shaw (2003) and Zuur et al. (2007), as well as in the third edition of Davis' classic text (Davis, 2001). Nonparametric procedures using permutation methods are presented in Clarke and Warwick (2001) as implemented in their software program Primer. Many of the examples in this chapter were computed using Primer (http://www.primer-e.com/), a nonparametric multivariate software package used primarily by ecologists, but also very useful for analysis of censored data.

13.1.1 The Structure of Multivariate Procedures

Table 13.1 illustrates the standard layout of multivariate data. Rows in the table present instances, locations, or times of observations. Columns represent variables, which may be chemical measures, counts of organisms, or physical characteristics. In Table 13.1, six columns contain concentrations for variants of the toxic chemical compounds DDT, DDE, and DDD found in fish. Use of these compounds was banned in the 1970s in the United States, but still found their way into fish and other biota decades later. The seventh column classes the fish into one of two age groups, Young or Mature. Of interest might be to determine groupings of stations with similar patterns of the six chemicals (Q mode), patterns of covariance between chemicals

TABLE 13.1   DDT and Related Compounds in Fish

Site	opDDD	ppDDD	opDDE	ppDDE	opDDT	ppDDT	Age
1	<5	<5	<5	14	<5	<5	Young
2	<5	42	8.4	130	<5	31	Mature
3	5.3	38	<5	250	<5	11	Mature
4	<5	12	<5	57	<5	<5	Mature
5	<5	<5	<5	16	<5	<5	Young
6	<5	<5	<5	<5	<5	<5	Young
7	<5	14	<5	52	<5	14	Mature
8	<5	15	<5	48	<5	<5	Mature
9	<5	12	<5	110	<5	20	Mature
10	5.1	39	<5	100	<5	24	Mature
11	<5	5.7	<5	87	18	<5	Mature
12	<5	9.4	<5	53	<5	7.3	Mature
13	<5	18	<5	210	<5	30	Mature
14	5.1	27	<5	140	<5	33	Mature
15	<5	10	<5	24	<5	5.8	Mature
16	<5	7.6	<5	15	<5	10	Mature
17	9	46	<5	110	<5	21	Mature
18	<5	22	<5	51	<5	7.4	Mature
19	9.8	41	6.9	50	<5	11	Mature
20	<5	13	<5	66	<5	8.6	Mature
21	<5	26	<5	110	<5	26	Mature
22	5.1	24	250	38	<5	11	Mature
23	8	100	8	160	<5	<5	Mature
24	<5	<5	<5	23	<5	<5	Young
25	<5	<5	<5	17	<5	<5	Young
26	<5	<5	<5	16	<5	<5	Young
27	5.7	27	<5	140	<5	14	Mature
28	<5	15	6	8.1	<5	<5	Mature
29	<5	20	7.5	22	<5	6.4	Mature
30	<5	22	<5	190	5.2	27	Mature
31	<5	31	<5	42	<5	25	Mature
32	<5	15	<5	23	<5	5.6	Mature

(R mode), or to test whether the pattern of chemicals differs between Young and Mature fish.

The first step is to construct a resemblance matrix, a concise summary of the similarities or distances between rows or between columns. A triangular-shaped matrix results from computing numerical measures of similarity (or conversely, the distance) between all pairs of rows or columns. To investigate similarities among sites (rows), a similarity measure would be computed between each pair of rows (Figure 13.1).

In Table 13.1, there are six columns of chemical variables, so six comparisons are made for each pair of rows, together resulting in one number that describes the resemblance between these two sites. Resemblance may either be a measure of

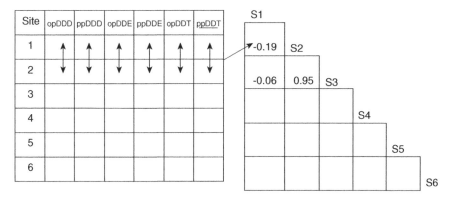

FIGURE 13.1   Construction of the top portion of a Q-mode resemblance matrix for DDT data.

similarity (larger numbers for more similar pairs of sites) or distance (larger numbers for less similar pairs of sites). The resemblance between Site 1 and Site 2 is computed and stored in the first cell of the triangular matrix. Values for each of the $n(n-1)/2$ similarities between the n rows of observations are entered into the matrix. Resemblance measures include correlation coefficients such as Pearson's r, Spearman's rho, or Kendall's tau, Euclidean distances between points in multidimensional space (where small distances indicate similar sites), or more specialized measures from the biological sciences. Resemblance measures particularly suited to censored data are listed in Table 13.2. A more complete discussion of resemblance measures is given in Zuur et al. (2007) and Clarke and Warwick (2001).

Once the resemblance matrix is complete, each multivariate procedure operates on that matrix in its own way. Cluster analysis links rows with higher similarities together in the same cluster, and rows with lower similarities into different clusters. ANOSIM tests for significantly higher similarities between rows within the same predefined group (e.g., Young or Mature fish in Table 13.1) than between rows in differing groups. Examples of these procedures are given in the following sections where binary, ordinal, or u-score resemblance matrices are computed.

An R-mode triangular-shaped matrix measures the similarity or differences between columns (Figure 13.2). Here the intent is to learn whether groups of variables cluster together in patterns. Operating on this matrix, exploratory factor analysis or a clustering procedure might be used to indicate groups of chemicals that covary

TABLE 13.2   Resemblance Measures Suited to Analysis of Censored Data

Resemblance Measure	Type of Data	Resulting Procedures
Simple matching	Presence/absence	Binomial methods
Euclidean distance	Ranks of data	Ordinal methods
Euclidean distance	u-Scores	Wilcoxon-type methods

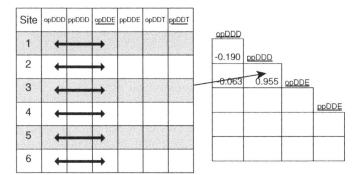

FIGURE 13.2 Construction of the top portion of an R-mode resemblance matrix for DDT data. Comparison of the first and third columns produces one of the similarity coefficients.

(factors), or groups of organisms with similar spatial patterns. Plots of these associations using either MDS or, if the relationships are linear, PCA can provide insight into which variables appear similar to and different from others.

The following sections illustrate several Q- and R-mode approaches using censored data. The data used in this chapter include a Date variable and Age variable that were created just to illustrate the types of analysis that can be performed with censored data.

13.2 NONPARAMETRIC METHODS AFTER CENSORING AT THE HIGHEST REPORTING LIMIT

13.2.1 Binary Method–Using Symmetrical Matching Coefficients

A simple procedure to employ multivariate methods for censored data is to recensor data into two groups, below versus greater than or equal to the highest reporting limit. A binary multivariate resemblance matrix can then be computed.

Several similarity coefficients are used in the biological sciences for presence/ absence (0/1) data. The commonly used asymmetric coefficients such as Jaccard and Sorensen (Clarke and Warwick, 2001) do not count occurrences that are absent/absent (0/0), so as not to heavily weight correlations between sites based on species that are absent in both cases. This is appropriate for species occurrence, so that (for example) the correlation between fauna of two streams in North America is not increased because water buffalo are absent at both locations! Yet coefficients disregarding joint absences are not appropriate for 0/1 data derived from concentrations reduced to being recorded as either below or above a reporting limit. Consider the case where a 1 indicates that the concentration of a particular chemical is below the highest reporting limit, and a 0 indicates the concentration is above that reporting limit. Two streams which both have high concentrations for that chemical, and therefore both have 0s recorded, should count toward the correlation between sites. Therefore a symmetric coefficient, one which counts both joint zeros and joint ones equally, gives the same similarity result regardless of whether high concentrations are assigned a 0 or 1. The "simple matching coefficient" between points j and k is a symmetric coefficient defined as

$$S_{jk} = 100 \times \frac{a+d}{a+b+c+d} \qquad (13.1)$$

where a equals the number of joint occurrences of a 1, d equals the joint occurrences of a 0, and b and c are mismatches where one site in the pair has a presence and one an absence of a detection at and above the highest limit. If a high value is switched from the designation of a 0 to that of a 1, a and d switch places, as do b and c, but the computed matching coefficient stays the same.

The highest (and only) reporting limit in Table 13.1 is 5. Table 13.3 shows Table 13.1 data when coded as a 1 for concentrations greater than or equal to 5, and a 0 for concentrations below 5.

The top six rows of the triangular resemblance matrix using the simple matching coefficient on the 0/1 values of Table 13.3 is given in Table 13.4. In a triangular matrix, the trivial 100% match of a cell with itself (Site 2 versus Site 2, Site 3 versus Site 3, etc.) is left blank. The top half of the square 32×32 matrix is also blank because it is a mirror image of the bottom half. In Table 13.4, the match between Sites 1 and 2 has an S of 50, and between 2 and 3 an S of 66.67, and so on. Therefore, the presence/absence pattern of DDT compounds at Site 2 is more similar to Site 3 than to Site 1. Note the 100% similarity between Sites 1 and 5, which in Table 13.1 are seen to have the same pattern of concentrations below and above 5.

Several operations can now be performed on this resemblance matrix to provide insight into the DDT data. First a plot or "map" of the similarities between sites (Figure 13.3) is provided by nonmetric multidimensional scaling (NMDS) (Clarke and Warwick, 2001). NMDS was first proposed by Kruskal (1964) and plots distances between points in the same rank order as distances (or 100-similarities) in the resemblance matrix. From Table 13.3 it is seen that Sites 1, 5, and 24 through 26 all have the same pattern–concentrations below 5 for all compounds except ppDDE. These five sites plot essentially on top of one another at the right of Figure 13.3. Site 6 has nondetects below 5 for all six compounds, and plots just to the right of the other five sites. These sites have triangles pointing up on the plot, showing they are the fish of Young age. Sites 4 and 8 have detected concentrations for ppDDE and one other compound, ppDDT, and plot to the left of the Young group. Sites with five compounds above 5 μg/L plot at the left side of the NMDS.

There are no axis names or scales on NMDS as the distances reported are just ranks of the (100-similarities). By investigating the pattern of where points plot it is often possible to interpret what the directions along the axes mean. In Figure 13.3, the horizontal axis represents the number of compounds with detections, with infrequent detections to the right. The vertical axis provides insight on which compounds are detected. At the top, Site 23 contains detections for DDD and DDE metabolite compounds, but not DDT. Site 30 at the bottom has detections above 5 for both DDT compounds. So the vertical axis is a gradation between detections for DDT and those for its metabolites DDE and DDD. Note that the direction of both axes is arbitrary–the

TABLE 13.3   DDT and Related Compounds in Fish, Coded as 0 for Concentrations Below 5 and 1 for Concentrations Greater than or Equal to 5

Site	opDDD	ppDDD	opDDE	ppDDE	opDDT	ppDDT	Age
1	0	0	0	1	0	0	Young
2	0	1	1	1	0	1	Mature
3	1	1	0	1	0	1	Mature
4	0	1	0	1	0	0	Mature
5	0	0	0	1	0	0	Young
6	0	0	0	0	0	0	Young
7	0	1	0	1	0	1	Mature
8	0	1	0	1	0	0	Mature
9	0	1	0	1	0	1	Mature
10	1	1	0	1	0	1	Mature
11	0	1	0	1	1	0	Mature
12	0	1	0	1	0	1	Mature
13	0	1	0	1	0	1	Mature
14	1	1	0	1	0	1	Mature
15	0	1	0	1	0	1	Mature
16	0	1	0	1	0	1	Mature
17	1	1	0	1	0	1	Mature
18	0	1	0	1	0	1	Mature
19	1	1	1	1	0	1	Mature
20	0	1	0	1	0	1	Mature
21	0	1	0	1	0	1	Mature
22	1	1	1	1	0	1	Mature
23	1	1	1	1	0	0	Mature
24	0	0	0	1	0	0	Young
25	0	0	0	1	0	0	Young
26	0	0	0	1	0	0	Young
27	1	1	0	1	0	1	Mature
28	0	1	1	1	0	0	Mature
29	0	1	1	1	0	1	Mature
30	0	1	0	1	1	1	Mature
31	0	1	0	1	0	1	Mature
32	0	1	0	1	0	1	Mature

TABLE 13.4   Top Six of 32 Rows of the Triangular Similarity Matrix for DDT and Related Compounds in Fish

Site	1	2	3	4	5	6
2	50					
3	50	66.67				
4	83.33	66.67	66.67			
5	100	50	50	83.33		
6	83.33	33.33	33.33	66.67	83.33	
7	66.67	83.33	83.33	83.33	66.67	50

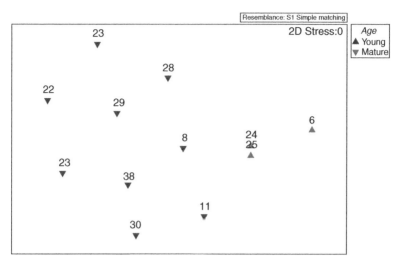

FIGURE 13.3   NMDS of 32 sites for the DDT in fish data.

orientation of the plots may be reversed left–right or up–down with no change in meaning.

The NMDS is computed in R as follows, after loading two libraries:

```
> library("vegan")
> library("MASS")
> data(DDT01)
> attach(DDT01)
> fish=data.frame(opDDD,ppDDD,opDDE,ppDDE,opDDT,ppDDT)
> ddt.symm=designdist(fish, method"1.00001-((a+d)/(a+b+c+d))",
  + terms = c("binary"),abcd=TRUE,"symm")
> ddt.mds=metaMDS(ddt.symm,zerodist="add", autotransform=FALSE)
> plot(ddt.mds,type="n")
> text(ddt.mds$points,labels=as.character(1:32))
```

The designdist command computes a user-defined distance matrix, which here is 1–similarity (with a very small constant added to ensure no distances are exactly zero).

It appears in Figure 13.3 that the pattern for the six chemicals differs between the Young and Mature fish groups. To test the hypothesis of no difference in the pattern between the two groups, the nonparametric ANOSIM test (Clarke, 1999) can be employed. ANOSIM determines whether the cells in the resemblance matrix are significantly different for sites within the same group in comparison to sites overall. It functions as a nonparametric multivariate analysis of variance. Because the test operates on the resemblance matrix, it uses all of the information in the entire pattern of chemical occurrences (Clarke and Warwick, 2001). In contrast, an older parametric approach might have been to compute an analysis of variance on the first principal component of multivariate data (after the unfortunate practice of replacing censored values with one-half the reporting limit). By using only the first component, only a portion of the information in the pattern of occurrences would be captured and used. ANOSIM instead for this data set uses all 6×32 dimensions of the multivariate information present in the data.

ANOSIM is computed by ranking the similarities in the matrix. Ranks are averaged for the sites within the same group (young fish to young fish, mature to mature), and for sites across groups (young to mature). The test statistic R (equation 13.2) divides the difference in these average ranks by one-half the number of cells in the triangular similarity matrix. When data within groups are self-similar the numerator becomes large, indicating a difference between the patterns in the two groups.

$$R = \frac{(\overline{\text{rank}}_b - \overline{\text{rank}}_w)}{n(n-1)/4} \tag{13.2}$$

where the b subscript indicates average ranks for similarities between different groups, and the subscript w indicates average ranks of similarities within the same group. ANOSIM is a permutation test, where p-values for the test are determined not by using an assumption of normality of data, but by computing a large number of permutations representing the null hypothesis and comparing their distribution to the one observed test statistic R. For even this moderate sized data set there are 906,192 different ways in which 32 rows can be assigned to the two groups. The null hypothesis of no difference between groups implies that each of these assignments is equally likely. ANOSIM randomly selects a subset of these possible assignments and compares the resulting R for each to the one observed R statistic for the original data. A histogram of the resulting test statistics is shown in Figure 13.4.

The histogram of R values in Figure 13.4 represents the variation in values expected when the null hypothesis is true—when observations called Young and those called Mature are just a chance occurrence and there is no difference in pattern between the groups. The observed ANOSIM R of 0.69 is larger than all of the 999 R statistics

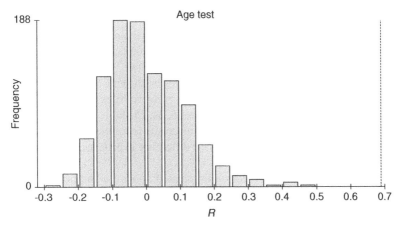

FIGURE 13.4 Observed ANOSIM R of 0.69 compared to test statistics (histogram) computed after randomly assigning the 32 sites to Young and Mature fish groups.

resulting from the random assignment of chemical patterns to groups. Therefore, the resulting probability of obtaining an R statistic of 0.69 or larger when the null hypothesis is true is estimated as 1/1000, or a p-value of 0.001 for the test. The null hypothesis that the patterns of the six chemicals is the same in both Young and Mature fish is rejected, and the two groups are determined (without substitution for nondetects of any kind) to have different patterns of DDT and metabolites.

ANOSIM can be computed within the vegan package of R with the following command:

```
> anosim(ddt.symm,Age)
```

with the resulting output:

```
Dissimilarity: symm
ANOSIM statistic R: 0.6876
     Significance: 0.001
Based on 999 permutations
```

A third procedure to extract information on groupings of sites based on a resemblance matrix for binary data is cluster analysis. Clustering involves using the pattern of similarities or differences to sort the sites into collections of sites, or clusters. The goal is to place sites into the same cluster that are more similar to one another than to sites in different clusters. Clustering is an unstructured classification scheme–no *a priori* information on groupings is used (such as Young versus Mature fish) to classify sites, but only the numbers within the resemblance matrix. In addition to the distance/similarity measure chosen (the simple matching coefficient for these detect/nondetect data), clustering methods require a choice of the type of linkage used to build up clusters from individual sites or smaller clusters (Shaw, 2003). Three methods are common in software: single linkage, complete linkage, and average linkage or Ward's. The first two methods may too easily put sites that are far away into the same cluster, or sites nearby into different clusters, depending on the arrangement of the data (Shaw, 2003). Therefore, average linkage or Ward's methods (two very similar methods) have become standard operating procedure in most cluster applications. Figure 13.5 presents the result of using average linkage clustering on the DDT fish similarity matrix, resulting in a dendrogram of sites.

The vertical axis is the similarity between sites or clusters. Starting at the top and working down, the strongest difference between groups of sites results in the first split into two clusters. The first cluster contains Sites 6, 24, 25, 26, 1, and 5, with remaining sites in the second cluster. Note that this six-site cluster is composed of the Young fish, so the unstructured classification discerns the difference in patterns observed between the two classifications of fish.

Using R, the distance matrix ddt.symm created above for the binary detection/ nondetection data can be clustered with the hclust command, and the dendrogram plotted:

```
> ddt.clust=hclust(ddt.symm,"average")
> plot(ddt.clust)
```

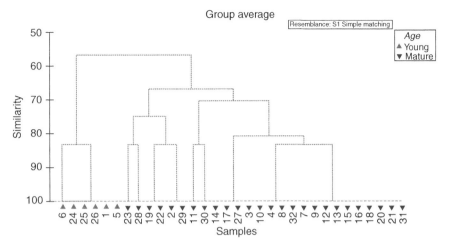

FIGURE 13.5   Dendrogram of clustering the binary low/high DDT-related data.

and the dendrogram may be reordered by the Age group to appear something like Figure 13.5 by

```
> ddt.den=as.dendrogram(ddt.clust)
> dage=reorder(ddt.den,Age)
> plot(dage)
```

Similar methods may be used to look at similarities among the variables, here the six DDT-related compounds. A new R-mode similarity matrix using the simple matching similarity coefficient is computed and shown in Table 13.5.

The NMDS map computed from this similarity matrix (Figure 13.6) illustrates the relationships among the six DDT-related compounds.

A separation between the op-compounds and the pp-compounds is seen along the horizontal axis. There is a strong difference in the pp- and op-compound patterns. The vertical axis increases from DDT at the bottom to the DDE and DDD metabolites further up.

TABLE 13.5   R-Mode Similarity Matrix for the Six DDT-Related Compounds

	opDDD	ppDDD	opDDE	ppDDE	opDDT
opDDD					
ppDDD	43.75				
opDDE	75	37.5			
ppDDE	28.125	84.375	21.875		
opDDT	68.75	25	75	9.375	
ppDDT	53.125	84.375	40.625	68.75	34.375

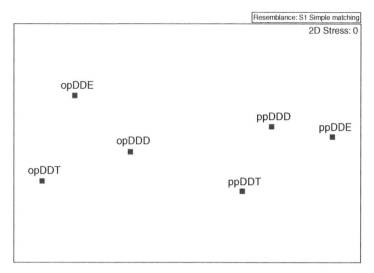

FIGURE 13.6   NMDS for the six DDT-related compounds.

A joint look at the pattern of sites and pattern of compounds is available with a biplot. Biplots are plots of the projections of sites and compounds onto the first two principal components. Biplots present a two-dimensional slice through multivariate space. They picture the two largest directions of data variation as defined by PCA. However, they leave out information in other dimensions (picture some points actually closer to you in front of the page, and other points behind the page). A biplot constructed from the Table 13.3 presence/absence data is shown in Figure 13.7. PCA employs a different similarity coefficient, Pearson's correlation coefficient r, to

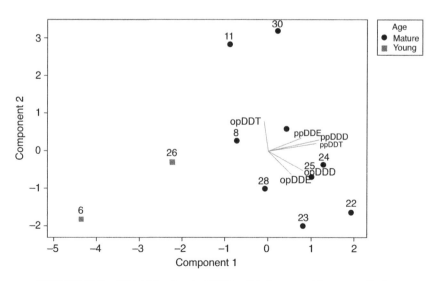

FIGURE 13.7   PCA biplot for the DDT in fish (presence/absence at 5) data.

measure similarities among rows and separately among columns. Pearson's r measures linear correlations, so patterns of variation should be linear to use PCA on untransformed data. Linearity for binary 0/1 patterns is difficult to evaluate, but the pattern observed in Figure 13.7 is a useful one, and similar to the patterns observed in the two earlier NMDS plots.

Component 1 is a contrast between higher concentrations of pp forms of the compounds to the right, and low concentrations to the left. This is evident from the direction the ppDDT, ppDDD and ppDDE vectors are pointing—vectors on a biplot point toward the direction of higher values. The sites for Young fish are to the left, showing lower concentrations for these compounds. The second component is based on the op forms of the compounds, with higher values of opDDT toward the top (high component 2 values) and higher values of opDDD and opDDE generally toward the bottom (low values of component 2). So this second component splits the patterns in the Mature fish by whether opDDT (Sites 11 and 30) or its metabolites (Site 23 and others) are more often present in concentrations above the reporting limit. As with NMDS, the assignment of these patterns to positive and negative values of the components, and so their directions on the biplot, is arbitrary. A biplot that is a mirror image in the horizontal or vertical directions is a totally equivalent plot, and could be produced by alternate software.

The PCA biplot is produced in R with the following commands:

```
> ddt.pca=princomp(fish,cor=TRUE)
> biplot(ddt.pca)
```

13.2.2   Ordinal Methods for One Reporting Limit

Portraying concentrations in a binary format above and below the highest reporting limit, following with multivariate procedures using the simple matching coefficient, provides great utility for analysis of censored data without substitution. There is however additional information in the relative order (ranks or percentiles) of values above the highest reporting limit. Ordinal nonparametric methods capture this additional information by assigning unique ranks to data at and above that limit, while ranks of all data below the highest limit are tied with one another. There is no reason not to go this second step if your data include detected concentrations. However as we will see, the binary approach of the previous section captures much of the information, the information in proportions that is contained in censored data.

The data of Table 13.1 are ranked within each compound separately, and presented in Table 13.6. Note that within each compound all values originally reported as <5 have the same average rank, which is below the ranks of concentrations detected at and above 5 for that compound. Many nonparametric methods can be thought of as tests on the average rank (Conover and Iman, 1981). The Kruskal–Wallis test can be approximated by whether the average rank is similar or different in each of k groups, as one example. So the ordinal methods of this section apply routine multivariate procedures to the ranks of censored data, rather than to the observations themselves. If the multivariate procedure is a nonparametric method, the results are identical

TABLE 13.6   Ranks of the DDT in Fish Data of Table 13.1

Site	opDDD	ppDDD	opDDE	ppDDE	opDDT	ppDDT	Age	Date
1	12.5	3.5	13.5	3	15.5	6	Young	1996
2	12.5	30	31	26	15.5	31	Mature	1990.5
3	28	27	13.5	32	15.5	20	Mature	1993.5
4	12.5	11.5	13.5	19	15.5	6	Mature	2001.5
5	12.5	3.5	13.5	5.5	15.5	6	Young	2000.5
6	12.5	3.5	13.5	1	15.5	6	Young	1999.5
7	12.5	14	13.5	17	15.5	22.5	Mature	1998
8	12.5	16	13.5	14	15.5	6	Mature	2002
9	12.5	11.5	13.5	24	15.5	24	Mature	1997
10	26	28	13.5	22	15.5	26	Mature	1994.5
11	12.5	7	13.5	21	32	6	Mature	1999
12	12.5	9	13.5	18	15.5	15	Mature	2000
13	12.5	18	13.5	31	15.5	30	Mature	1995.5
14	26	24.5	13.5	27.5	15.5	32	Mature	1992.5
15	12.5	10	13.5	11	15.5	13	Mature	2002.5
16	12.5	8	13.5	4	15.5	18	Mature	2003
17	31	31	13.5	24	15.5	25	Mature	1991.5
18	12.5	20.5	13.5	16	15.5	16	Mature	1998.5
19	32	29	28	15	15.5	20	Mature	1992
20	12.5	13	13.5	20	15.5	17	Mature	1999.5
21	12.5	23	13.5	24	15.5	28	Mature	1996
22	26	22	32	12	15.5	20	Mature	1995
23	30	32	30	29	15.5	6	Mature	1991
24	12.5	3.5	13.5	9.5	15.5	6	Young	1998.5
25	12.5	3.5	13.5	7	15.5	6	Young	1997
26	12.5	3.5	13.5	5.5	15.5	6	Young	1999.5
27	29	24.5	13.5	27.5	15.5	22.5	Mature	1994
28	12.5	16	27	2	15.5	6	Mature	2000.5
29	12.5	19	29	8	15.5	14	Mature	1997.5
30	12.5	20.5	13.5	30	31	29	Mature	1993
31	12.5	26	13.5	13	15.5	27	Mature	1996.5
32	12.5	16	13.5	9.5	15.5	12	Mature	2000.5

whether applied to the original values or the ranks. If the multivariate procedure is a parametric procedure, computation on the ranks produces an approximate nonparametric procedure. Nonparametric methods are useful with multivariate-censored data because a <5 is known to be less than a detected 10, but the exact distance between them used by standard parametric procedures (other than MLE) is not known.

For i rows or columns of data, the Euclidean distance measure E (equation 13.3) can be computed on the ranks of censored data. Euclidean distance is the straight-line distance in i dimensions between two points.

$$E = \sqrt{\sum_i (y_{i1} - y_{i2})^2} \tag{13.3}$$

TABLE 13.7   The Top Six Rows of a Q-Mode Resemblance (Euclidean Distance) Matrix of Ranked DDT Concentrations

Site	1	2	3	4	5	6
2	46.50					
3	42.77	26.69				
4	17.89	36.37	29.08			
5	2.5	45.32	41.12	15.69		
6	2	47.52	44.15	19.70	4.5	
7	24.05	26.75	25.31	16.81	22.69	25.27

In two dimensions it is the familiar hypotenuse of a triangle. Table 13.7 presents the top six rows of the triangular resemblance matrix formed by computing Euclidean distances between pairs of sites for the ranks of the fish DDT data in Table 13.6. With Euclidean distances, smaller values in the matrix represent sites that are closer together in six-dimensional space. Now the same multivariate methods used in the previous section on the resemblance matrix computed from binary data can be applied to this new resemblance matrix to better understand and test patterns in the data. Following a short presentation of the same multivariate methods used previously with binary data, a multivariate trend test using the Date values in Table 13.6 will be presented. Called the "test of seriation" (Clarke and Warwick, 2001), the test is in essence a multivariate analog of the popular univariate Mann–Kendall test for trend (Helsel and Hirsch, 2002).

Note that Site 1 shows a small distance to Sites 5 and 6, which are very similar in ranks of concentration. Figure 13.8 presents an NMDS plot for the DDT rank data. Note the similarity in patterns of sites to that for the binary results in Figure 13.4. The Young and Mature groups of fish stand out separately on different areas of Figure 13.8, illustrating their difference.

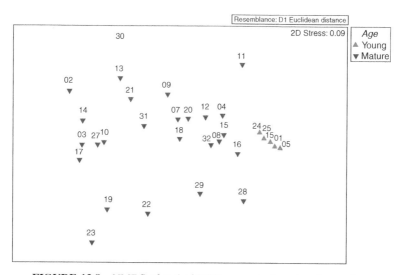

FIGURE 13.8   NMDS of ranked DDT concentrations by site number.

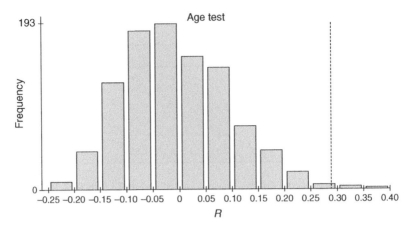

FIGURE 13.9 Observed ANOSIM R of 0.288 compared to test statistics (histogram) computed after randomly assigning the 32 sites to Young and Mature fish groups.

In R, the NMDS plot is computed on the Euclidean distance matrix of the ranks by

```
load("DDTrank.rda")
attach(DDTrank)
> frank=data.frame(opDDD,ppDDD,opDDE,ppDDE,opDDT,ppDDT)
> f.mds=metaMDS(frank,distance="euclidean",zerodist="add",
autotransform=FALSE)
> plot(f.mds, type="n")
> text(f.mds$points,labels=as.character(1:32))
```

An ANOSIM test of the difference in the two groups produces a p-value of 0.005, showing that the two groups have differing patterns of ranked concentrations (without substituting any values for censored observations). Figure 13.9 pictures the observed test statistic of 0.288 along with the histogram of 999 test statistics from randomized group assignments characteristic of the null hypothesis.

The ANOSIM test on the ranks is computed using:

```
> rankfish=data.frame(opDDD,ppDDD,opDDE,ppDDE,opDDT,ppDDT)
> anosim(rankfish,Age,distance="euclidean")
```

with the resulting output:

```
Dissimilarity: euclidean
ANOSIM statistic R: 0.2884
      Significance: 0.008
Based on 999 permutations
```

A PCA biplot of the ranks (Figure 13.10) illustrates the same patterns. PCA computes Pearson's r correlation as its similarity measure, here applied to the ranks of the data and so in essence a PCA using Spearman's rho correlation coefficient on the original data. It finds the two primary linear directions through multidimensional space that

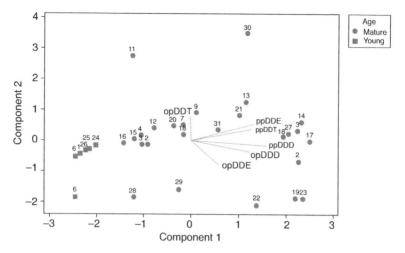

FIGURE 13.10   PCA biplot of the ranked DDT data.

best describe the variability of the data. In essence, these two components maximize the spread of the data on the plot in comparison to any other viewpoint. A biplot is more restrictive than NMDS in that it assumes linear axes, so a constant rate of change between points in space. But the benefit of a biplot over NMDS is that it produces axes interpretable as linear combinations of the original variables. The first principal component captures the variation in ppDDE and ppDDT with lesser contributions from ppDDD. High ranked concentrations are found for Sites 2, 3, 14, and 17, among others, while low ranked concentrations are found at the Young fish sites. The second principal component combines the effects of opDDT with its metabolites opDDE and to a much lesser extent, opDDD. Sites 11 and 30 have high ranks for opDDT, while Sites 19, 22, 23, 28, and 29 have low ranks for opDDT but high ranks for the metabolites.

The PCA biplot on ranks is computed in R with the commands:

```
> f.pca=princomp(rankfish,cor=TRUE)
> biplot(f.pca)
```

Now we will look at only the Mature fish group to determine whether concentrations in these fish are changing over time. The term "trend test" is usually applied to testing changes over time. However, the mechanics are applicable to change in concert with any explanatory variable. Clarke and Warwick (2001) used the more general term "seriation" to describe a consistent change in multivariate pattern along an axis of one external explanatory variable not used to construct the multivariate resemblance matrix. Here the external variable is Date, a time sequence, but it need not be so. As a nonparametric test, a change in pattern is tested against the ranks of the explanatory variable, and so is applicable to any ordinal or continuous variable, including time.

The test for seriation is computed by constructing an aggregate rank correlation coefficient such as Kendall's tau between each element of two triangular matrices

(Clarke and Warwick, 2001). The first matrix is the resemblance matrix of multivariate patterns partly shown in Table 13.7. The second matrix is the rank of distances between pairs of points in the direction of the explanatory variable. Two points close together in time have a small rank, while comparisons between an early point and late point will have larger rank distance. If there is a correlation between the resemblance pattern and the pattern of rank distances along the explanatory axis, the null hypothesis of no seriation is rejected and a significant association between resemblance pattern and explanatory variable is established. In essence this is much like a Mann–Kendall test for trend (Helsel and Hirsch, 2002) in the joint multivariate pattern of all variables used to construct the resemblance matrix. As with ANOSIM, the test for seriation uses all of the multivariate information in the data, a distinct advantage over using only a subset by regressing the first principal component scores as the response variable versus time.

The results of the test for seriation in the pattern of DDT and metabolites over time (Date) are given in Figure 13.11. The Kendall's correlation between elements of the two matrices is 0.49, with a permutation significance of 0.001. Kendall's tau between the two matrices was higher than all the randomly generated tau values computed by first randomizing the order of dates. A visual picture of the trend is shown in Figure 13.12, an NMDS plot where the Date variable has been split into three groups of approximately 4 years each. A progression from the left to the right illustrates that Date is strongly correlated with the left to right change on the NMDS map.

The test for seriation is one of a class of tests called Mantel tests that are available in R. The test of seriation on the Mature age fish can be computed by first subsetting the data into only the Mature fish (first three commands), then computing a distance matrix between Dates (time.dist) and a separate distance matrix between rows of the six chemical compounds (ddt.dist). Finally, the test of seriation is performed by computing Kendall's tau correlation between the elements of the two distance matrices. This is accomplished using the mantel command with method = "Kendall".

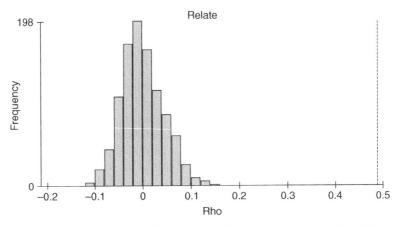

FIGURE 13.11   Observed Kendall's tau of 0.49 compared to tau values (histogram) computed after randomly assigning Dates to data at the 32 sites.

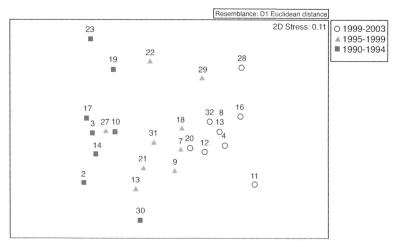

FIGURE 13.12   NMDS for ranks of the DDT in fish data, classed by Date category.

```
> mat=subset(DDTrank,subset=Age=="Mature")
> matddt=data.frame(mat$opDDD,mat$ppDDD,mat$opDDE,mat$ppDDE,
mat$opDDT,mat$ppDDT)
> matdate=data.frame(mat$Date)
> time.dist=dist(matdate,method="manhattan")
> ddt.dist=dist(matddt,method="euclidean")
> seriation=mantel(time.dist,ddt.dist,method="kendall")
> seriation
```

producing the test results:

```
Mantel statistic based on Kendall's rank correlation tau
Mantel statistic r: 0.4889
       Significance: 0.001
Empirical upper confidence limits of r:
90%     95%     97.5% 99%
0.0610 0.0787 0.0992 0.1283
Based on 999 permutations
```

Other multivariate procedures can be performed by ranking data censored at the highest reporting limit and computing a Euclidean resemblance matrix on the ranks. R-mode analyses can investigate relationships among the variables (chemicals). Factor analysis could be computed on the ranks to determine if groups of chemicals show a similar pattern, and so are part of the same factor. Hren et al. (1984) performed discriminant analysis on the ranks of several chemical variables to discern groupings of sites in order to prioritize cleanup efforts. The principle should be clear–rank censored data with all values below the highest reporting limit given tied ranks. Compute the Euclidean distance between rows, or between columns, and apply the same multivariate procedures that are suitable for uncensored data.

13.3 MULTIVARIATE METHODS FOR DATA WITH MULTIPLE REPORTING LIMITS

At least three types of multivariate methods have been applied to censored environmental data without recensoring values to the common highest limit. Software for the first two, maximum likelihood and multiple imputation methods, is somewhat difficult to find. Multiple imputation methods are available in R, but both types of methods are fairly complex to understand and implement. As the focus in this book is on procedures that can be understood and used by scientists with the typical training in statistics, most of this section focuses on the third procedure, methods based on u-scores. u-scores undergird many nonparametric procedures including the rank-sum test. While simple, they are a powerful tool in the analysis of censored multivariate data.

13.3.1 Multivariate Maximum Likelihood Methods for Data with Multiple Reporting Limits

If assumptions about the distributional shape of data for each variable can be realistically made, censored maximum likelihood procedures could be used to model the variables, and then determine whether there are any significant patterns in the joint distributions of variables. Maximum likelihood methods have been used with uncensored data to produce exploratory factor analysis and principal component solutions (Wentzell and Lohnes, 1999). Those procedures are found in a few statistics software programs. Andrews and Wentzell (1997) applied maximum likelihood PCA to censored data by assigning large variances to the estimated censored values. Kamakura and Wedel (2001) performed a multivariate-censored likelihood factor analysis using Tobit models, censored non-negative distributions, in solving for a factor analysis solution describing inter-relationships among variables. However, MLE that accounts for both censoring and the joint distributions of multiple variables is sufficiently complex to be a computational challenge, even with today's computers. It is especially difficult with data that have a high degree of censoring. While restrictions and simplifications can be used to cut down the complexity of computations, this advanced modeling approach is currently not generally applicable by the environmental scientist. Instead, imputation methods have become more widely used by numerical modelers.

13.3.2 Multivariate Multiple Imputation and EM Methods for Data with Multiple Reporting Limits

The concept of imputation was introduced in Chapter 6, where ROS was used to estimate summary statistics. Imputation is model-based estimation of values for missing or censored observations. Models for the context of environmental data include a distributional assumption and/or a correlation structure with other variables. The resulting imputed values are "placeholders" representing the general expectation for data in aggregate, rather than values actually expected for that particular observation, unless the correlation with other variables is strong. The use of the imputed value for one particular observation in addition to its use in representing the overall pattern of interaction between variables and sites is a judgment call based on science, rather than a numerical decision from the statistical output. If there is sufficient strength in the correlation between the censored variable and a suitable explanatory variable, the imputed value might be considered a plausible estimate for that specific observation. Otherwise it is only used collectively with other imputed values to fill in the overall distribution.

Maul and El-Shaarawi (1993) used the expectation maximization (EM) algorithm to impute values for censored halocarbon concentrations in water. After doing this they used cluster analysis to determine which locations in Lake Ontario were similar to others. EM uses an assumed distribution (such as normal or lognormal) and assumed parameters (mean and standard deviation) for each variable to estimate expected values for censored observations. With EM usually the mean of the distribution for that variable is used as the placeholder. Then a statistic that computes the fit between the set of observed data plus imputed means to the assumed distribution is evaluated, and the parameter estimates adjusted. The process is repeated until the fit between the set of observed data plus imputed means and the assumed distribution cannot be improved. This becomes the EM solution for fitting distributions to each variable.

Francis et al. (2009) used a similar procedure called Markov Chain Monte Carlo (MCMC) to impute values for censored disinfection by-product concentrations in drinking waters. MCMC uses both the distributional assumptions and cross-correlations between variables to estimate imputed values. By taking the correlations among variables into account, MCMC has a much greater chance of imputing realistic values for censored observations. As an interactive procedure, MCMC repeatedly evaluates the match between imputed plus observed data and the initial assumptions, adjusting the generating function until the best match is obtained. The product is a probability distribution of concentrations for each variable. From these the mean and other summary statistics can be reported, clustering or other multivariate procedures performed, or as Francis et al. did, sum individual variables that were components of similar chemicals to arrive at a total concentration. MCMC methods applicable to censored data are available in the R package MCMCglmm, where general linear models follow the MCMC imputation to perform parametric analysis such as a multivariate analysis of variance.

The two references just cited applied single imputation methods, where one solution is found and one individual value is imputed for each censored observation. As Hopke et al. (2001) observe, "no matter how carefully it is done and how knowledgeable the imputer may be, results from the analysis of singly imputed completed data are generally misleading because the single values being imputed cannot reflect sampling variability about the actual values . . . ." The one estimated number is not the "truth," and cannot in itself provide an estimate of how close to the truth it might be. This is the rationale for multiple imputation, ". . . which replaces each censored value with two or more plausible values, where each is drawn from the joint distribution of possible values" (Hopke et al., 2001). The same equations are combined with models of the variability of data to generate multiple possible values for each censored observation. Standard multivariate methods are applied to each equivalent possible outcome to provide the range of plausible results for a PCA, clustering, or other procedure. If the original assumptions about distributions and correlations among variables are imperfect (which they are), multiple imputation can produce an evaluation of how sensitive the outcomes are to these assumptions.

While the software to perform multiple imputation is becoming more easily available, it is a complex method requiring a number of parameters to be chosen. There is a niche remaining for simpler methods that capture the information in multivariate data, do not use substitution, and can computed and understood relatively easily. The multivariate scoring methods of the next section fill that niche.

13.3.3   Multivariate Score and Ranking Methods for Data with Multiple Reporting Limits

Ranks place observations in order from low to high. When there is no censoring and no ties, each observation is given a unique rank and so a unique place in the order from low to high. Ranks form the basis of nonparametric methods, where observations are judged to be higher or lower than other observations, but the numeric scale of those differences may not be constant, or may not even be known. Censoring at multiple thresholds complicates the issue. While a <1 should be ranked lower than a detected 3, how should it be ranked in comparison to a 0.8? Multiple variables further complicate the picture. How can information for six different measures be combined to form one overall analysis?

Working within one variable for the moment, one of the simplest scoring statistics is the u-score (equation 13.4). The u-score is the sum of the algebraic sign of differences comparing the ith observation to all other observations within the same variable:

$$u_i = \sum_{i \neq k} \text{sign}(x_i - x_k) \tag{13.4}$$

The u-score forms the basis for the Mann–Whitney test, and is related to Kendall's tau and other nonparametric methods. Kaplan–Meier percentiles for censored observations use this score. It is the number of observations known to be lower than x_i, minus the number of observations known to be higher. With censoring, where $x_i = 7$ cannot be determined to be higher or lower than $x_k = <10$, the sign of the difference is zero. The median observation will have a score u_i of zero, with negative scores for observations below the median and positive scores above the median.

Multivariate μ-score (called "mu-score") procedures have been developed by Wittkowski et al. (2008) and applied to ranking sports teams or individual players across multiple measures. The same methods are applied in genomics (Morales et al., 2008) to look at gene trait relationships to observable outcomes. In these applications, the goal has been to produce one overall ranking of observations. If the interest here were in ranking all 32 observed fish to determine where each was located on a ranked scale of contamination, the u-score methods could be applied directly. Here instead u-scores (or their ranks) will be constructed variable by variable. Using a Euclidean distance measure on the u-scores or their ranks, NMDS, clustering, PCA, and other procedures can be applied to find and illustrate patterns in the data. Similar in concept to what was done after censoring at the highest reporting limit, the difference now is that the scoring process accommodates censoring at multiple reporting limits.

If u-scores are computed for data with one reporting limit, with the lower end of that interval being a 0 (0 to RL), then the ranks of the u-scores will be identical to the ranks computed in the previous section for ordinal methods with one reporting limit. But u-scores can also be computed for data with multiple limits. Table 13.8 is an

TABLE 13.8   Fish DDT Data from Table 13.1 Altered to be Censored at Two Reporting Limits

Site	opDDD	ppDDD	opDDE	ppDDE	opDDT	ppDDT	Age	Date
1	0–5	0–5	0–5	14	0–5	0–5	Young	1996
2	0–5	42	8.4	130	0–5	31	Mature	1990.5
3	5.3	38	0–5	250	0–5	11	Mature	1993.5
4	0–2	12	0–2	57	0–2	0–2	Mature	2001.5
5	0–2	0–2	0–2	16	0–2	0–2	Young	2000.5
6	0–2	0–2	0–2	0–2	0–2	0–2	Young	1999.5
7	2–5	14	2–5	52	2–5	14	Mature	1998
8	0–2	15	0–2	48	0–2	0–2	Mature	2002
9	2–5	12	2–5	110	0–2	20	Mature	1997
10	5.1	39	0–5	100	0–5	24	Mature	1994.5
11	0–2	5.7	2–5	87	18	0–2	Mature	1999
12	0–2	9.4	0–2	53	0–2	7.3	Mature	2000
13	0–5	18	0–5	210	0–5	30	Mature	1995.5
14	5.1	27	0–5	140	0–5	33	Mature	1992.5
15	0–2	10	0–2	24	0–2	5.8	Mature	2002.5
16	0–2	7.6	0–2	15	0–2	10	Mature	2003
17	9	46	0–5	110	0–5	21	Mature	1991.5
18	0–2	22	2–5	51	0–2	7.4	Mature	1998.5
19	9.8	41	6.9	50	0–5	11	Mature	1992
20	0–2	13	2–5	66	0–2	8.6	Mature	1999.5
21	0–5	26	0–5	110	0–5	26	Mature	1996
22	5.1	24	250	38	0–5	11	Mature	1995
23	8	100	8	160	0–5	0–5	Mature	1991
24	0–2	0–2	0–2	23	0–2	0–2	Young	1998.5
25	0–2	0–2	0–2	17	0–2	0–2	Young	1997
26	0–2	0–2	0–2	16	0–2	0–2	Young	1999.5
27	5.7	27	0–5	140	0–5	14	Mature	1994
28	0–2	15	6	8.1	0–2	0–2	Mature	2000.5
29	0–2	20	7.5	22	0–2	6.4	Mature	1997.5
30	0–5	22	0–5	190	5.2	27	Mature	1993
31	2–5	31	2–5	42	0–2	25	Mature	1996.5
32	0–2	15	0–2	23	0–2	5.6	Mature	2000.5

alteration of Table 13.1 to mimic a situation where before mid-1996, data were censored below a reporting limit of 5—concentrations reported as (0–5). In mid-1996, the laboratory began reporting values as either below the method detection limit (0–2) or as remarked values between 2 and 5 (3.1J, etc.) below the quantitation limit of 5. These in-between data are here coded to be between the detection and quantitation limits (2–5), respecting the signal-to-noise issue that caused the remark—see Chapter 3. (Scientists should speak with their laboratory chemist to determine whether or not remarked data should be treated as intervals. Can a 3.1J be considered reliably smaller than 3.5J, for example? If not, record them as interval censored.) The purpose of altering the data is to demonstrate the multivariate scoring process for data censored at two different levels, here the detection and quantitation limits.

Table 13.9 presents the u-scores of the ordering within each column of chemical compounds computed using the Minitab® macro %u-score (the NADA for R function u-score performs the same computation). Each observation is compared to all the others within the same column. Comparisons between two entries recorded as (0–2) would be tied and the sign of that difference zero. A zero sign also results from comparing (0–5) with (2–5), and (0–5) with a detected 4. A sign of + 1 results from comparing a (2–5) cell to a (0–2) observation elsewhere in the column. Summing the signs within a cell, higher observations have a higher score. Note that in Table 13.9, the 16 values for opDDD at (0–2) end up tied with each other at a score of −11 while the five (0–5) values receive a score of −8.Why? The (0–2) values are known to be below the (2–5) values, and so are below more observations than the (0–5) values. Wider intervals are less precise measurements, and will result in less extreme u-scores than those with more precise analytical results.

Table 13.9 scores were ranked to bring the values back to a familiar and non-negative scale (Table 13.10). Either the u-score itself or its rank could be used in subsequent computations such as NMDS, clustering, ANOSIM, and tests for seriation.

Euclidean distances were computed on the Table 13.10 ranks and stored in a triangular resemblance matrix. Figure 13.13 shows the NMDS plot computed on that resemblance matrix. Recall that the altered concentrations are now censored at two reporting limits, 2 and 5, and that no substitution has been done. The pattern is very similar to the one in Figure 13.8 for the data censored at the reporting limit of 5. The only changes in concentration were to shift some 0–5 censored values to either (0–2) or (2–5). Yet the change in reporting limit has had a measurable effect. Site 1 was a pre-1996 observation with a reporting limit of 5. All other Young data used a reporting limit of 2. Separation of Site 1 from the other Young fish is easily visible in Figure 13.13 because of this. Sites 13 and 31 were not too far apart in Figure 13.8, both Mature fish with moderate concentrations. Site 13 was pre-1996 and kept the (0–5) level for censored values. Site 31 was after mid-1996 and low concentrations were recorded as (0–2). In Figure 13.13, the two sites are much farther apart.

The R commands for producing the equivalent NMDS plot to Figure 13.13 are

TABLE 13.9   u-Scores for the Altered Fish DDT Data from Table 13.8

Site	u-Score opDDD	u-Score ppDDD	u-Score opDDE	u-Score ppDDE	u-Score opDDT	u-Score ppDDT	Age	Date
1	−8	−26	−6	−27	−2	−21	Young	1996
2	−8	27	29	19	−2	29	Mature	1990.5
3	23	21	−6	31	−2	7	Mature	1993.5
4	−11	−10	−12	5	−3	−21	Mature	2001.5
5	−11	−26	−12	−22	−3	−21	Young	2000.5
6	−11	−26	−12	−31	−3	−21	Young	1999.5
7	8	−5	5	1	15	12	Mature	1998
8	−11	−1	−12	−5	−3	−21	Mature	2002
9	8	−10	5	15	−3	15	Mature	1997
10	19	23	−6	11	−2	19	Mature	1994.5
11	−11	−19	5	9	31	−21	Mature	1999
12	−11	−15	−12	3	−3	−3	Mature	2000
13	−8	3	−6	29	−2	27	Mature	1995.5
14	19	16	−6	22	−2	31	Mature	1992.5
15	−11	−13	−12	−11	−3	−7	Mature	2002.5
16	−11	−17	−12	−25	−3	3	Mature	2003
17	29	29	−6	15	−2	17	Mature	1991.5
18	−11	8	5	−1	−3	−1	Mature	1998.5
19	31	25	23	−3	−2	7	Mature	1992
20	−11	−7	5	7	−3	1	Mature	1999.5
21	−8	13	−6	15	−2	23	Mature	1996
22	19	11	31	−9	−2	7	Mature	1995
23	27	31	27	25	−2	−21	Mature	1991
24	−11	−26	−12	−14	−3	−21	Young	1998.5
25	−11	−26	−12	−19	−3	−21	Young	1997
26	−11	−26	−12	−22	−3	−21	Young	1999.5
27	25	16	−6	22	−2	12	Mature	1994
28	−11	−1	21	−29	−3	−21	Mature	2000.5
29	−11	5	25	−17	−3	−5	Mature	1997.5
30	−8	8	−6	27	29	25	Mature	1993
31	8	19	5	−7	−3	21	Mature	1996.5
32	−11	−1	−12	−14	−3	−9	Mature	2000.5

```
>load("FishDDTalt.rda")
> attach(FishDDTalt)
> lo=data.frame(opDDD,ppDDD,opDDE,ppDDE,opDDT,ppDDT)
> hi=data.frame(opDDD_Hi,ppDDD_Hi,opDDE_Hi,ppDDE_Hi,opDDT_Hi,
ppDDT_Hi)
> urk=uscores(lo,hi,out="ranks")
> u.mds=metaMDS(urk,distance="euclidean",zerodist="add",
autotransform=FALSE)
> plot(u.mds, type="n")
> text(u.mds$points,labels=as.character(1:32))
```

TABLE 13.10   Ranks of the Table 13.9 u-Scores for the Altered Fish DDT Data

Site	Rank opDDD	Rank ppDDD	Rank opDDE	Rank ppDDE	Rank opDDT	Rank ppDDT	Age	Date
1	19	3.5	16	3	23.5	6	Young	1996
2	19	30	31	26	23.5	31	Mature	1990.5
3	28	27	16	32	23.5	20	Mature	1993.5
4	8.5	11.5	6	19	9	6	Mature	2001.5
5	8.5	3.5	6	5.5	9	6	Young	2000.5
6	8.5	3.5	6	1	9	6	Young	1999.5
7	23	14	23.5	17	30	22.5	Mature	1998
8	8.5	16	6	14	9	6	Mature	2002
9	23	11.5	23.5	24	9	24	Mature	1997
10	26	28	16	22	23.5	26	Mature	1994.5
11	8.5	7	23.5	21	32	6	Mature	1999
12	8.5	9	6	18	9	15	Mature	2000
13	19	18	16	31	23.5	30	Mature	1995.5
14	26	24.5	16	27.5	23.5	32	Mature	1992.5
15	8.5	10	6	11	9	13	Mature	2002.5
16	8.5	8	6	4	9	18	Mature	2003
17	31	31	16	24	23.5	25	Mature	1991.5
18	8.5	20.5	23.5	16	9	16	Mature	1998.5
19	32	29	28	15	23.5	20	Mature	1992
20	8.5	13	23.5	20	9	17	Mature	1999.5
21	19	23	16	24	23.5	28	Mature	1996
22	26	22	32	12	23.5	20	Mature	1995
23	30	32	30	29	23.5	6	Mature	1991
24	8.5	3.5	6	9.5	9	6	Young	1998.5
25	8.5	3.5	6	7	9	6	Young	1997
26	8.5	3.5	6	5.5	9	6	Young	1999.5
27	29	24.5	16	27.5	23.5	22.5	Mature	1994
28	8.5	16	27	2	9	6	Mature	2000.5
29	8.5	19	29	8	9	14	Mature	1997.5
30	19	20.5	16	30	31	29	Mature	1993
31	23	26	23.5	13	9	27	Mature	1996.5
32	8.5	16	6	9.5	9	12	Mature	2000.5

ANOSIM can again test for whether the Young and Mature age groups have significantly different concentration patterns. The ANOSIM on Table 13.10 ranks produces a test statistic of 0.362 and a p-value of 0.001 (Figure 13.14). The groups are strongly significantly different. The R command to perform the ANOSIM test is

```
> anosim(urk,Age,distance="euclidean")
```

Figure 13.15 shows the result of a cluster analysis on Table 13.10 ranks of u-scores. The Young fish are clustered with one another except for Site 1 with the higher reporting limit. Figure 13.5 previously showed the clustering of data classed as simply

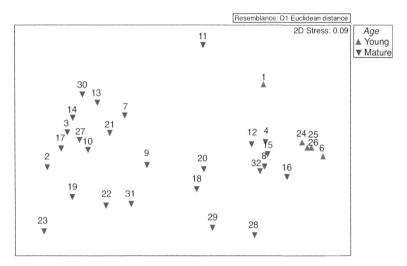

FIGURE 13.13   NMDS for Table 13.10 ranks of u-scores from data with multiple reporting limits.

below versus greater than or equal to a concentration of 5. The new clustering shows that several clusters can be considered significantly different from one another using the SIMPROF test (Clarke and Warwick, 2001). Dark black lines are significant separators between clusters in Figure 13.15, so that Site 1 is by itself and significantly different from other clusters, as is Site 11. Sites 18, 20, 28, and 29 cluster together and differ from sites further to the right on the dendrogram. There were no significant differences between clusters shown in Figure 13.5 for the binary data. The increased numerical detail provided by the multiple reporting limit procedure provides greater discrimination between patterns at the sites than with the binary procedure.

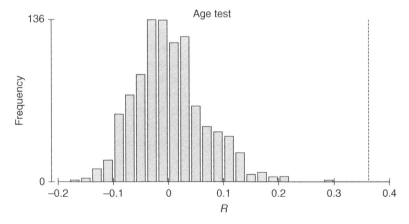

FIGURE 13.14   ANOSIM results for Table 13.10 ranked u-scores from data with multiple reporting limits.

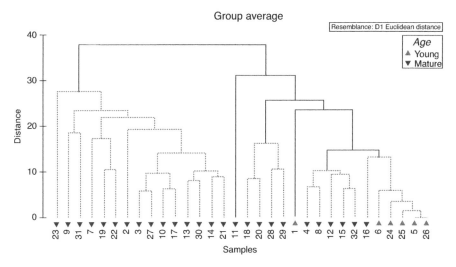

FIGURE 13.15 Clustered sites for Table 13.10 ranks of u-scores from data with multiple reporting limits.

The cluster analysis in R is performed by

```
> u.dist=dist(urk,method="euclidean")
> u.clust=hclust(u.dist,"average")
> plot(u.clust)
```

13.4 SUMMARY OF MULTIVARIATE METHODS FOR CENSORED DATA

Analysis methods are available for multivariate data that are censored at multiple reporting limits, or delineated at both method detection and quantitation limits. Much can be done with data classified simply as below or above a single reporting limit, and the results do not contain invasive patterns placed into the data by fabricating values with substitutions. Even more detail in the patterns of multiple chemicals, community measures, or other attributes is available using ranks of data censored at the highest reporting limit, or ranks of u-scores for data censored at multiple reporting limits. Nonparametric methods have been emphasized because ranks can be computed without fabricating values for censored observations, and without assuming a distributional shape that is difficult to test for in multiple dimensions.

14 The NADA for R Software

14.1 A BRIEF OVERVIEW OF R AND THE NADA SOFTWARE

R is a free statistics programming language and environment available online under the General Public License (GPL). It brings both basic and advanced statistical capabilities to anyone having a need to use them. R was developed as an aid to teach statistics by Ross Ihaka and Robert Gentleman at the University of Auckland in New Zealand. In 1997, the R Core Team of international statisticians and computer scientists was established to guide and quality control the development of R. The R Core Team maintains and quality assures the base R program, and oversees the format and virus-free status of the thousands of user-written programs available as add-on R packages. The content and accuracy of the user-written programs are the responsibility of packages' authors. Compiled versions of R are available for the Macintosh, Windows, and Unix operating systems. You may use and distribute R freely, as long as all receiving it maintain the same rights.

R will read and write data in a variety of formats. The standard format is as a set of data columns, with the first row a text header providing the name for each column, stored as a text file. There are user-written functions to read data sets output by commercial statistics software and by Excel, as well. More information is available on the sites listed below. The base R software includes basic graphics capabilities, but more colorful, complex, and sometimes interactive graphics are available in user-written packages. R can link with other software so that analysis results and complex color graphics can be automatically output to reports in Microsoft Word or pdf formats. A number of online presentations of data on Internet sites run R code behind the scenes.

If you are completely new to R, the best place to learn about it is at the R Project web site (http://www.r-project.org). The Manuals section provides several free user manuals for specific purposes, including the *Introduction to R* manual. The *R Journal* is an online journal that includes applications of R to topics of interest, including some that might interest you! The R Project web site also maintains a list of commercially available books on using R—click the "Books" link. Base R and each add-on package

Statistics for Censored Environmental Data Using Minitab® and R, Second Edition.
Dennis R. Helsel.
© 2012 John Wiley & Sons, Inc. Published 2012 by John Wiley & Sons, Inc.

include a help system that provides detailed guidance on available options and required formats.

The location for downloading base R and user-written packages is at the Comprehensive R Archive Network (CRAN) http://cran.r-project.org/.

Mirror sites are located across the globe, and the preference is that a mirror site near you be used when downloading software. Clicking on the Packages link at CRAN will bring up an alphabetical list of the many user-contributed software additions to R. These packages cover a multitude of areas where statistics is applied, including GIS and geostatistics, biostatistics, and many other specialized applications. Packages can be installed from within R using

```
Packages > Install Packages
```

in the pull-down menu, or by typing install.packages on the command line, followed by the package name in parentheses and double quotes:

```
install.packages ("NADA")
```

Installing a package downloads the software files onto your computer. The modular form of R means that for any given session, only those packages necessary for the work at that time need to be active. So at the beginning of each session, the user loads the packages needed, often only a subset of the ones installed, using

```
Packages > Load Packages
```

on the pull-down menu, or by typing the library command

```
library("NADA")
```

When you quit R you will need to reload the package again the next time. Scripts can be set up to automate repeated procedures in R, including loading in a set of several packages for performing your usual and regular tasks.

The Nondetects And Data Analysis (NADA) user-written package was authored and is maintained by Lopaka Lee. It is a collection of data sets and R-language implementations of methods described in this book. In writing the NADA package, Mr. Lee has made a concerted effort to make function names and usage consistent with the base R system. Almost all functions begin with the prefix "cen" for procedures on censored data—for example, "cenfit" and "cenmle." Also, generic functions such as "mean," "quantile," and "plot" can be used with output objects from any of the NADA for R functions. For more detail on usage and capabilities, see the usage in examples throughout the book or download the user manual at

```
http://cran.r-project.org/web/packages/NADA/index.html
```

Keep in mind that both R and the NADA for R package are rapidly improving. What is missing today could be added very shortly. Additionally, development is completely open. If there is some functionality that is missing or something you have found to be wrong, you are welcomed to suggest, or contribute the code, for a solution—all contributors are properly recognized within the documentation. For suggestions email us at <mailto:nada@practicalstats.com> for additions to or comments on the NADA for R package.

14.1.1 Some Basic Commands to Get Started

Most of the commands below are loaded with the NADA package

```
library("NADA")
```

The ictest command is part of the interval package, and the ANOSIM and other multivariate commands are part of the vegan package. The MASS package is also needed for performing NMDS.

```
library("interval")
library("vegan")
library("MASS")
```

One of the data sets that comes with NADA can be loaded (made active and available) by first identifying it with the data command, and then making it active:

```
> data (ShePyrene)
> attach(ShePyrene)
```

Attaching the data set notifies R that these are the data you are working on. Variables can then be referred to using only the variable name rather than preceding the variable name with the name of the data set. To find out which variables are in the data set, type either the names or head command.

```
> names(ShePyrene)
  [1] "Pyrene" "PyreneCen"
```

Names lists the short names of variables available in this data set. Pyrene is a vector of pyrene concentrations where censored observations are represented by their reporting limit. PyreneCen is an indicator variable, with values of 0 (False) for uncensored observations and a 1 (True) for censored observations.

```
> head(ShePyrene)
  Pyrene PyreneCen

1    28    TRUE
2    31    FALSE
3    32    FALSE
4    34    FALSE
```

```
5   35   TRUE
6   35   TRUE
```

Head prints the names plus the first six lines of data. Records where PyreneCen = True, such as the first observation of <28, have the reporting limit value listed in the variable Pyrene. The second row is a detected concentration (PyreneCen=FALSE) of 31.

An important command is the help command. Typing the name of a function within the parentheses, such as the cenfit function in NADA, brings up an html page describing the command and its options.

```
help(cenfit)
```

For general help on help in R, type

```
help(help)
```

If you are not sure of a command name, but want to search the R help files for items related to Kendall's tau, for example, use the help.search function:

```
> help.search("tau")
```

14.2   SUMMARY OF THE COMMANDS AVAILABLE IN NADA

14.2.1   Plotting Censored Data

```
cenboxplot(Y,Yc)              plots a censored boxplot of Y on a log scale
cenboxplot(Y,Yc, log=F)       plots a censored boxplot of Y, no log transform
cenboxplot(Y,Yc, Gp)          plots censored boxplots of Y for each group in Gp
cenxyplot(X,Xc,Y,Yc)          censored scatterplot of Y by X. Requires censoring
                                 indicators
Yobj=cenfit(Y,Yc)
Plot(Yobj)                    censored edf of Y
Yros=cenros(Y,Yc)
Plot(Yros)                    censored lognormal ROS probability plot of Y
Yros=cenros(Y,Yc, dist="gaussian")
Plot(Yros)                    censored normal ROS probability plot of Y
Ymle=cenmle(Y,Yc)
Plot(Ymle)                    censored lognormal MLE probability plot of Y
```

14.2.2   Descriptive Statistics for Censored Data

```
Ymle=cenmle(Y,Yc)
Ymle                          MLE summary statistics for a lognormal distribution
Ymle=cenmle(Y,Yc, dist="gaussian")
Ymle                          MLE summary statistics for a normal distribution
```

```
Ykm=cenfit(Y,Yc)
Ykm                        Kaplan–Meier summary statistics
Yros=cenros(Y,Yc, dist="gaussian")
Yros                       ROS summary statistics for a normal distribution
Yros=cenros(Y,Yc)
Yros                       ROS summary statistics for a lognormal distribution
censtats(Y,Yc)             computes mean, median with all three methods.
                           Assumes lognormal distribution.
```

14.2.3 Confidence Intervals for Censored Data

```
Ymle=cenmle(Y,Yc)
mean(Ymle)                 95% conf interval assuming a lognormal dist. (Cox method)
Ymle=cenmle(Y,Yc, conf.int=0.90, dist="gaussian")
mean(Ymle)                 90% conf interval assuming a normal distribution
```

14.2.4 Group Tests for Censored Data

```
wilcox.test(y~Gp)          rank-sum test for 1 RL. <RL value lower than RL
kruskal.test(y,Gp)         Kruskal–Wallis test for 1 RL. <RL value lower than RL
A=cenmle(Y,Yc,Gp)          MLE test of mean logs of Y by Gp
plot(A)                    MLE probability plot of residuals of logarithms
B=cenmle(Y,Yc,Gp, dist="gaussian")       MLE test of mean of Y by Gp
plot(B)                    MLE probability plot of residuals
C=cendiff(Y,Yc,Gp)
C                          nonparametric Wilcoxon test on Y by Gp
plot(cenfit(Y,Yc,Gp)       edfs for the groups being tested
D=ictest(Ylow,Yhi,Gp,scores="wmw",Lin=Lt,Rin=Rt)
D                          Wilcoxon test on interval-censored Y by Gp
```

14.2.5 Correlation and Regression for Censored Data

```
cor.test(X,Y,method="spearman")          rho and test for data with 1 RL
cor.test(X,Y,method="kendall")           tau and test for data with 1 RL
A=cenreg(Cen(Y,Yc)~X, dist="gaussian")   MLE censored regression
A                          also reports likelihood r correlation coeff
B=cenken(Y,Yc,X,Xc)        tau and ATS line for data with 1 or more RLs
B                          the Xc is optional
C=ictest(Surv(Ylo,Yhi,type="interval2")~X,scores="wmw",
  Lin=Lt,Rin=Rt)
C                          censored regression for interval-censored Y.
D=glm(formula=GTrl~X1+X2,family=binomial(logit))
D                          logistic regression for data with 1 RL
E=glm(formula=GTrl~1,family=binomial(logit))
```

```
E                                          null logistic regression for data with 1 RL
anova(D,E,test="Chisq")                    test to compare two logistic regression models
```

14.2.6   Multivariate Methods for Censored Data

```
library("vegan")                           loads the vegan and MASS packages, which include the
library("MASS")                            commands needed for multivariate analysis
A=data.frame(C1,C2,C3,C4)                  places data from four columns into one data frame
D=designdist(A, method= "1.00001-((a+d)/(a+b+c+d))", terms=c("binary"),
+ abcd=TRUE, "dist")                       stores symmetric distance matrix into D
D.mds=metaMDS(D,zerodist="add", autotransform=FALSE)          NMDS on D
plot(D.mds,type="n")                       plots the NMDS without points
text(D.mds$points,labels=as.character(1:32))          adds row number as point locator
anosim(A,Gp,distance="euclidean")          ANOSIM on data frame A by group Gp
X=dist(T,method="manhattan")               distance matrix of explanatory variable X
Y=dist(A,method="euclidean")               distance matrix of response variables
ser=mantel(X,Y, method ="kendall")         test of seriation
ser                                        prints test of seriation results
u.clust=hclust(Y,"average")                cluster analysis of distance matrix Y with average linkage
plot(u.clust)                              plots the cluster dendrogram
lo=data.frame(C1,C2,C3,C4)                 data frame from columns at low ends of intervals
hi=data.frame(C11,C12,C13,C14)             data frame from columns at upper ends of intervals
urk=uscores(lo,hi,out="ranks")             ranks of u-scores from interval-censored data
u.mds=metaMDS(urk,distance="euclidean",zerodist="add",
  autotransform=FALSE)
plot(u.mds, type="n")                      NMDS from u-scores of interval-censored data
text(u.mds$points,labels=as.character(1:32))          adds row labels to NMDS
```

Appendix: Datasets

The author's appreciation is extended to each of the scientists who provided their data for use in this book, especially to those who directly provided data not available in published reports. All data sets listed can be found at http://www.practicalstats.com/nada in both Microsoft Excel (.xls) and Minitab® worksheet (.mtw) formats. Many of the data sets are also included within the NADA for R package. Type

```
data()
```

to list the data sets that are available within the packages you have already loaded in R.

AsExample	Artificial numbers representing arsenic concentrations in a drinking water supply.
File Name:	AsExample.xls
Reference:	None. Generated.
Objective:	Determine what can be done with data where all values are below the reporting limit.
Censoring:	A reporting limit at 1, and a reporting limit at 3 µg/L.
Used in:	Chapter 8
Atra	Atrazine concentrations in a series of Nebraska wells before (June) and after (September) the growing season.
File Name:	Atra.xls
Reference:	Junk et al., 1980, *Journal of Environmental Quality* 9, pp. 479–483.
Objective:	Determine if concentrations increase from June to September.
Censoring:	One reporting limit, at 0.01 µg/L.
Used in:	Chapters 4, 5, and 9

Statistics for Censored Environmental Data Using Minitab® and R, Second Edition.
Dennis R. Helsel.
© 2012 John Wiley & Sons, Inc. Published 2012 by John Wiley & Sons, Inc.

AtraAlt	Atrazine concentrations altered from the Atra data set so that there are more censored observations, adding a second reporting limit at 0.05.
File Name:	AtraAlt.xls
Reference:	Altered from the data of Junk et al., 1980, *Journal of Environmental Quality* 9, pp. 479–483.
Objective:	Determine if concentrations increase from June to September.
Censoring:	Two reporting limits, at 0.01 and 0.05 μg/L.
Used in:	Chapters 5 and 9

Atrazine	The same atrazine concentrations as in Atra, stacked into one column (col.1). Column 2 indicates the month of collection. Column 3 indicates which data are below the reporting limit—those with a value of 1.
File Name:	Atrazine.xls
Reference:	Junk et al., 1980, *Journal of Environmental Quality* 9, pp. 479–483.
Objective:	Determine if concentrations increase from June to September.
Censoring:	One reporting limit, at 0.01 μg/L.
Used in:	Chapter 9

Bloodlead	Lead concentrations in the blood of herons in Virginia.
File Name:	Bloodlead.xls
Reference:	Golden et al., 2003, *Environmental Toxicology and Chemistry* 22, pp. 1517–1524.
Objective:	Compute interval estimates for lead concentrations.
Censoring:	One reporting limit, at 0.02 μg/g.
Used in:	Chapter 7

Cd	Cadmium concentrations in fish for two regions of the Rocky Mountains.
File Name:	Cd.xls
Reference:	none. Data modeled after several reports.
Objective:	Determine if concentrations are the same or different in fish livers of the two regions.
Censoring:	Four reporting limits, at 0.2, 0.3, 0.4, and 0.6 μg/L.
Used in:	Chapter 9

ChlfmCA	Chloroform concentrations in groundwaters of California.
File Name:	ChlfmCA.xls
Reference:	Squillace et al., 1999, *Environmental Science and Technology* 33, pp. 4176–4187.
Objective:	Determine if concentrations differ between urban and rural areas.

Censoring:	Three reporting limits, at 0.05, 0.1, and 0.2 µg/L.
Used in:	Chapter 9

CuZn	Copper and zinc concentrations in groundwaters from two zones in the San Joaquin Valley of California. The zinc concentrations were used.
File Name:	CuZn.xls
Reference:	Millard and Deverel, 1988, *Water Resources Research* 24, pp. 2087–2098.
Objective:	Determine if zinc concentrations differ between the two zones.
Censoring:	Zinc has two reporting limits, at 3 and 10 µg/L.
Used in:	Chapters 4, 5, and 9

CuZnAlt	Zinc concentrations of the CuZn data set; concentrations in the Alluvial Fan zone have been altered so that there are more censored observations. This produces a greater signal, even with more censored observations.
File Name:	CuZnAlt.xls
Reference:	Altered from the data of Millard and Deverel, 1988, *Water Resources Research* 24, pp. 2087–2098.
Objective:	Determine if zinc concentrations differ between the two zones.
Censoring:	Zinc has two reporting limits, at 3 and 10 µg/L.
Used in:	Chapter 9

DFe	Dissolved iron concentrations over several years in the Brazos River, Texas. Summer concentrations were used.
File Name:	DFe.xls
Reference:	Hughes and Millard, 1988, *Water Resources Bulletin* 24, pp. 521–531.
Objective:	Determine if there is a trend over time.
Censoring:	Iron has two reporting limits, at 3 and 10 µg/L.
Used in:	Chapters 5, 11, and 12

Doc	Dissolved organic carbon concentrations in groundwaters of irrigated and nonirrigated areas.
File Name:	DOC.xls
Reference:	Junk et al., 1980, *Journal of Environmental Quality* 9, pp. 479–483.
Objective:	Determine if concentrations differ between irrigated and non-irrigated areas.
Censoring:	One reporting limit at 0.2 µg/L.
Used in:	Chapter 9

Golden	Lead concentrations in the blood and several organs of herons in Virginia.

File Name:	Golden.xls
Reference:	Golden et al., 2003, *Environmental Toxicology and Chemistry* 22, pp. 1517–1524.
Objective:	Determine the relationships between lead concentrations in the blood and various organs. Do concentrations reflect environmental lead concentrations, as represented by dosing groups?
Censoring:	One reporting limit, at 0.02 µg/g.
Used in:	Chapters 10 and 11
Hatchery	Proportions of detectable concentrations of antibiotics (µg/L) in drainage from fish hatcheries across the United States.
File Name:	Hatchery.xls
Reference:	Thurman et al., 2002, Occurrence of antibiotics in water from fish hatcheries. USGS Fact Sheet FS 120-02.
Objective:	Compute confidence intervals and tests on proportions.
Censoring:	One reporting limit for each compound, all at 0.05 µg/L.
Used in:	Chapters 8 and 9
HgFish	Mercury concentrations in fish across the United States.
File Name:	HgFish.xls
Reference:	Brumbaugh et al., 2001, USGS Biological Science Report BSR-2001-0009.
Objective:	Do mercury concentrations differ by land use of the watershed? Can concentrations be related to water and sediment characteristics of the streams?
Censoring:	Three reporting limits, at 0.03, 0.05, and 0.10 µg/g wet weight.
Used in:	Chapters 10, 11, and 12
MDCu +	Copper concentrations in groundwater from the Alluvial Fan zone in the San Joaquin Valley of California. One observation was altered to become a <21, larger than all of the uncensored observations (the largest detected observation is a 20).
File Name:	MDCu + .xls
Reference:	Millard and Deverel, 1988, *Water Resources Research* 24, pp. 2087–2098.
Objective:	Calculation of summary statistics when the largest observation is censored.
Censoring:	Five reporting limits, at 1, 2, 5, 10, and 20 µg/L. An additional artificial reporting limit of 21 was added to illustrate a point.
Used in:	Chapter 6
Oahu	Arsenic concentrations (µg/L) in an urban stream, Manoa Stream at Kanewai Field, on Oahu, Hawaii.
File Name:	OahuAs.xls

Reference:	Tomlinson, 2003, Effects of Groundwater/Surface-Water Interactions and Land Use on Water Quality. Written communication (draft USGS report).
Objective:	Characterize conditions by computing summary statistics.
Censoring:	Three reporting limits, at 0.9, 1, and 2 µg/L. Uncensored values reported below the lowest reporting limit indicate that insider censoring may have been used, and so the results are likely biased high.
Used in:	Chapter 6
Recon	Atrazine concentrations in streams throughout the Midwestern United States.
File Name:	Recon.xls
Reference:	Mueller et al., 1997, *Journal of Environmental Quality* 26, pp. 1223–1230.
Objective:	Develop a regression of model for atrazine concentrations using explanatory variables.
Censoring:	One reporting limit, at 0.05 µg/L.
Used in:	Chapter 12
Roach	Lindane concentrations in fish from tributaries of the Thames River, England.
File Name:	Roach.xls
Reference:	Yamaguchi et al., 2003, *Chemosphere* 50, pp. 265–273.
Objective:	Determine whether lindane concentrations are the same at all sites.
Censoring:	One reporting limit at 0.08 µg/kg.
Used in:	Chapter 9
SedPb	Lead concentrations in stream sediments before and after wildfires.
File Name:	SedPb.xls
Reference:	Eppinger et al., 2003, USGS Open-File Report 03-152.
Objective:	Determine whether lead concentrations are the same pre- and postfire.
Censoring:	One reporting limit at 4 µg/L.
Used in:	Chapter 9
Silver	Silver concentrations in a standard solution sent to 56 laboratories as part of a quality assurance program.
File Name:	Silver.xls
Reference:	Helsel and Cohn, 1988, *Water Resources Research* 24, pp. 1997–2004.

Objective:	Estimate summary statistics for the standard solution. The median or mean might be considered the "most likely" estimate of the concentration.
Censoring:	Twelve reporting limits, the largest at 25 µg/L.
Used in:	Chapter 6

Tbl1_1	Contaminant concentrations in test and a control group.
File Name:	Tbl1_1.xls
Reference:	None. Generated data.
Objective:	Determine whether a test group has higher concentrations than a control group.
Censoring:	Three reporting limits, at 1, 2, and 5 µg/L.
Used in:	Chapter 1, Table 1.1

TCE	TCE concentrations (µg/L) in groundwaters of Long Island, New York. Categorized by the dominant land use type (low, medium, or high density residential) surrounding the wells.
File Name:	TCE.xls
Reference:	Eckhardt et al., 1989, USGS Water Resources Investigations Report 86-4142.
Objective:	Determine if concentrations are the same for the three land use types.
Censoring:	Four reporting limits, at 1,2,4, and 5 µg/L.
Used in:	Chapter 10

TCEReg	TCE concentrations (µg/L) in groundwaters of Long Island, New York, along with several possible explanatory variables.
File Name:	TCEReg.xls
Reference:	Eckhardt et al., 1989, USGS Water Resources Investigations Report 86-4142.
Objective:	Determine if concentrations are related to one or more explanatory variables.
Censoring:	Four reporting limits, at 1, 2, 4, and 5 µg/L. One column indicates whether concentrations are above or below 5.
Used in:	Chapter 12

Thames	Dieldrin, lindane and PCB concentrations in fish of the Thames River and tributaries, England.
File Name:	Thames.xls
Reference:	Yamaguchi et al., 2003, *Chemosphere* 50, pp. 265–273.
Objective:	Determine if concentrations differ among sampling sites. Are dieldrin and lindane concentrations correlated?
Censoring:	One reporting limit per compound.
Used in:	Chapters 11 and 12

References

Aitchison, J. and I.A.C. Brown, 1957, *The Lognormal Distribution*. Cambridge University Press, Cambridge, 176 pp.

Akritas, M.G., 1986, Bootstrapping the Kaplan–Meier estimator. *Journal of the American Statistical Association* **81**, 1032–1038.

Akritas, M.G., 1992, Rank transform statistics with censored data. *Statistics and Probability Letters* **13**, 209–221.

Akritas, M.G., 1994, Statistical analysis of censored environmental data, Chapter 7 in G.P. Patil and C.R. Rao,eds., *Handbook of Statistics*,Vol. 12. North-Holland, Amsterdam, 927 pp.

Akritas, M.G. and J. Siebert, 1996, A test for partial correlation with censored astronomical data. *Monthly Notices of the Royal Astronomical Society* **278**, 919–924.

Akritas, M.G., S.A. Murphy, and M.P. LaValley, 1995, The Theil-Sen estimator with doubly-censored data and applications to astronomy. *Journal of the American Statistical Association* **90**, 170–177.

Allison, P.D., 1995, *Survival Analysis Using the SAS System: A Practical Guide*.SAS Institute, Inc., Cary, NC, 292 pp.

Andrews, D.T. and P.D. Wentzell, 1997, Applications of maximum likelihood principal component analysis: incomplete data sets and calibration transfer. *Analytica Chimica Acta* **350**, 341–352.

Antweiler, R.C. and H.C. Taylor, 2008, Evaluation of statistical treatments of left-censored environmental data using coincident uncensored data sets: I. Summary statistics. *Environmental Science and Technology* **42**, 3732–3738.

Aruga, R., 1997, Treatment of responses below the detection limit: some current techniques compared by factor analysis on environmental data. *Analytica Chimica Acta* **354**, 255–262.

ASTM, 1983, American Society of Testing Materials, Sec. D4210.

ASTM, 1991, Standard practice for 99%/95% interlaboratory detection estimate (IDE) for analytical methods with negligible calibration error. American Society of Testing Materials, Sec. D6091.

ASTM, 2000, Standard practice for interlaboratory quantitation estimate (IQE). American Society of Testing Materials, Sec. D 6512.

Baccarelli, A., R. Pfeiffer, D. Consonni, et al., 2005, Handling of dioxin measurement data in the presence of non-detectable values: overview of available methods and their application in the Seveso chloracne study. *Chemosphere* **60**, 898–906.

Statistics for Censored Environmental Data Using Minitab® and R, Second Edition.
Dennis R. Helsel.
© 2012 John Wiley & Sons, Inc. Published 2012 by John Wiley & Sons, Inc.

Barr, D.R. and T. Davidson, 1973, A Kolmogorov–Smirnov test for censored samples. *Technometrics* **15**, 739–757.

Boos, D.D. and J.M. Hughes-Oliver, 2000, How large does n have to be for Z and t intervals? *American Statistician* **54**, 121–126.

Borgan, O. and K.A. Liestøl, 1990, A note on the confidence intervals and bands for the survival curve based on transformations. *Scandinavian Journal of Statistics* **17**, 35–41.

Brookmeyer, R. and J. Crowley, 1982, A confidence interval for the median survival time. *Biometrics* **38**, 29–41.

Brown, B.W., M. Hollander, and R.M. Korwar, 1974, Nonparametric tests of independence for censored data, with applications to heart transplant studies. *Reliability and Biometry* 327–354.

Brüggemann, L., P. Morgenstern and R. Wennrich, 2010, Comparison of international standards concerning the capability of detection for analytical methods. *Accreditation and Quality Assurance* **15**, 99–104.

Brumbaugh, W.G., D.P. Krabbenhoft, D.R. Helsel, J.G. Wiener, and K.R. Echols, 2001, A national pilot study of mercury contamination of aquatic ecosystems along multiple gradients—bioaccumulation in fish. U.S. Geological Survey Biological Science Report BSR–2001–0009, 25 pp.

Buckley, J. and I. James, 1979, Linear regression with censored data. *Biometrika* **66**, 429–436.

Cade, B.S. and B.R. Noon, 2003, A gentle introduction to quantile regression for ecologists. *Ecological Environment* **1**, 412–420.

California Environmental Protection Agency, 2005, California Ocean Plan, 45 p. Available at http://www.swrcb.ca.gov/water_issues/programs/ocean/docs/oplans/oceanplan2005.pdf.

Chaloupka, M., D. Parker, and G. Balazs, 2004, Modeling post-release mortality of loggerhead sea turtles exposed to the Hawaii-based pelagic longline fishery. *Marine Ecology Progress Series* **280**, 285–293.

Chay, K.Y. and B.E. Honore, 1998, Estimation of censored semiparametric regression models: an application to changes in Black–White earnings inequality during the 1960s. *Journal of Human Resources* **33**, 4–38.

Chung, C.F., 1990, Regression analysis of geochemical data with observations below detection limits, in G. Gaal and D.F. Merriam,eds., *Computer Applications in Resource Estimation*. Pergammon Press, New York, pp. 421–433.

Chung, C.F. and W.A. Spirito, 1989, Estimation of distribution parameters from data with observations below detection limit with an example from South Nahanni River area, District of Mackenzie, in F.P. Agterberg and G.F. Bonham-Carter, eds., *Statistical Applications in the Earth Sciences*. Geological Survey of Canada Paper 89-9, p. 233–242.

Clarke, J.U., 1998, Estimation of censored data methods to allow statistical comparisons among very small samples with below reporting limit observations. *Environmental Science and Technology* **32**, 177–183.

Clarke, K.R., 1999, Nonmetric multivariate analysis in community-level ecotoxicology. *Environmental Toxicology and Chemistry* **18**, 118–127.

Clarke, K.R. and R.M. Warwick, 2001, *Change in Marine Communities: An Approach to Statistical Analysis and Interpretation*, 2nd edition. Primer-E, Ltd., Plymouth, UK, 169 pp.

Cohen, A.C., 1957, On the solution of estimating equations for truncated and censored samples from normal populations. *Biometrika* **44**, 225–236.

Cohen, A.C., 1959, Simplified estimators for the normal distribution when samples are singly censored or truncated. *Technometrics* **1**, 217–237.

Cohen, A.C., 1961, Tables for maximum likelihood estimates singly truncated and singly censored samples. *Technometrics* **3**, 535–541.

Cohn, T.A., 1988, Adjusted maximum likelihood estimation of the moments of lognormal populations from type I censored samples. U.S. Geological Survey Open-File Report 88-350, 34 pp.

Collett, D., 2003, *Modeling Survival Data in Medical Research*, 2nd edition. Chapman and Hall/CRC, London, 391 pp.

Conover, W.J., 1968, Two *k*-sample slippage tests. *Journal of the American Statistical Association* **63**, 614–626.

Conover, W.J., 1999, *Practical Nonparametric Statistics*, 3rd edition. Wiley, New York, 584 pp.

Conover, W.J. and R.L. Iman, 1981, Rank transformations as a bridge between parametric and nonparametric statistics. *American Statistician* **35**, 124–129.

Cook P., J. Robbins, D. Endicott, K. Lodge, P. Guiney, M. Walker, E. Zabel, and R. Peterson, 2003, Effects of aryl hydrocarbon receptor-mediated early life stage toxicity on lake trout populations in Lake Ontario during the 20th Century. *Environmental Science and Technology* **37**, 3864–3877.

Currie, L.A., 1968, Limits for qualitative detection and quantitative determination. *Analytical Chemistry* **48**, 586–593.

Davis, J.C., 2001, *Statistics and Data Analysis in Geology*, 3rd edition. Wiley, New York, NY, 656 pp.

Davis, C.B. and N.E. Grams, 2006, When laboratories should not censor analytical data, and why. Presented at the 25th National Conference on Managing Environmental Systems, April 24–27, 2006, Austin, TX. Available at http://www.epa.gov/quality/qs-docs/25-3.pdf.

Dietz, E.J., 1987, A comparison of robust estimation in simple linear regression. *Communications in Statistical Simulation* **16**, 1209–1227.

Dietz, E.J., 1989, Teaching regression in a nonparametric statistics course. *American Statistician* **43**, 35–40.

Dietz, E.J. and T.J. Killeen, 1981, A nonparametric multivariate test for monotone trend with pharmaceutical applications. *Journal of the American Statistical Association* **76**, 169–174.

Eckhardt, D.A., W.J. Flipse, and E.T. Oaksford, 1989, Relation between land use and ground-water quality in the upper glacial aquifer in Nassau and Suffolk Counties, Long Island NY. U.S. Geological Survey Water Resources Investigations Report 86–4142, 26 pp.

Efron, B., 1981, Censored data and the bootstrap. *Journal of the American Statistical Association* **374**, 312–319.

Efron, B. and R.J. Tibshirani, 1986, Bootstrap methods for standard errors, confidence intervals and other measures of statistical accuracy. *Statistical Science* **1**, 54–77.

El-Makarim, A. and A. Aboueissa, 2009, Maximum likelihood estimators of population parameters from multiply censored samples. *Environmetrics* **20**, 312–330.

El-Shaarawi, A.H. and S.R. Esterby, 1992, Replacement of censored observations by a constant: an evaluation. *Water Research* **26**, 835–844.

Emerson, J.D., 1982, Nonparametric confidence intervals for the median in the presence of right censoring. *Biometrics* **38**, 17–27.

Eppinger, R.G., P.H. Briggs, B. Rieffenberger, C. Van Dorn, Z.A. Brown, J.G. Crock, P.H. Hagemann, A. Meier, S.J. Sutley, P.M. Theodorakos, and S.A. Wilson, 2003, Geochemical data for stream sediment and surface water samples from Panther Creek, the middle fork of the Salmon River, and the main Salmon River, collected before and after the Clear Creek, Little Pistol, and Shellrock wildfires of 2000 in Central Idaho. U.S. Geological Survey Open-File Report 2003-152. Available on CD at http://pubs.er.usgs.gov.

Everitt, B.S. and G. Dunn, 2001, *Applied Multivariate Data Analysis,*2nd edition.Arnold, London, 342 pp.

Farewell, V.T., 1989, Some comments on analysis techniques for censored water quality data. *Environmental Monitoring and Assessment* **12**, 285–294.

Farnham, I.M., A.K. Singh, K.J. Stetzenbach, and K.H. Johannesson, 2002, Treatment of nondetects in multivariate analysis of groundwater geochemistry data. *Chemometrics and Intelligent Laboratory Systems* **60**, 265–281.

Fay, M.P. and P.A. Shaw, 2010, Exact and asymptotic weighted logrank tests for interval censored data: the interval R package. *Journal of Statistical Software* **36**(2), 34.

Feigelson E. and P. Nelson, 1985, Statistical methods for astronomical data with upper limits. I. Univariate distributions. *The Astrophysical Journal* **293**, 192–206.

Finkelstein, M.M., 2008, Asbestos fibre concentrations in the lungs of brake workers: another look. *Annals of Occupational Hygiene* **52**, 455–461.

Flynn, M., 2010, Analysis of censored exposure data by constrained maximization of the Shapiro-Wilk W statistic. *Annals of Occupational Hygiene* **54**, 263–271.

Fong, D.Y.T., C.W. Kwan, K.F. Lam, and K.S.L. Lam, 2003, Use of the sign test for the median in the presence of ties. *American Statistician* **57**, 237–240.

Francis, R.A., M.J. Small, and J.M. VanBriesen, 2009, Multivariate distributions of disinfection by-products in chlorinated drinking water. *Water Research* **43**, 3453–3468.

Ganser, G.H. and P. Hewitt, 2010, An accurate substitution method for analyzing censored data. *Journal of Occupational and Environmental Hygiene* **7**, 233–244.

Gehan, E.A., 1965, A generalized Wilcoxon test for comparing arbitrarily singly censored samples. *Biometrika* **52**, 203–223.

Geosyntec Consultants and Wright Water Engineers, 2009, Urban stormwater BMP performance monitoring manual. A report to the U.S. Environmental Protection Agency updating EPA-821-B-02-001. Available at http://www.bmpdatabase.org/MonitoringEval. htm#MonitoringGuidance

Gibbons, R.D., 1995, Some statistical and conceptual issues in the detection of low level environmental pollutants. *Environmental and Ecological Statistics* **2**, 125–145.

Gibbons, R.D. and D.E. Coleman, 2001, *Statistical Methods for Detection and Quantification of Environmental Contamination.*Wiley, New York, 384 pp.

Gibbons, R.D., D.E. Coleman, and R.F. Maddalone, 1997, An alternative minimum level definition for analytical quantification. *Environmental Science and Technology* **31**, 2071–2077.

Gilbert, R.O., 1987, *Statistical Methods for Environmental Pollution Monitoring.*Wiley, New York, 320 pp.

Gilbert, R.O. and R.R. Kinnison, 1981, Statistical methods for estimating the mean and variance from radionuclide data sets containing negative, unreported or less-than values. *Health Physics* **40**, 377–390.

Gilliom, R.J., and D.R. Helsel, 1986, Estimation of distributional parameters for censored trace level water quality data, 1. Estimation techniques. *Water Resources Research* **22**, 135–146.

Gilliom, R.J., R.M. Hirsch, and E.J. Gilroy, 1984, Effect of censoring trace-level water-quality data on trend-detection capability. *Environmental Science and Technology* **18**, 530–535.

Gleit, A., 1985, Estimation for small normal data sets with reporting limits. *Environmental Science and Technology* **19**, 1201–1206.

Golden, N.H., B.A. Rattner, J.B. Cohen, D.J. Hoffman, E. Russek-Cohen, and M.A. Ottinger, 2003, Lead accumulation in feathers of nestling black-crowned night herons (*Nycticorax nycticorax*) experimentally treated in the field. *Environmental Toxicology and Chemistry* **22**, 1517–1524.

Griffith, M.B., B.H. Hill, A.T. Herlihy, and P.R. Kaufmann, 2002, Multivariate analysis of periphyton assemblages in relation to environmental gradients in Colorado Rocky Mountain streams. *Journal of Phycology* **38**, 83–95.

Grünfeld, K., 2005, Dealing with outliers and censored values in multi-element geochemical data—a visualization approach using XmdfTool. *Applied Geochemistry* **20**, 341–352.

Haas, C.N., and P.A. Scheff, 1990, Estimation of averages in truncated samples. *Environmental Science and Technology* **24**, 912–919.

Hahn, G.J. and Meeker, W.Q., 1991, *Statistical Intervals: A Guide for Practitioners.* Wiley, New York, 392 pp.

Harrell, F.E., 2001, *Regression Modeling Strategies, with Applications to Linear Models, Logistic Regression, and Survival Analysis.*Springer, New York, 568 pp.

Harrington, D.P. and Fleming, T.R., 1982, A class of rank test procedures for censored survival data. *Biometrika* **69**, 553–566.

Hawkins, D.M. and G.W. Oehlert, 2000, Characterization using normal or log-normal data with multiple censoring points. *Environmetrics* **11**, 167–181.

Helsel, D.R. 1990, Less than obvious: statistical treatment of data below the detection limit. *Environmental Science and Technology* **24**, 1766–1774.

Helsel, D.R., 2005, Insider censoring: distortion of data with censored observations. *Human and Ecological Risk Assessment* **11**, 1127–1137.

Helsel, D.R., 2006, Fabricating data: how substituting values for censored observations can ruin results, and what can be done about it. *Chemosphere* **65**, 2434–2439.

Helsel, D.R. and T.A. Cohn, 1988, Estimation of descriptive statistics for multiply censored water quality data. *Water Resources Research* **24**, 1997–2004.

Helsel, D.R. and R.M. Hirsch, 2002, *Statistical Methods in Water Resources.* U.S. Geological Survey Techniques of Water Resources Investigations, Book 4, Chapter A3, 512 pp. Available at http://water.usgs.gov/pubs/twri/twri4a3/.

Hettsmansperger, T.P., J.W. McKean, and S.J. Sheather, 1997, Rank-based analyses of linear models, Chapter 7 in G. S. Maddala and C. R. Rao, *Handbook of Statistics,*Vol. 15. North-Holland, Amsterdam, 716 pp.

Hewett P. and G.H. Ganser, 2007, A comparison of several methods for analyzing censored data. *Annals of Occupational Hygiene* **51**, 611–632.

Hinton, S.W., 1993, Delta log-normal statistical methodology performance. *Environmental Science and Technology* **27**, 2247–2249.

Hirsch, R.M. and J.R. Slack, 1984, A nonparametric trend test for seasonal data with serial dependence. *Water Resources Research* **20**, 727–732.

Hirsch, R.M. and J.R. Stedinger, 1987, Plotting positions for historical floods and their precision. *Water Resources Research* **23**, 715–727.

Hollander, M. and D.A. Wolfe, 1999, *Nonparametric Statistical Methods*, 2nd edition. Wiley, New York, 787 pp.

Hopke, P.K., C. Liu and D.B. Rubin, 2001, Multiple imputation for multivariate data with missing and below-threshold measurements: time-series concentrations of pollutants in the Arctic. *Biometrics* **57**, 22–33.

Hornung, R.W. and L.D. Reed, 1990, Estimation of average concentration in the presence of nondetectable values. *Applied Occupational Environmental Hygiene* **5**, 46–51.

Hosmer, D.W. and S. Lemeshow, 2000, *Applied Logistic Regression*, 2nd edition. Wiley, New York, 375 pp.

Hren, J., K.S. Wilson and D.R. Helsel, 1984, A statistical approach to evaluate the relation of coal mining, land reclamation, and surface-water quality in Ohio. U.S. Geological Survey Water-Resources Investigations Report 84–4117, 325 pp.

Hughes, J.P. and S.P. Millard, 1988, A tau-like test for trend in the presence of multiple censoring points. *Water Resources Bulletin* **24**, 521–531.

Huybrechts, T., O. Thas, J. Dewulf, and H. Van Langenhove, 2002, How to estimate moments and quantiles of environmental data sets with non-detected observations? A case study on volatile organic compounds in marine water samples. *Journal of Chromatography* **975**, 123–133.

Hyslop, N.P. and W.H. White, 2008, An empirical approach to estimating detection limits using collocated data. *Environmental Science and Technology* **42**, 5235–5240.

Ireson, M.J, and P.V. Rao, 1985, Interval estimation of slope with right-censored data. *Biometrika* **72**, 601–608.

Isobe, T., E.D. Feigelson, and P.I. Nelson, 1986, Statistical methods for astronomical data with upper limits. II. Correlation and regression. *Astrophysical Journal* **306**, 490–507.

IUPAC, 1997, *Detection and quantification capabilities*,Chapter 18, Section 4.3.7, in *Compendium of Analytical Nomenclature*, 3rd edition.Definitive Rules 1997. International Union of Pure and Applied Chemistry. Available at http://old.iupac.org/publications/analytical_compendium/.

Jain, R.B., S.P. Caudill, R.Y. Wang and E. Monsell, 2008, Evaluation of maximum likelihood procedures to estimate left censored observations. *Analytical Chemistry* **80**, 1124–1132.

Jeng, S.L. and W.Q. Meeker, 2001, Parametric simultaneous confidence bands for cumulative distributions from censored data. *Technometrics* **43**, 450–461.

Junk, G.A., R.F. Spalding, and J.J. Richard, 1980, Areal, vertical, and temporal differences in ground-water chemistry: II. Organic constituents. *Journal of Environmental Quality* **9**, 479–483.

Kahn, H.D., W.A. Telliard, and C.E. White, 1998, Comment on "An alternative minimum level definition for analytical quantification." *Environmental Science and Technology* **32**, 2346–2348.

Kalbfleisch, J.D. and R.L. Prentice, 2002, *The Statistical Analysis of Failure Time Data*, 2nd edition.Wiley, New York, 439 pp.

Kamakura, W.A. and M. Wedel, 2001, Exploratory tobit factor analysis for multivariate censored data. *Multivariate Behavioral Research* **36**, 53–82.

Kaus, R. 1998, Detection limits and quantitation limits in the view of international harmonization and the consequences for analytical laboratories. *Accreditation and Quality Assurance* **3**, 150–154.

Keith, L.H., 1992, *Environmental Sampling and Analysis: A Practical Guide*. Lewis Publishers, Chelsea, MI, 143 pp.

Kendall, M.G., 1955, *Rank Correlation Methods*, 2nd edition. Charles Griffin and Company, London, 196 pp.

Klein, J.P. and M.L. Moeschberger, 2003, *Survival Analysis: Techniques for Censored and Truncated Data*, 2nd edition. Springer, New York, 536 pp.

Kolpin, D.W., J.E. Barbash, and R.J. Gilliom, 2002 Atrazine and metolachlor occurrence in shallow ground water of the United States, 1993 to 1995. Relations to explanatory factors. *Journal of the American Water Resources Association* **38**, 301–311.

Krishnamoorthy K, A Mallick, and T. Matthew, 2009, Model-based imputation approach for data analysis in the presence of non-detects. *Annals of Occupational Hygiene* **4**, 249–263.

Kroll, C.N. and J.R. Stedinger, 1996, Estimation of moments and quantiles using censored data. *Water Resources Research* **32**, 1005–1012.

Kroll, C.N. and J.R. Stedinger, 1999, Development of regional regression relationships with censored data. *Water Resources Research* **35**, 775–784.

Kruskal, J.B., 1964, Multidimensional scaling by optimizing goodness of fit to a nonmetric hypothesis. *Psychometrika* **29**, 1–27.

Land, C.E., 1972, An evaluation of approximate confidence interval estimation methods for lognormal means. *Technometrics* **14**, 145–158.

Latta, R.B., 1981, A Monte Carlo study of some two-sample rank tests with censored data. *Journal of the American Statistical Association* **76**, 713–719.

Law, C. and R. Brookmeyer, 1992, Effects of mod-point imputation on the analysis of doubly censored data. *Statistics in Medicine* **11**, 1569–1578.

Lee, E.T., and J.W. Wang, 2003, *Statistical Methods for Survival Data Analysis*, 3rd edition. Wiley, New York, 534 pp.

Libiseller, C.L. and A. Grimvall, 2002, Performance of partial Mann–Kendall tests for trend detection in the presence of covariates. *Environmetrics* **13**, 71–84.

Lindstrom, R.M., 2001, Limits for qualitative detection and quantitative determination, in D.R. Lide,ed., *A Century of Excellence in Measurements, Standards, and Technology*, NIST Special Publication 958, pp. 164–166. Available at nvl.nist.gov/pub/nistpubs/sp958-lide/164-166.pdf.

Liu, S., S.T. Yen, and D.W. Kolpin, 1996, Atrazine concentrations in near-surface aquifers: a censored regression approach. *Journal of Environmental Quality* **25**, 992–999.

Looney, S.W. and T.R. Gulledge, 1985, Use of the correlation coefficient with normal probability plots. *American Statistician* **39**, 75–79.

Lubin, J.H., J.S. Colt, D. Camann, S. Davis, J.R. Cerhan, R.K. Severson, L. Bernstein, and P. Hartge, 2004, Epidemiologic evaluation of measurement data in the presence of detection limits. *Environmental Health Perspectives* **112**, 1691–1696.

Martin, J.J., S.D. Winslow and D.J. Munch, 2007, A New Approach to Drinking-Water Quality Data: Lowest-Concentration Minimum Reporting Level. *Env. Sci. and Technol.* **41**, 677–681.

Maul A. and A.H. El-Shaarawi, 1993, Stochastic models applied to cluster analysis of censored water quality data. *Water Resources Research* **29**, 2705–2711.

Meeker, W.O. and L.A. Escobar, 1998, *Statistical Methods for Reliability Data*.Wiley, New York, 680 pp.

Meier, P., T. Karrison, R. Chappell, and H. Xie, 2004, The price of Kaplan–Meier. *Journal of the American Statistical Association* **99**, 890–897.

Miesch, A., 1967, Methods of computation for estimating geochemical abundance. U.S. Geological Survey Professional Paper 574-B, 15 pp.

Millard, S.P. and S.J. Deverel, 1988, Nonparametric statistical methods for comparing two sites based on data with multiple nondetect limits. *Water Resources Research* **24**, 2087–2098.

Minnesota Pollution Control Agency, 1999, Data Analysis Protocol for the Ground Water Monitoring and Assessment Program. St. Paul, MN, 42 pp. Available at http://www.pca.state.mn.us/index.php/water/water-types-and-programs/groundwater/groundwater-monitoring-and-assessment/publications-ambient-groundwater-monitoring-and-assessment.html.

Morales, J.F., T.Song, A.D. Auerbach, and K.M. Wittkowski, 2008, Phenotyping genetic diseases using an extension of μ-scores for multivariate data. *Statistical Applications in Genetics and Molecular Biology* **7**(1), 19.

Mueller, D.K., B.C. Ruddy, and W.A. Battaglin, 1997, Logistic model of nitrate in streams of the Upper Midwestern United States. *Journal of Environmental Quality* **26**, 1223–1230.

Murphy, S.A., 1995, Likelihood ratio-based confidence intervals in survival analysis. *Journal of the American Statistical Association* **90**, 1399–1405.

Nair, V.N., 1984, Confidence bands for survival functions with censored data: a comparative study. *Technometrics* **26**, 265–275.

Nehls, G.J. and G.G. Akland, 1973, Procedures for handling aerometric data. *Journal of the Air Pollution Control Association* **23**, 180–184.

Oakes, D., 1982, A concordance test for independence in the presence of censoring. *Biometrics* **38**, 451–455.

Oblinger Childress, C.J., W.T. Foreman, B.F. Connor, and T.J. Maloney, 1999, New reporting procedures based on long-term method detection levels and some considerations for interpretations of water-quality data provided by the U.S. Geological Survey National Water Quality Laboratory. USGS Open-File Report 99-193, 19 pp.

O'Brien, P.C. and T.R. Fleming, 1987, A paired Prentice–Wilcoxon test for censored paired data. *Biometrics* **43**, 169–180.

O'Connell, S.G., M. Arendt, A. Segars, T. Kimmel, J. Braun-McNeill, L. Avens, B. Schroeder, L. Ngai, J.R. Kucklick, and J.M. Keller, 2010, Temporal and spatial trends of perfluorinated compounds in juvenile loggerhead sea turtles (*Caretta caretta*) along the east coast of the United States. *Environmental and Science Technology* **44**, 5202–5209.

Olsson, U., 2005, Confidence intervals for the mean of a log-normal distribution. *Journal of Statistics Education* **13**(1). Available online at http://www.amstat.org/publications/se/v13n1/Olsson.html.

Ontario Ministry of the Environment, 2010, Pesticides in Ontario's Treated Municipal Drinking Water 1986–2006. Ottawa, Ontario, 27 pp. Available at http://www.ene.gov.on.ca/environment/en/resources/STD01_078819.html.

Owen, W., and T. DeRouen, 1980, Estimation of the mean for lognormal data containing zeros and left-censored values, with applications to the measurement of worker exposure to air contaminants. *Biometrics* **36**, 707–719.

Perkins, J.L., G.N. Cutter, and M.S. Cleveland, 1990, Estimating the mean, variance, and confidence limits from censored (<limit of detection), lognormally-distributed exposure data. *American Industrial Hygiene Association Journal* **51**, 416–419.

Peto, R. and J. Peto, 1972, Asymptotically efficient rank invariant test procedures (with discussion). *Journal of the Royal Statistical Society, Series A* **135**, 185–206.

Pettitt, A.N., 1976, Cramer–von Mises statistics for testing normality with censored samples. *Biometrika* **63**, 475–481.

Porter, P.S., R.C. Ward, and H.F. Bell, 1988, The reporting limit. *Environmental Science and Technology* **22**, 856–861.

Pratt, J.W., 1959, Remarks on zeros and ties in the Wilcoxon signed rank procedures. *Journ. Amer. Statistical Assoc.* **54**, 655–667.

Prentice, R.L., 1978, Linear rank tests with right-censored data. *Biometrika* **65**, 167–179.

Prentice, R.L. and P. Marek, 1979, A qualitative discrepancy between censored data rank tests. *Biometrics* **35**, 861–867.

Rao, S.T., J.Y. Ku, and K.S. Rao, 1991, Analysis of toxic air contaminant data containing concentrations below the limit of detection. *Journal of Air and Waste Management Association* **41**, 442–448.

Reimann, C., P. Filzmoser, and R.G. Garrett, 2002, Factor analysis applied to regional geochemical data: problems and possibilities. *Applied Geochemistry* **17**, 185–206.

Ren, J., 2003, Goodness of fit tests with interval censored data. *Scandinavian Journal of Statistics* **30**, 211–226.

Rigo, H.G., 1999, Comment on "An alternate minimum level definition for analytical quantification." *Environmental Science and Technology* **33**, 1311–1312.

Rogers Commission, 1986, Report to the President by the Presidential Commission on the Space Shuttle Challenger Accident. June 6th, 1986. Washington, D.C.

Royston, P., 1993, A toolkit for testing non-normality in complete and censored samples. *The Statistician* **42**, 37–43.

Royston, P., 1995, Remark AS R94: a remark on algorithm AS 181: the W-test for normality. *Journal of the Royal Statistical Society, Series C (Applied Statistics)* **44**, 547–551

Ryan, T.P., 1997, *Modern Regression Methods*. Wiley, New York, 515 pp.

Self, S.G. and E.A. Grossman, 1986, Linear rank tests for interval-censored data with application to PCB levels in adipose tissue of transformer repair workers. *Biometrics* **42**, 521–530.

Sen, P.K., 1968, Estimates of the regression coefficient based on Kendall's tau. *Journal of the American Statistical Association* **63**, 1379–1389.

Shapiro, S.S. and M.B. Wilk, 1965, An analysis of variance test for normality (complete samples). *Biometrika* **52**, 591–611.

Shaw, P.J.A., 2003, *Multivariate Statistics for the Environmental Sciences*. Arnold, London, 233 pp.

She, N., 1997, Analyzing censored water quality data using a non-parametric approach. *Journal of the American Water Resources Association* **33**, 615–624.

Shumway, R.H., A.S. Azari, and P. Johnson, 1989, Estimating mean concentrations under transformation for environmental data with detection limits. *Technometrics* **31**, 347–356.

Shumway, R.H., R.S. Azari, and M. Kayhanian, 2002, Statistical approaches to estimating mean water quality concentrations with detection limits. *Environmental Science and Technology* **36**, 3345–3353.

Simon, R. and Y.J. Lee, 1982, Nonparametric confidence limits for survival probabilities and median survival time. *Cancer Treatment Reports* **66**, 37–42.

Singh, A. and J. Nocerino, 2002, Robust estimation of mean and variance using environmental data sets with below detection limit observations. *Chemometrics and Intelligent Laboratory Systems* **60**, 69–86.

Singh, A.K., A. Singh, and M. Engelhardt, 1997, The lognormal distribution in environmental applications. U.S. Environmental Protection Agency Report EPA/600/R–97/006, 19 pp.

Singh, A., R. Maichle, and S.E. Lee, 2006, On the computation of a 95% upper confidence limit of the unknown population mean based upon data sets with below detection limit observations. U.S. Environmental Protection Agency Report EPA/600/R-06/022.

Slyman, D.J., A. de Peyster, and R.R. Donohoe, 1994, Hypothesis testing with values below detection limit in environmental studies. *Environmental Science and Technology* **28**, 898–902.

Smith, D.E. and K.C. Burns, 1998, Estimating percentiles from composite environmental samples when all observations are nondetectable. *Environmental and Ecological Statistics* **5**, 227–243.

Squillace, P.J., M.J. Moran, W.W. Lapham, C.V. Price, R.M. Clawges, and J.S. Zogorski, 1999, Volatile organic compounds in untreated ambient groundwater of the United States, 1985–1995. *Environmental Science and Technology* **33**, 4176–4187.

Stephens, M.A., 1974, EDF statistics for goodness of fit and some comparisons. *Journal of the American Statistical Association* **69**, 730–737.

Stetzenbach, K.J., I.M. Farnham, V.F. Hodge, and K.H. Johannesson, 1999, Using multivariate statistical analysis of groundwater major cation and trace element concentrations to evaluate groundwater flow in a regional aquifer. *Hydrological Processes* **13**, 2655–2673.

Succop, P.A., S. Clark, and M. Chen, 2004, Imputation of data values that are less than a detection limit. *Journal of Occupational and Environmental Health* **1**, 436–441.

Sun, J., 2006, *The Statistical Analysis of Interval-Censored Failure Time Data*. Springer, New York, 302 pp.

Tajimi, M., R. Uehara, M. Watanabe, I. Oki, T. Ojima, and Y. Nakamura, 2005, Correlation coefficients between the dioxin levels in mother's milk and the distances to the nearest waste incinerator which was the largest source of dioxins from each mother's place of residence in Tokyo, Japan. *Chemosphere* **61**, 1256–1262.

Tempelman, A.A. and M.G. Akritas, 1996, Model testing for multivariate censored data. Part 1: Simple null hypotheses. *Probability Theory and Related Fields* **106**, 351–369.

Theil, H., 1950, A rank-invariant method of linear and polynomial regression analysis. *Nederl. Akad. Wetensch, Proceed.*, **53**, 386–392.

Thompson, M.L. and K.P. Nelson, 2003, Linear regression with Type I interval- and left-censored response data. *Environmental and Ecological Statistics* **10**, 221–230.

Thomsen, V., D. Schatzlein, and D. Mercuro, 2003, Limits of detection in spectroscopy. *Spectroscopy* **18**(12), 112–114.

Thurman, E.M., J.E. Dietze, and E.A. Scribner, 2002, Occurrence of antibiotics in water from fish hatcheries. USGS Fact Sheet FS 120-02, 4 pp. Available at http://pubs.water.usgs.gov/fs-120-02/.

Tobin, J., 1958, Estimation of relationships for limited dependent variables. *Econometrica* **26**, 24–26.

Tomlinson, M.S., 2003, Effects of ground-water/surface-water interactions and land use on water quality. Written communication in advance of becoming a USGS report.

Travis, C.C. and M.L. Land, 1990, Estimating the mean of data sets with nondetectable values. *Environmental Science and Technology* **24**, 961–962.

Tressou, J., 2006, Nonparametric modeling of the left censorship of analytical data in food risk assessment. *Journal of the American Statistical Association* **101**, 1377–1386.

Turnbull, B.W., 1976, The empirical distribution function with arbitrarily grouped, censored and truncated data. *Journal of the Royal Statistical Society, Series B* **38**, 290–295.

U.S. Environmental Protection Agency, 1982, Definition and procedure for the determination of the method detection limit—revision 1.11. Code of Federal Regulations 40, Part 136, Appendix B, pp. 565–567.

U.S. Environmental Protection Agency, 1989, *Risk Assessment Guidance for Superfund (RAGS)*, Vol. I. Human Health Evaluation Manual (Part A). USEPA Office of Solid Waste, EPA/540/1-89/002. Available at http://www.epa.gov/oswer/riskassessment/ragsa/.

U.S. Environmental Protection Agency, 1991, Technical support document for water-quality based toxics control, EPA/505/2-90-001. USEPA Office of Water, Washington, DC. Available at http://www.epa.gov/npdes/pubs/owm0264.pdf.

U.S. Environmental Protection Agency, 1998a, Guidance for data quality assessment. Practical methods for data analysis, EPA/600/R-96/084. Available at http://www.epa.gov/swerust1/cat/epaqag9.pdf.

U.S. Environmental Protection Agency, 1998b, Guidelines for ecological risk assessment, EPA/630/R-95/002F. Available at http://oaspub.epa.gov/eims/eimscomm.getfile?p_download_id=36512.

U.S. Environmental Protection Agency, 2001, Workshop report on the application of 2,3,7,8-TCDD toxicity equivalence factors to fish and wildlife, EPA/630/R-01/002. Available at http://cfpub.epa.gov/ncea/raf/pdfs/tefworkshopforum.pdf.

U.S. Environmental Protection Agency, 2002a, Development document for proposed effluent limitations guidelines and standards for the concentrated aquatic animal production industry point source category, EPA-821-R-02-016. Available at http://www.epa.gov/waterscience/guide/aquaculture/tdd/complete.pdf.

U.S. Environmental Protection Agency, 2002b, Guidance for comparing background and chemical concentrations in soils for CERCLA sites, EPA-540-R-01-003. Available at http://www.epa.gov/superfund/programs/risk/background.pdf.

U.S. Environmental Protection Agency, 2002c, Development document for the proposed effluent limitations guidelines and standards for the meat and poultry products industry point source category (40 CFR 432), EPA-821-B-01-007. Office of Water, U.S. Environmental Protection Agency.

U.S. Environmental Protection Agency, 2003, Technical support document for the assessment of detection and quantitation approaches, EPA-821-R-03-005, 71 pp. Available at http://www.epa.gov/waterscience/methods/det/dgch1–3.pdf.

U.S. Environmental Protection Agency, 2007, Report of the federal advisory committee on detection and quantitation approaches and uses in Clean Water Act Programs. Final report 12/28/07, 64 pp. plus appendices. Available at http://www.epa.gov/waterscience/methods/det/faca/final-report-200712.pdf.

U.S. Environmental Protection Agency, 2008, Framework for application of the toxicity equivalence methodology for polychlorinated dioxins, furans, and biphenyls in ecological risk assessment, EPA 10/R-08/004, USEPA, Washington, DC. Available at http://www.epa.gov/raf/tefframework/pdfs/tefs-draft-052808.pdf.

U.S. Environmental Protection Agency, 2009, Statistical analysis of groundwater monitoring data at RCRA facilities, Unified guidance, EPA 530-R-09-007. Available at http://www.epa.gov/waste/hazard/correctiveaction/resources/guidance/sitechar/gwstats/unified-guid.pdf.

Van den Berg, M., L. Birnbaum, B.T.C. Bosveld, B. Brunstrom, P.M. Cook, M. Feeley, J.P. Giesy, A. Hanberg, R. Hasegawa, S.W. Kennedy, T. Kubiak, J.C. Larsen, F.X.R. van Leewen, A.K. Djien Liem, C. Nolt, R.E. Peterson, R.E. Poellinger, S. Safe, D. Schrenk, D. Tillitt, M. Tysklind, F. Waern, and T. Zacherewski, 1998, Toxic equivalency factors (TEFs) for PCBs, PCDDs, PCDFs for humans and wildlife. *Environmental Health Perspectives* **106**, 775–792.

Velleman, P.F. and D.C. Hoaglin, 1981, *Applications, Basics, and Computing of Exploratory Data Analysis.* Duxbury Press, Boston, 354 pp.

Verrill, S. and R.A. Johnson, 1988, Tables and large-sample distribution theory for censored-data correlation statistics for testing normality. *Journal of the American Statistical Association* **83**, 1192–1197.

Wahlin, K. and A. Grimvall, 2010, Roadmap for assessing regional trends in groundwater quality. *Environmental Monitoring and Assessment* **165**, 217–231.

Waller, L.A. and B.W. Turnbull, 1992, Probability plotting with censored data. *The American Statistician* **46**, 5–12

Ware, J.H. and D.L. DeMets, 1976, Reanalysis of some baboon descent data. *Biometrics* **32**, 459–463.

Wentzell, P.D. and M.T. Lohnes, 1999, Maximum likelihood principal component analysis with correlated measurement errors: theoretical and practical considerations. *Chemometrics and Intelligent Laboratory Systems* **45**, 65–85.

Weston, S.A. and W.Q. Meeker, 1991, Coverage probabilities of nonparametric simultaneous confidence bands for a survival function. *Journal of Statistical Computing Simulation*, **38**, 83–97.

White, C.E. and H.D. Kahn, 1995, Discussion on the paper by R.D. Gibbons, "Some statistical and conceptual issues in the detection of low level environmental pollutants." *Environmental and Ecological Statistics* **2**, 149–154.

Wilcox, R.R., 1998, Simulations on the Theil–Sen regression estimator with right-censored data. *Statistics and Probability Letters* **39**, 43–47.

Winslow, S.D., B.V. Pepich, J.J. Martin, G.R. Hallberg, D.J. Munch, C.P. Frebis, and E.J. Hedrick, 2006, Statistical Procedures for Determination and Verification of Minimum Reporting Levels for Drinking Water Methods, *Env. Sci. and Technol.* **40**, 281–288.

Wittkowski, K.M., T. Song, K. Anderson, and J.E. Daniels, 2008, U-scores for multivariate data in sports. *Journal of Quantitative Analysis in Sports* **4**(3), article #7.

Yamaguchi, N., D. Gazzard, G. Scholey, and D.W. MacDonald, 2003, Concentrations and hazard assessment of PCBs, organochlorine pesticides and mercury in fish species from the upper Thames—River pollution and its potential effects on top predators. *Chemosphere* **50**, 265–273.

Zar, J.H., 1999, *Biostatistical Analysis*, 4th edition.Prentice-Hall, Upper Saddle River, 875 pp.

Zuur, A.F., E.N. Ieno, and G.M. Smith, 2007, *Analyzing Ecological Data.*Springer, New York, NY, 672 pp.

INDEX

accelerated failure time models *249*
air quality *14, 62, 90*
Akritas test *188, 190*
Akritas-Theil-Sen slope 20, *See*
   Theil-Sen slope
all nondetects *142, 143, 144, 150, 151*
analysis of variance *194, 195, 216*
ANOSIM 20, 270, 272, 276, 277, 278, 284,
   286, 292, 294, 295, 299, 302
ANOVA *202*
arbitrary censoring *67*, 200, 249, *254*
astronomy *234, 309*
average linkage 278, 296

balanced errors *27*
between the limits *28, 29*
bias *29, 30, 31, 91, 238*
binomial probability 127, 129, 135, *142*,
   144, 152
binomial test *149, See* quantile test
biplot 280, 281, 284, 285
Bonferroni adjustment *208, 209, 212*
bootstrapping *90, 102, 103, 116, 131, 136*
boxplot 44, 45, 153
Buckley-James regression *258, 265, 266*

calibration-based limits *35*
canonical correlation 270
cdf *15, 47, See* cumulative distribution
   function
censored data 2, 10, 15, 18, 19, 20,
   21, 315
censored regression *20, 161, 162, 165, 200,
   201, 204, 252*
Central Limit Theorem *102, 105*
Challenger accident *xii, xv*

cluster analysis 269, 272, 278, 289, 294,
   296, 302, 316
coefficient of determination *224*
Cohen's method *64, 66, 67, 88, 89, 90*
Cohen's 2, 64, 66, 85, 87, 88, 89, 90
compliance *191, See* legal standard
confidence bound 99, 102, 106, 107, 108,
   118, 119, 121, 135
confidence interval *99, 101, 102, 115, 186*
confidence interval for percenitles *120*
confidence interval for the median *119, 128,
   132, 134*
confidence intervals for percentiles *135*
constant standard deviation *23*
contingency table test *196, 215*
correlation 20, 218
correspondence analysis 269
coverage *110*
critical value *23*
cumulative distribution function *15, 16, 47,
   72, 148, 151, 168, 250*

database *37*
decision level *23*
deleting nondetects *xv, 44*
delta-lognormal method 3
detection limit *8, 12, 15, 22, 23*
Discriminant function 270
doubly-censored data *258, 309*

edf *47, 48, 49, 61, 171, See* empirical
   distribution function
EM algorithm 288, 289
empirical distribution function 44, 47, 72
Euclidean distance 272, 282, 283, 284,
   287, 291

Statistics for Censored Environmental Data Using Minitab® and R, Second Edition.
Dennis R. Helsel.
© 2012 John Wiley & Sons, Inc. Published 2012 by John Wiley & Sons, Inc.

Printed and bound by CPI Group (UK) Ltd, Croydon, CR0 4YY

16/04/2025

14658532-0003